Textbooks in Electrical and Elect[...]gineering

Books are to be returned on or before the last date below.

~~111202~~

LIVERPOOL
JOHN MOORES UNIVERSITY
AVRIL ROBARTS LRC
TEL. 0151 231 4022

LIBREX

LIVERPOOL JMU LIBRARY

3 1111 00910 2755

Series Editors

G. Lancaster J.K. Sykulski E.W. Williams

---

1. Introduction to fields and circuits
   GORDON LANCASTER
2. The CD-ROM and optical recording systems
   E.W. WILLIAMS
3. Engineering electromagnetism: physical processes and computation
   P. HAMMOND and J.K. SYKULSKI
4. Integrated circuit engineering
   L.J. HERBST
5. Electrical circuits and systems: an introduction for engineers and physical scientists
   A.M. HOWATSON
6. Applied numerical modelling for engineers
   D. DE COGAN and A. DE COGAN
7. Electromagnetism for engineers: an introductory course (Fourth edition)
   P. HAMMOND
8. Instrumentation for engineers and scientists
   JOHN TURNER and MARTYN HILL
9. Introduction to error control codes
   SALVATORE GRAVANO
10. Materials science for electrical and electronic engineers
   IAN P. JONES
11. Integrated converters: D to A and A to D architectures, analysis and simulation
   PAUL G.A. JESPERS

Materials Science for Electrical and Electronic Engineers

Ian P. Jones

University of Birmingham

OXFORD
UNIVERSITY PRESS

Great Clarendon Street, Oxford OX2 6DP

Oxford University Press is a department of the University of Oxford.
It furthers the University's objective of excellence in research, scholarship,
and education by publishing worldwide in

Oxford New York
Athens Auckland Bangkok Bogotá Buenos Aires Calcutta
Cape Town Chennai Dar es Salaam Delhi Florence Hong Kong Istanbul
Karachi Kuala Lumpur Madrid Melbourne Mexico City Mumbai
Nairobi Paris São Paulo Singapore Taipei Tokyo Toronto Warsaw

with associated companies in Berlin Ibadan

Oxford is a registered trade mark of Oxford University Press
in the UK and in certain other countries

Published in the United States
by Oxford University Press Inc., New York

© Ian P. Jones 2001

The moral rights of the author have been asserted
Database right Oxford University Press (maker)

First published 2001

All rights reserved. No part of this publication may be reproduced,
stored in a retrieval system, or transmitted, in any form or by any means,
without the prior permission in writing of Oxford University Press,
or as expressly permitted by law, or under terms agreed with the appropriate
reprographics rights organization. Enquiries concerning reproduction
outside the scope of the above should be sent to the Rights Department,
Oxford University Press, at the address above

You must not circulate this book in any other binding or cover
and you must impose this same condition on any acquirer

A catalogue record for this book is available from the British Library

Library of Congress Cataloging in Publication Data
(Data applied for)

ISBN 0 19 856294 2

Typeset by EXPO Holdings, Malaysia
Printed in Great Britain
on acid-free paper by T. J. International Ltd.,
Padstow, Cornwall

To my parents

I like to think my father would have approved of parts of this

Preface

One evening in the mid-80s, just before going to bed, I started idly to flick through a new introductory book on materials aimed at structural engineering students. It was called *Engineering materials* and the authors were M. F. Ashby and D. R. H. Jones. Four hours later I put the book down, having read it from cover to cover, and my ideas about what constituted a good textbook had been turned upside down. *Engineering materials* is well written, witty and superbly scientific. One cannot wait to turn the page to find out what happens next. It has been my inspiration and guiding star in writing *this* book.

My central theme is that the type of bonding in a solid not only controls its electrical properties but also, and just as directly, its mechanical properties and how things are made from it. Thus the reason why a copper wire can conduct electricity is exactly the same reason it can be drawn into wire in the first place. The reason why a piece of porcelain does *not* conduct electricity is the same as why it cannot be rolled into its final shape—as copper could—and has thus to be made directly. This common origin of electrical and mechanical properties dictates the structure of the book.

In Part I the different types of bonds are described and conductors and insulators are introduced.

Part II is devoted to electrical conductors: metals. In Chapter 2, the properties characteristic of a metal are explained in terms of the metallic bond. Mechanical properties are the subject of Chapter 3. Elemental metal conductors, which account for such a large proportion of the electrical market, the way in which conductors are made strong and the consequences for their electrical conductivity, are described in some detail in Chapter 4. Steel is not, *per se*, an electrical engineering material, but its use is so universal that anyone wanting to call themselves an engineer should have some understanding of it. Chapter 5 is therefore devoted to steel. The final chapter in this section, Chapter 6, deals with the important topic of electrochemistry: electroplating and its evil first cousin, corrosion.

Part III is about electrical insulators. There are two main types, ceramics and plastics, and these respectively are the subjects of Chapters 7 and 8. Again, the consequences of the bonding (here ionic and covalent) for electrical and mechanical properties are emphasised, as are the implications for processing and manufacturing routes.

Part IV deals with the half-way house of semiconductors—so crucial to electronic engineers—and with some specific topics which fall across the borders defined earlier in the book and do not fit easily into that scheme of things. These topics are magnetic materials, superconductors and optical materials.

This is a book for electrical and electronic engineers, not for materials scientists. I have tried to render every explanation in its simplest and clearest form and especially I have included as many relevant examples as possible. If the reader finds him- or herself wondering why on earth an electrical or electronic engineer should need to know this or that, then I shall have failed. I have tried to make clear at every point the direct relevance of every topic to the reader's main course of study: electrical or electronic engineering. I hope, however, that some of the excitement I felt when I was researching this book—coming a complete innocent (in electrical terms) to the subject, and discovering so many fascinating things—I hope some of this excitement will communicate itself to you.

Although the structure of the book is my own, the main topics were taken from the IEE suggested syllabus, which satisfies component E1A of an Electronic and Electrical first degree in the U.K. For those readers outside this country, E1A is the materials related part of the general engineering background which students receive in their first year.

It is a pleasure here to thank many colleagues for reading parts of this book. Mike Wise and Dennis Milner between them checked Chapter 4. Stan Glover dictated most of Chapter 5 *extempore et verbatim* during a car journey to Nottingham and I am grateful to him and also to Claire Davis who checked that I had written it down correctly. Peter Farr cast a broody but tolerant eye over Chapter 6. He also supplied me with the examples. Clive Ponton, who, unfortunately for him, occupied the office next to mine for most of the time it took to write this book, took on an increasingly haggard appearance during the composition of Chapter 7 and his relief at its completion was hardly less than the author's. Nigel Mills made some characteristically terse contributions to Chapter 8 and Mark Aindow had three heroic attempts at explaining p–n junctions to me (Chapter 9). I was very lucky as far as Chapter 10 was concerned to be in the same building as Rex Harris's magnetic materials research group. Andy Williams very kindly read and corrected Chapter 10 and Stuart Abell checked the superconductivity part of Chapter 11. Finally, John Knott and Ray Smallman subjected themselves to the whole of the first part and I am very grateful to them for their care and for their helpful suggestions. (Chris Hardy drew the cartoons for Chapter 2 with her customary easy inspiration.)

I would also like to thank my very patient and courteous hosts on numerous industrial visits.

Richard Lawrence was my original editor. Richard must be the most patient man in the world. I cannot sufficiently thank him for his encouragement and neverfailing good humour.

Birmingham I.P.J.
June 2

Contents

LIVERPOOL
JOHN MOORES UNIVERSITY
AVRIL ROBARTS LRC
TEL. 0151 231 4022

Part I **Introduction**

1

Conductors, insulators and semiconductors

Chapter objectives

- The differences between electrical conductors and insulators
- How atoms in solids bond together
- The structure of the atom

(but not in that order!)

1.1 Introduction

This book is about the *materials* which are used by electrical and electronic engineers. I will be dealing with their *electrical* properties and with their *mechanical* properties. Their electrical properties are *why* an electrical or electronic engineer might wish to use them, but their mechanical properties control *how* they are used and how they are made. We shall see that both electrical and mechanical properties depend very directly on the *bonding* between the atoms in the material. Nearly all the materials described in this book are *solids*. It is the bonds between the atoms in a solid which hold it together. Bonds are of different types (three strong, two weak in fact). The type of bond which forms between atoms in a solid depends on which atoms are present.

Take copper, for example. Copper is a metal, and the most commonly used electrical conductor. It is easily bent or squashed and so when things are made from copper they are usually made by *working* (rolling, or extrusion, or wiredrawing, etc.). Copper is also an *element*—that is, all the atoms in a piece of copper are of the same type: they are copper atoms. They are held together by a type of bond called (not surprisingly) the *metallic bond*.

Now consider a common insulator: alumina. Alumina is aluminium oxide, a *compound* of aluminium and oxygen and so alumina contains two types of atoms: aluminium atoms and oxygen atoms. These are held together by *ionic bonds*. Anyone who has handled alumina knows that it is a hard brittle substance whose shape cannot be changed (except by breaking pieces off it). Alumina insulators have to be made in their final shape (precisely how will come much later in this book—in Chapter 7).

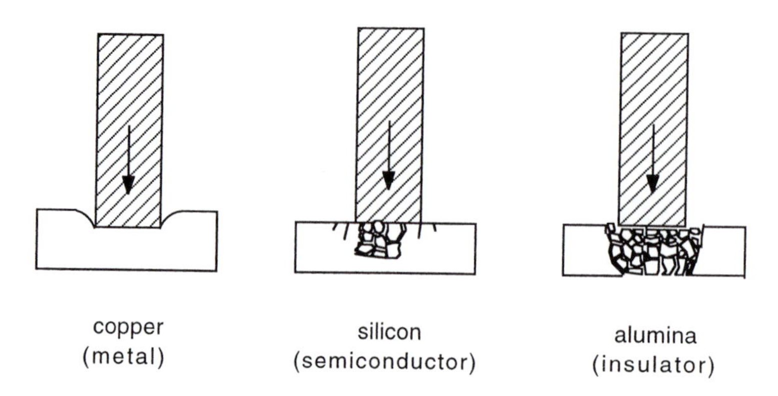

Fig. 1.1 Metals can be squashed, but semiconductors and insulators cannot—they just crumble.

A third and final example. Silicon, like copper, is an element. A piece of silicon contains just silicon atoms. Silicon is very different from copper, however. The atoms are bound together via the *covalent bond*. Silicon is a poor conductor of electricity (it is called a *semiconductor*: halfway between a conductor and an insulator). It is also very brittle. As with alumina, articles made from silicon have to be made in their final shape.

We will discover later on that it is possible, however, to make *flexible* insulators, by exploiting especially weak bonds in plastics.

Why? Why do copper atoms join together via metallic bonding to form a metal, whereas silicon atoms join together via covalent bonds to form a semiconductor? Why do aluminium atoms and oxygen atoms join together to form an insulator? Why can we draw wires from copper, but ionically bonded alumina and covalently bonded silicon will not, *can*not, suffer a shape change?

The answer to all of these questions lies in the bonding between the atoms. To understand why this is so, and how and why bonds form, we need to understand a little bit about the structure of the atom. I will start this chapter, therefore, by describing how atoms are constructed.

1.2 The structure of the atom

An atom consists of a small dense positively charged *nucleus* surrounded by a number of light, negatively charged *electrons*. These occupy orbits of different sizes and shapes which, together with the energies of the electrons in them, make up the *electronic structure* of the atom. I shall return to this in more detail in the next section. The largest electronic orbits (see Fig. 1.2) define the 'size' of the atom, just as the orbit of the planet

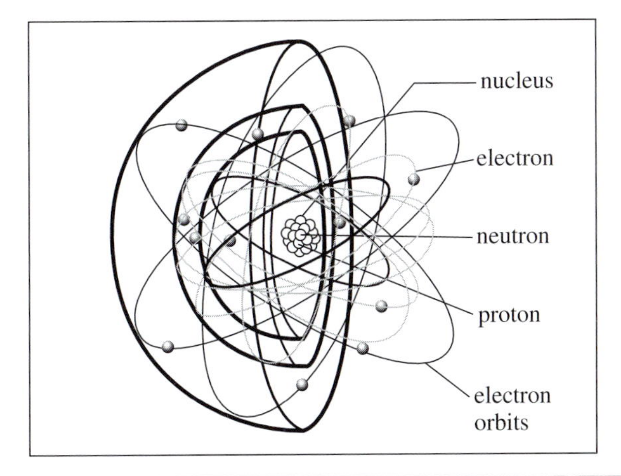

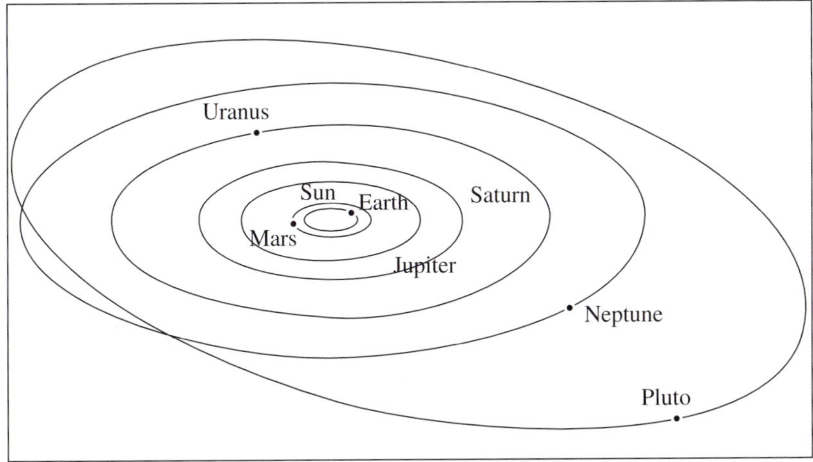

Fig. 1.2 (a) An atom consists of a small positively charged nucleus surrounded by a cloud of light negatively charged electrons. (Note: the nucleus is much, much smaller than I have shown it.) (b) The solar system (for comparison).

Pluto might be said to define the 'size' of our solar system. As with the solar system, however, most of the atom is free space, or at least space containing a small density of electrons; the vast majority of the weight of an atom resides in the small central nucleus. Thus the diameter of the smallest atom, hydrogen, is $\sim$100 pm ($= 10^{-10}$ m), while the diameter of its nucleus, which contains over 99.9% of the weight, is 1 fm ($= 10^{-15}$ m). So if an atom were the size of an average living room, the nucleus would be one tenth of a millimetre across! One difference between an atom and the solar system derives from the very different sizes of the two objects. The planets in the solar system have well defined sizes and positions. Because of its small mass the electrons in an atom do not. A single electron in an atom is not defined as being in a

Fig. 1.3 The atomic nucleus.

particular place at a particular time and with a particular speed. Instead, the electron is described by a *probability density function*, which defines the probability of the electron occupying any particular position in space. This is not because it is too small to measure conveniently—it is a fundamental property of small objects. I will come back to this in the next section.

The nucleus consists of a roughly equal number of positively charged *protons* and electrically neutral *neutrons*. Because the whole atom must be electrically neutral there must be as many electrons as protons. This number is usually given the symbol 'Z' and is called the *atomic number*. There are 92 naturally occurring elements, with atomic numbers ranging from 1 (hydrogen) to 92 (uranium).[†] These elements occur with very different frequencies and this is one of the reasons why iron, for example, is so cheap while gold is relatively expensive. The neutrons act to glue the protons together via the so-called *strong nuclear force*—otherwise the positive charges of the protons would cause them to repel each other and blow the nucleus apart. The number of neutrons in a nucleus can sometimes vary by one or two; this gives rise to different *isotopes* of the same element. The *relative atomic mass* (what used to be the atomic weight) of an element is a weighted average of the masses of the different naturally occurring isotopes of that element, according to the relative frequencies with which they occur on earth, ratioed roughly to the mass of the hydrogen atom. Thus, for example:

Cu^{63} is an isotope of copper whose relative atomic mass is 62.93. It constitutes 69% of terrestrial copper.

Cu^{65} is the other isotope of copper. Its relative atomic mass is 64.93 and it constitutes 31% of terrestrial copper.

The relative atomic mass of terrestrial copper is therefore $62.93 \times 0.69 + 64.93 \times 0.31 = 63.55$.

1.3 The electronic structure of atoms

In this section, I return to the *electronic structure* of atoms, which I mentioned above. Most of the volume of an atom is occupied by electrons which are in orbits about the atom's nucleus. Thus the electron orbits define the size of the atom and when two

[†] A few of these are unstable (radioactive) and therefore uncommon. Prometheum and technetium effectively do not occur.

atoms approach each other, it is the electrons of each atom which 'see' each other, or interact with each other, or whatever. Thus *chemistry* is about electronic structure, not about protons and neutrons.

We are interested in two things concerning the electrons: energy and position. We want to know *where* they are relative to the nucleus, and, for example, *how much energy* it would require to detach them from the atom.

The electrons fall into *shells* around the nucleus. Each shell contains a certain number of electrons, all of which have roughly the same energy. The smallest shell has the lowest energy and is closest to the nucleus. It can contain 2 electrons. The electrons in the next shell have a somewhat larger energy and they are further away from the nucleus. This second shell can contain 8 electrons. And so on. The maximum capacities of the shells go as follows:

Shell number	1	2	3	4	...	n
Maximum population of electrons	2	8	18	32	...	$2n^2$

The number of each shell is called the *principal quantum number*. This is because the mathematical theory which describes atomic phenomena like the structure of the atom is called 'quantum mechanics'. *Quantum* derives from the fact that it is a natural consequence of this theory that the energy associated with bound particles, like an electron orbiting a nucleus, is *quantised*, i.e. can only take certain fixed values.

Within each shell the electrons do not have precisely the same energy. There are three further quantum numbers and so each electron has four quantum numbers in all attached to it and these four quantum numbers define where the electron is likely to be and its energy. The four quantum numbers are as follows:

1.	Principal quantum number	n	The number of the shell. Defines roughly the *energy* of the electron.
2.	Orbital quantum number	l	This defines the *shape* of the orbit and the *angular momentum* of the electron about the nucleus. It has a small effect on the energy of the electron.
3.	Magnetic quantum number	m	This has no effect on the energy of the electron, unless the atom is placed in a magnetic field.
4.	Spin quantum number	s	This describes the *spin* of the electron about its own centre (not about the nucleus). It has a very small effect on the electron's energy.

The rules for n, l, m and s, which can be deduced from quantum mechanics, are as follows:

	n	can take the values	1, 2, 3 etc. (i.e. any positive integer)
	l	can then take the values	$0, 1 \ldots n-1$
	m	can then take the values	$-l$ to $+l$ (including 0)
and	s	can take the values	$-\frac{1}{2}$ or $+\frac{1}{2}$

If you combine these rules together, you will find that the total number of electrons in any shell is $2n^2$, as I described earlier.

It is usual to label l using letters instead of numbers. Thus:

l	=	0	1	2	3	...
		s	p	d	f	...

The letters come from spectroscopy (s = sharp; p = principal; d = diffuse; f = fine).

We can imagine an atom looking as in Fig. 1.4, where the shells are filled with electrons and complete shells (e.g. shell number 2 = 2s + 2p) are overall spherically symmetric about the nucleus. Instead of *orbits*, which are precise trajectories, we talk about *orbitals*, which are fuzzy. We cannot say precisely *where* the electron is at any given time. The blacker the orbital in Fig. 1.4, the greater is the chance of finding an electron there. The individual orbitals within each shell are not spherical—just their

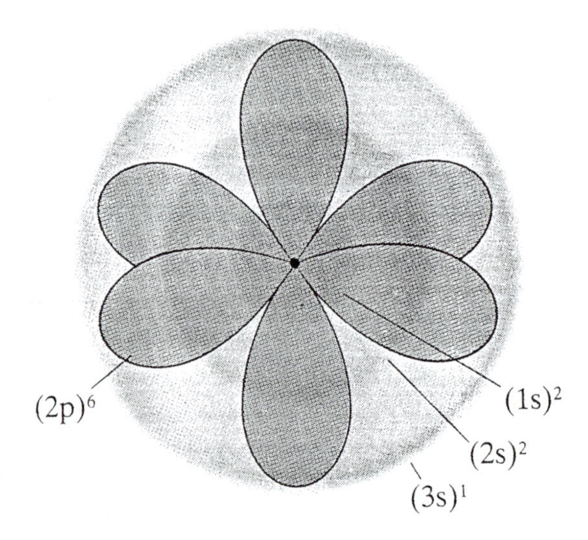

Fig. 1.4 Electron orbitals in a sodium atom. Whereas orbits are definite trajectories, orbitals are fuzzy. The electron is more likely to be where the orbital is blackest.

combined, averaged-out effect. Figure 1.4 shows some individual orbitals for $l = 0$ (i.e. 's') and 1 ('p').

The electron energy levels are shown schematically in Fig. 1.5. The exact values of the energy levels, and the order in which they occur, depend on how many electrons are actually in the atom. In the hydrogen atom, the energy of an electron depends solely on its principal quantum number. On the left of the diagram is the scheme for a light atom and on the right for a heavy atom. Here I have ignored the small effect which the fourth quantum number, s, has on the energy. Since the third quantum number, m, has no effect in the absence of a magnetic field, Fig. 1.5 involves only the first two quantum numbers, n and l.

Notice that, although the principal quantum number n does largely dictate the overall energy of a shell, there is considerable overlap between the energy levels of the shells as we go towards the higher numbered shells in the heavier atoms.

We can now use Fig. 1.5 to build up the electronic structures of the naturally occurring elements. As we start from the lightest element, hydrogen, we introduce electrons to each energy level in turn. At this stage we come to a very important question. *Why do not all the electrons go into the lowest energy level?* There is a fundamental law of physics, called the *Pauli Exclusion Principle*, which dictates that every electron state, defined by its unique combination of four quantum numbers, *will hold only one electron*. Thus as we add more and more electrons, they have to occupy higher and higher energy levels. The Pauli Exclusion Principle is quite general and does not apply just to atoms. It has very important consequences for metals, too, as we shall see later.

Table 1.1 shows a list of all the naturally occurring elements, from hydrogen up to uranium. The table shows the electronic configuration of each element, as far as the second quantum number l, and its chemical symbol and atomic number. Roughly, the relative atomic masses, measured on a scale where carbon is 12, are twice the atomic numbers. (On this scale, a proton or neutron weighs ~ 1.)

A very useful pictorial representation of the 92 elements is the *periodic table*, shown in Fig. 1.6. This picture will be very useful later in this book, telling us a tremendous amount about how the various elements are used in electrical engineering, so I shall spend a little time here interpreting it.

Let us start at the top with hydrogen and helium. The hydrogen atom contains one electron, which is in the first principal quantum shell and necessarily has orbital quantum number $l = 0$, i.e. it is an 's' electron. Thus we write the electronic configuration of hydrogen as $1s^1$. With helium, the first quantum shell is full, and we write the electronic configuration as $1s^2$. Helium is an *inert gas*. Inert gases have full quantum shells and are very stable (which is why neon, for example, is used in fluorescent tubes). *Chemical bonding* (see Section 1.4) is largely how assemblies of atoms achieve an inert gas structure.

At lithium, the third electron goes into the second principal quantum shell and we are on to the second period of the periodic table. The s electrons have the least energy

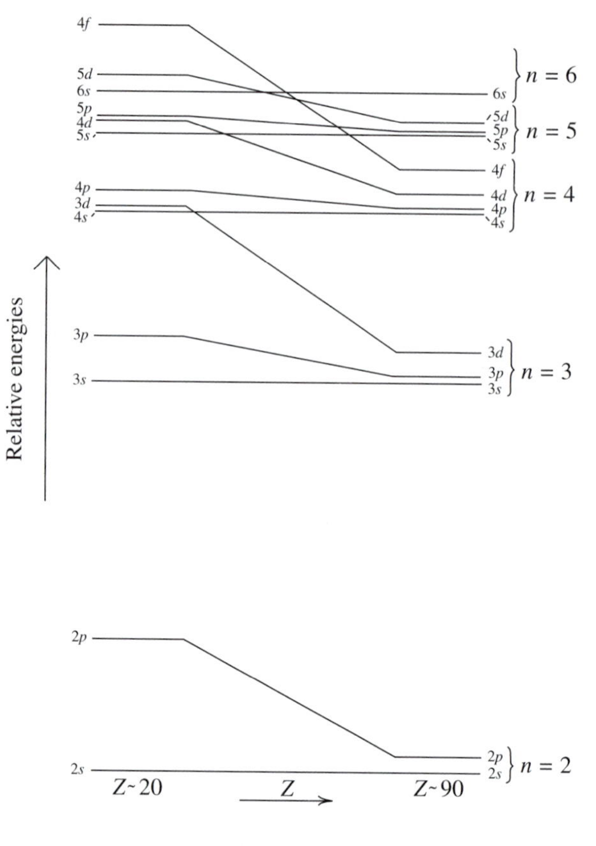

Fig. 1.5 Energy levels in an atom. On the left is a light atom and on the right a heavy atom. The number of electrons in the atom, Z, increases from left to right. As Z increases, so the individual energy levels separate and grow apart. Some of them also cross over.
(Remember in all of this that these energy levels are below zero, if we define zero energy as belonging to an electron at infinity. As the electron approaches the atom, it loses potential energy and its total energy goes down. In fact it recoups *half* the loss of potential energy as kinetic energy — it starts to orbit about the nucleus — so the drop in total energy is only *half* the drop in potential energy. As the electron gets closer to the nucleus — i.e. as the principal quantum number goes from 4 to 3 to 2 to 1 — so it goes faster and faster, but its potential energy drops twice as quickly as its kinetic energy rises, and so its *total* energy still falls.) (From *Atomic theory for students of metallurgy* by W. Hume-Rothery, courtesy of the Institute of Materials.)

in this shell (see Fig. 1.5) and we write the configuration as $1s^2\ 2s^1$. With beryllium the s sub-shell is full and in boron the fifth electron enters the 2p sub-shell. As we pass from boron to carbon, to nitrogen, oxygen and fluorine the 2p sub-shell fills until at neon it is complete. Neon is the second inert gas. The elements where the 'p' sub-shell is being filled are placed in a block on the right side of the table. We have now completed the second period of the periodic table, i.e. $n = 1$ and 2, $l = 0$ and

1, $m = -1$, 0 and 1 and $s = \pm 1/2$: $2 + 2 + 6 = 10$. With sodium, the outermost electron enters the 3s sub-shell and by argon ($Z = 18$) the 3s and 3p sub-shells are full. At this stage one might expect the next electron, the nineteenth, to enter the 3d shell (remember, we have available 1s; 2s, 2p; 3s, 3p, 3d; 4s, 4p, 4d, 4f; etc.). In fact the next two electrons (19: potassium, K and 20: calcium, Ca) enter the 4s sub-shell, and a quick glance at Fig. 1.5 shows why. The 4s sub-shell has slightly lower energy than the 3d: the principal quantum shells 3 and 4 overlap slightly (in *energy*, not in space). (Remember that this is because there are many electrons in the atom: in hydrogen all the 3 shell electrons—the 3s, 3p and 3d—would have the same energy, less than that of the 4th principal quantum shell, albeit there would not normally be any electrons in these levels.) Next, at scandium (Sc, $Z = 21$) the 3d sub-shell is started and the following nine elements through to zinc see it being filled. The elements from scandium to zinc are called *transition* elements and are placed in a block in the middle of the table between the 's' and 'p' blocks (hence 'transition'). Because the 3d and 4s levels are so close there is some movement of electrons between the 3d and 4s sub-shells. Thus we might expect copper ($Z = 29$) to have an electronic configuration $1s^2\ 2s^2\ 2p^6\ 3s^2\ 3p^6\ 3d^9\ 4s^2$, but in fact one of the 's' electrons drops down to the 'd' shell and the true configuration is $1s^2\ 2s^2\ 2p^6\ 3s^2\ 3p^6\ 3d^{10}\ 4s^1$ (remember I am talking here about *isolated* atoms). The very close energy levels in 4s and 3d give the transition elements some of their most characteristic properties—they form highly coloured compounds and are used in lasers. The common *ferromagnets* (Chapter 10) are also found in the first row of transition elements.

Using Fig. 1.5 and Table 1.1 you should be able to see how the rest of the Periodic table develops. The only additional complications arise when the 4f and 5f shells are being filled. Those elements where the 4f sub-shell is filling are called the 'rare earths' (a very inappropriate name, in fact) or the 'lanthanides' after lanthanum ($Z = 57$) which initiates the series. These are normally listed separately at the bottom of the Periodic table, as they are here, with a big 'f' on top to show that it is an 'f' sub-shell which is being filled. The equivalent series for 5f is called 'the actinides' after actinium ($Z = 89$). The lanthanides, with which you may be unfamiliar, will also turn up later in this book, in Chapter 10 on magnetic materials.

Although I do not want to anticipate what I shall say later on about bonding, you can see how important the periodic table is to how materials are used in electrical engineering from the fact that all the electrically conducting elements appear at the left and to the bottom of the table, all the insulating elements to the top right-hand corner and the semiconductors in between. In Fig. 1.6 I have shaded the insulating elements heavily and the semiconducting elements lightly.

So far I have been talking about isolated atoms. Electrical and electronic engineers, however, deal largely with solid materials, where the atoms are bound together with *chemical bonds*. In the next section I wish to explain why bonds form and why some types of bond make materials into electrical conductors while others make them into insulators.

Table 1.1 The electronic structures of the 92 naturally occurring elements. Full shells are shaded.

	n		1	2		3			4				5				6			7
	l		0	0	1	0	1	2	0	1	2	3	0	1	2	3	0	1	2	0
	l		s	s	p	s	p	d	s	p	d	f	s	p	d	f	s	p	d	s
Z	Element name	Symbol																		
1	Hydrogen	H	1																	
2	Helium	He	2																	
3	Lithium	Li		1																
4	Beryllium	Be	2	2																
5	Boron	B	2	2	1															
6	Carbon	C	2	2	2															
7	Nitrogen	N	2	2	3															
8	Oxygen	O	2	2	4															
9	Fluorine	F	2	2	5															
10	Neon	Ne	2	2	6															
11	Sodium	Na	2	2	6	1														
12	Magnesium	Mg	2	2	6	2														
13	Aluminium	Al	2	2	6	2	1													
14	Silicon	Si	2	2	6	2	2													
15	Phosphorus	P	2	2	6	2	3													
16	Sulphur	S	2	2	6	2	4													
17	Chlorine	Cl	2	2	6	2	5													
18	Argon	Ar	2	2	6	2	6													
19	Potassium	K	2	2	6	2	6		1											
20	Calcium	Ca	2	2	6	2	6		2											
21	Scandium	Sc	2	2	6	2	6	1	2											
22	Titanium	Ti	2	2	6	2	6	2	2											
23	Vanadium	V	2	2	6	2	6	3	2											
24	Chromium	Cr	2	2	6	2	6	5	1											
25	Manganese	Mn	2	2	6	2	6	5	2											
26	Iron	Fe	2	2	6	2	6	6	2											
27	Cobalt	Co	2	2	6	2	6	7	2											
28	Nickel	Ni	2	2	6	2	6	8	2											
29	Copper	Cu	2	2	6	2	6	10	1											
30	Zinc	Zn	2	2	6	2	6	10	2											
31	Gallium	Ga	2	2	6	2	6	10	2	1										
32	Germanium	Ge	2	2	6	2	6	10	2	2										
33	Arsenic	As	2	2	6	2	6	10	2	3										
34	Selenium	Se	2	2	6	2	6	10	2	4										
35	Bromine	Br	2	2	6	2	6	10	2	5										
36	Krypton	Kr	2	2	6	2	6	10	2	6										
37	Rubidium	Rb	2	2	6	2	6	10	2	6			1							
38	Strontium	Sr	2	2	6	2	6	10	2	6			2							
39	Yttrium	Y	2	2	6	2	6	10	2	6	1		2							
40	Zirconium	Zr	2	2	6	2	6	10	2	6	2		2							
41	Niobium	Nb	2	2	6	2	6	10	2	6	4		1							
42	Molybdenum	Mo	2	2	6	2	6	10	2	6	5		1							
43	Technetium	Tc	2	2	6	2	6	10	2	6	6		1							
44	Ruthenium	Ru	2	2	6	2	6	10	2	6	7		1							
45	Rhodium	Rh	2	2	6	2	6	10	2	6	8		1							
46	Palladium	Pd	2	2	6	2	6	10	2	6	10									

Table 1.1 (*cont.*)

n			1	2		3			4				5				6			7
l			0	0	1	0	1	2	0	1	2	3	0	1	2	3	0	1	2	0
Z	**Element name**	**Symbol**	s	s	p	s	p	d	s	p	d	f	s	p	d	f	s	p	d	s
47	Silver	Ag	2	2	6	2	6	10	2	6	10		1							
48	Cadmium	Cd	2	2	6	2	6	10	2	6	10		2							
49	Indium	In	2	2	6	2	6	10	2	6	10		2	1						
50	Tin	Sn	2	2	6	2	6	10	2	6	10		2	2						
51	Antimony	Sb	2	2	6	2	6	10	2	6	10		2	3						
52	Tellurium	Te	2	2	6	2	6	10	2	6	10		2	4						
53	Iodine	I	2	2	6	2	6	10	2	6	10		2	5						
54	Xenon	Xe	2	2	6	2	6	10	2	6	10		2	6						
55	Caesium	Cs	2	2	6	2	6	10	2	6	10		2	6			1			
56	Barium	Ba	2	2	6	2	6	10	2	6	10		2	6			2			
57	Lanthanum	La	2	2	6	2	6	10	2	6	10		2	6	1		2			
58	Cerium	Ce	2	2	6	2	6	10	2	6	10	2	2	6			2			
59	Praeseodymium	Pr	2	2	6	2	6	10	2	6	10	3	2	6			2			
60	Neodymium	Nd	2	2	6	2	6	10	2	6	10	4	2	6			2			
61	Prometheum	Pm	2	2	6	2	6	10	2	6	10	5	2	6			2			
62	Samarium	Sm	2	2	6	2	6	10	2	6	10	6	2	6			2			
63	Europeum	Eu	2	2	6	2	6	10	2	6	10	7	2	6			2			
64	Gadolinium	Gd	2	2	6	2	6	10	2	6	10	7	2	6	1		2			
65	Terbium	Tb	2	2	6	2	6	10	2	6	10	8	2	6	1		2			
66	Dysprosium	Dy	2	2	6	2	6	10	2	6	10	10	2	6			2			
67	Holmium	Ho	2	2	6	2	6	10	2	6	10	11	2	6			2			
68	Erbium	Er	2	2	6	2	6	10	2	6	10	12	2	6			2			
69	Thulium	Tm	2	2	6	2	6	10	2	6	10	13	2	6			2			
70	Ytterbium	Yb	2	2	6	2	6	10	2	6	10	14	2	6			2			
71	Lutetium	Lu	2	2	6	2	6	10	2	6	10	14	2	6	1		2			
72	Hafnium	Hf	2	2	6	2	6	10	2	6	10	14	2	6	2		2			
73	Tantalum	Ta	2	2	6	2	6	10	2	6	10	14	2	6	3		2			
74	Tungsten	W	2	2	6	2	6	10	2	6	10	14	2	6	4		2			
75	Rhenium	Re	2	2	6	2	6	10	2	6	10	14	2	6	5		2			
76	Osmium	Os	2	2	6	2	6	10	2	6	10	14	2	6	6		2			
77	Iridium	Ir	2	2	6	2	6	10	2	6	10	14	2	6	7		2			
78	Platinum	Pt	2	2	6	2	6	10	2	6	10	14	2	6	8		2			
79	Gold	Au	2	2	6	2	6	10	2	6	10	14	2	6	10		2			
80	Mercury	Hg	2	2	6	2	6	10	2	6	10	14	2	6	10		2			
81	Thallium	Tl	2	2	6	2	6	10	2	6	10	14	2	6	10		2	1		
82	Lead	Pb	2	2	6	2	6	10	2	6	10	14	2	6	10		2	2		
83	Bismuth	Bi	2	2	6	2	6	10	2	6	10	14	2	6	10		2	3		
84	Polonium	Po	2	2	6	2	6	10	2	6	10	14	2	6	10		2	4		
85	Astatine	At	2	2	6	2	6	10	2	6	10	14	2	6	10		2	5		
86	Radon	Rn	2	2	6	2	6	10	2	6	10	14	2	6	10		2	6		
87	Francium	Fr	2	2	6	2	6	10	2	6	10	14	2	6	10		2	6		1
88	Radium	Ra	2	2	6	2	6	10	2	6	10	14	2	6	10		2	6		2
89	Actinium	Ac	2	2	6	2	6	10	2	6	10	14	2	6	10		2	6	1	2
90	Thorium	Th	2	2	6	2	6	10	2	6	10	14	2	6	10		2	6	2	2
91	Protoactinium	Pa	2	2	6	2	6	10	2	6	10	14	2	6	10	2	2	6	1	2
92	Uranium	U	2	2	6	2	6	10	2	6	10	14	2	6	10	3	2	6	1	2

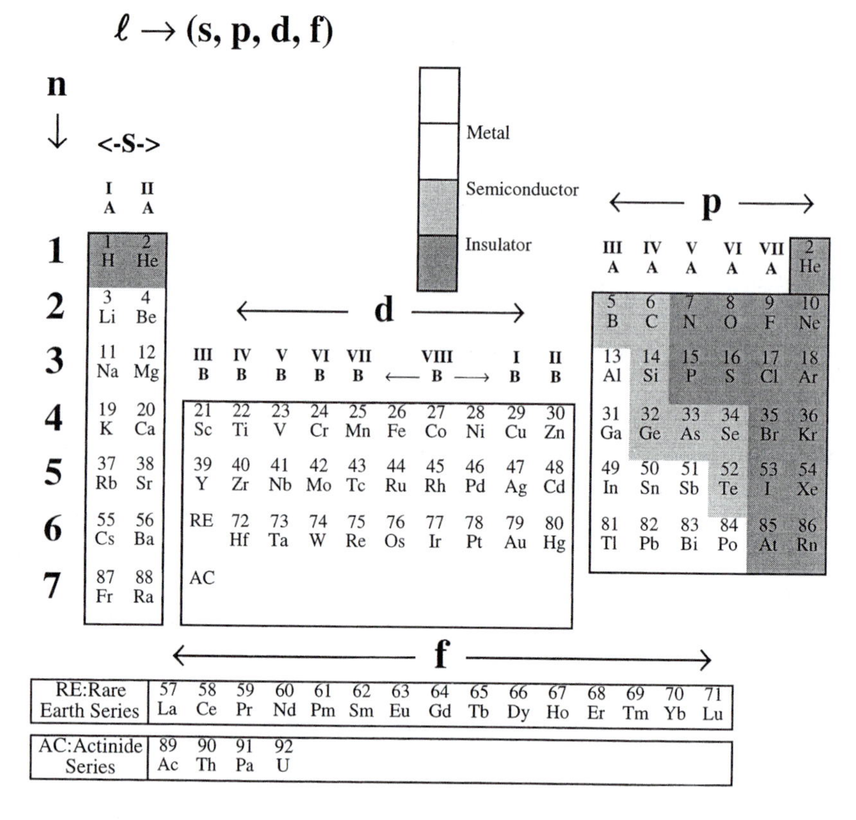

Fig. 1.6 The *periodic table* of the elements is a two-dimensional pictorial representation of all the elements. The rows correspond to the Principal Quantum Number, n, and each of the blocks corresponds to a different value of the Orbital Quantum Number, l. The rows are called *periods* because early chemists realised that the chemical properties of the elements repeated themselves ('periodically'). The columns are called *groups*; their conventional labels (IA, IIA etc.) appear at the tops of the columns. (Notice I have included He (helium) *twice* so that *all* the 'inert gases' appear down the right-hand side of the table.)

1.4 Chemical bonding in solids

First of all, what is a solid? This may seem an obvious question, but we need to have a clear idea of the *result* before it is useful to discuss *how* chemical bonding achieves it. There are various ways in which we can approach such a definition. One way is an *operational* approach. If I give you a single particle of something to hold and ask you whether it is a solid or not, if it runs between your fingers you are liable to tell me it is *not* a solid, but a liquid or heavy gas. This is an operational definition based on *viscosity*: solids are much (much, much ...) more viscous than liquids and so we might choose to define a solid in terms of its viscosity. Here, however, I prefer a *structural* definition, because it fits my immediate purpose better. **A solid is a state of matter where the constituent atoms retain their relative positions and distances for long periods of time**. Now this is not absolutely true, and a lawyer could very soon pick enough holes in this definition to leave it in shreds. For example, we shall see that atoms do move around in solids—and indeed this is quite important as it is one of the important mechanisms for electrical conductivity in insulators (yes, I do mean this!). Also, atoms vibrate about their normal positions, so their distances apart are always varying slightly. The definition above, however, captures the *essence* of what a solid is—atomic movement in the other commonly acknowledged states of matter—liquids and gases—is infinitely faster and more extensive.

We are looking, then, for a strong physical reason why, at lower temperatures, atoms choose to congregate close to each other in relatively permanent arrangements. This physical reason, of course, must incorporate a lowering of the energy of the whole system. When atoms can lower their energy by joining up with other atoms, we say that the atoms *bond* together. A *chemical bond* has formed between the atoms. This is not quite the same thing as saying they form a solid. For example, a finite group of atoms may bond together to form a *molecule* and the molecules themselves may constitute a gas, a liquid, or bond one to another to form a solid. As I explain bonding, I shall also explain how solids are formed.

There are three types of strong bond:

- the *ionic* bond
- the *covalent* bond

and
- the *metallic* bond

There are also a number of weaker bonds. I shall first of all describe and explain the strong bonds.

Strong chemical bonds

1.5 The ionic bond

Consider an element on the left hand side of the periodic table (Fig. 1.6) like sodium, and one on the right, like chlorine. Sodium has an electronic configuration (see Table 1.1):

$$\mathrm{Na} \quad 1s^2 2s^2 2p^6 3s^1$$

while chlorine is:

$$\mathrm{Cl} \quad 1s^2 2s^2 2s^6 3s^2 3p^5$$

Now, if you examine Fig. 1.5 you will notice that filled s and p sub-shells are very stable—there is a considerable gap between the energy of an s level and that of the level below it, belonging to the preceding principal quantum level (where this exists) (e.g. 2p « 3s). (This stability of filled s + p sub-shells explains, ultimately, why the periodic table is constructed around it.)

Thus, for example, the energy required to remove the 3s electron from a sodium atom is 5.14 eV.[†] To remove a second electron (from the 2p shell), however, would require 47.27 eV. This is a very high energy and it is not surprising that sodium can lose one electron easily, but not two. The removal of the second electron requires so much more energy because the full second principal quantum shell has been disturbed. The positively charged atom which results is called an *ion*, specifically a *cation* (*cat* from *cat*hode).

Turning now to chlorine, exactly the opposite situation exists. The chlorine atom is just one electron short of a full third principal quantum s+p sub-shell. The chlorine atom is very happy to receive one electron—in fact there is an energy *drop* of 3.62 eV when an electron is added to a chlorine atom—chlorine is said to have a positive *electron affinity*. The negatively charged chlorine ion is called an *anion* (from *an*ode). The chlorine atom would be far less pleased to receive a *second* electron, since it would have to occupy the much higher 4s energy level.

Putting the two operations together means that to transfer an electron from a sodium atom to a chlorine atom costs overall 5.14 eV–3.62 eV = 1.52 eV per ion pair created. The ion pair is Na^+ (positively charged) and Cl^- (negatively charged).

So what?

Well, so imagine if the ions created were close to one another. There would be an electrostatic attraction between the two ions, whose force $F = e^2/(4\pi\varepsilon_0 r^2)$ where e is the electronic charge, ε_0 is the permittivity of free space and r is the distance between the ions. More importantly, there would be a drop in potential—and total—energy as the ions are (imaginarily) brought together from ± infinity. This drop in energy

$$= \int_{\infty}^{r_0} \frac{e^2}{4\pi\varepsilon_0 r^2}\,dr = \frac{e^2}{4\pi\varepsilon_0 r_0} \qquad (1.1)$$

where r_0 is the final (equilibrium) separation of the ions.

If this drop in energy is greater than the energy expended in creating the pair of ions, then overall we have gained and the total energy of the system will drop. *This is the essence of the ionic bond.*

[†] eV stands for 'electron Volt'. It is a unit of energy whose size is very convenient for atomic phenomena. An eV is the energy an electron would gain via a potential difference of 1 volt. 1 eV = 1.6×10^{-19} J (see Question 1.8 at the end of this chapter.)

In solid NaCl, the ion separation is 0.282 nm. For a pair of ions at this separation,[†] then the energy balance sheet is as follows:[‡]

Paid		**Received**	
To ionise Na → Na+	5.14 eV	When Cl → Cl^-	3.62 eV
		Electrostatic potential energy	5.11 eV
Total paid	5.14 eV	**Total received**	8.73 eV
'Profit'	3.59 eV		

Therefore we 'receive' more than we 'paid' and a stable bond is formed. In more scientific terms, if the total energy falls, a stable bond will form.

So far we have made a *molecule* of sodium chloride, NaCl, consisting of two atoms, or ions. There are two ways in which we can conceive of a solid forming (see Fig. 1.7). In the left hand diagram, a sort of 'supermolecule' is formed, where the 'molecule' actually extends over the whole solid. In this case it is not useful to talk about molecules as such and we simply talk about an *ionic solid*. In the right-hand diagram the molecules of sodium chloride are held together by some new, as yet undefined, bond. This we call a *molecular solid*. For the ionic bond it turns out that the situation depicted in the diagram on the left has a lower energy than that in the diagram on the right, and is preferred. This is because the drop in electrostatic potential energy is even greater for an *array* of ions than for just two ions. For a 3-D array the drop in electrostatic potential energy per ion is written as

$$\left(\frac{e^2}{4\pi\varepsilon_0 r_0}\right)\mathbf{M} \qquad (1.1a)$$

where the expression in brackets is for a single pair of ions. **M** is called the Madelung constant; it describes how efficient a particular structure is at minimising the electrostatic potential energy.

The diagrams above are two dimensional, of course. Figure 1.8 shows the three dimensional structure of sodium chloride. Notice that the structure of this solid is very regular: it is called a *crystal*. Although we think of there being three main states of matter—solid, liquid and gas—in fact there is more than one solid state of matter. The *crystalline state* is one. Later on we will encounter the *glassy state*, which is another.

[†] Really, we should use the separation for an isolated NaCl molecule, but this calculation is just by way of being a demonstration, so we needn't worry too much about this.

[‡] This 'balance sheet' does not include the ion–ion repulsion described later.

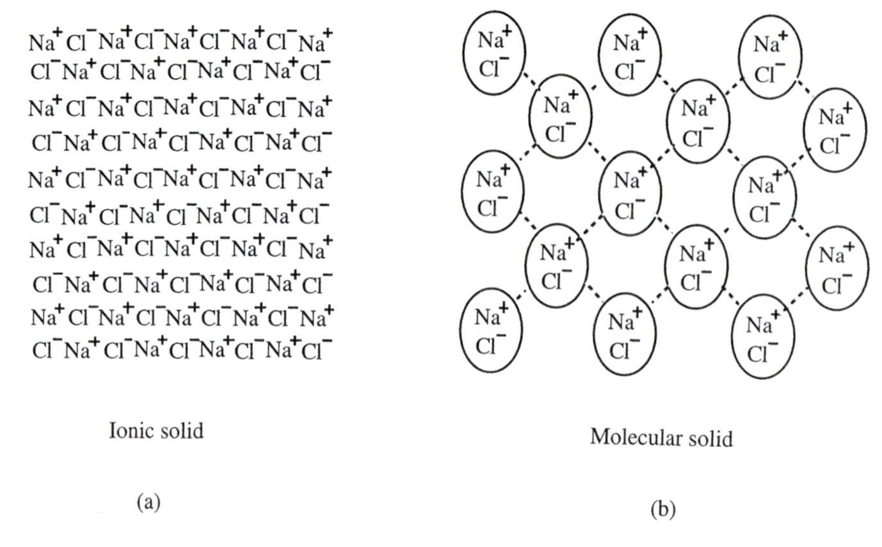

Fig. 1.7 Two possible ways of forming a solid from ionically bonded Na^+Cl^-. On the left (a) a 'supermolecule' forms. This is called an *ionic solid*. On the right (b) *molecules* of NaCl are loosely bound together by some, as yet unspecified, force.

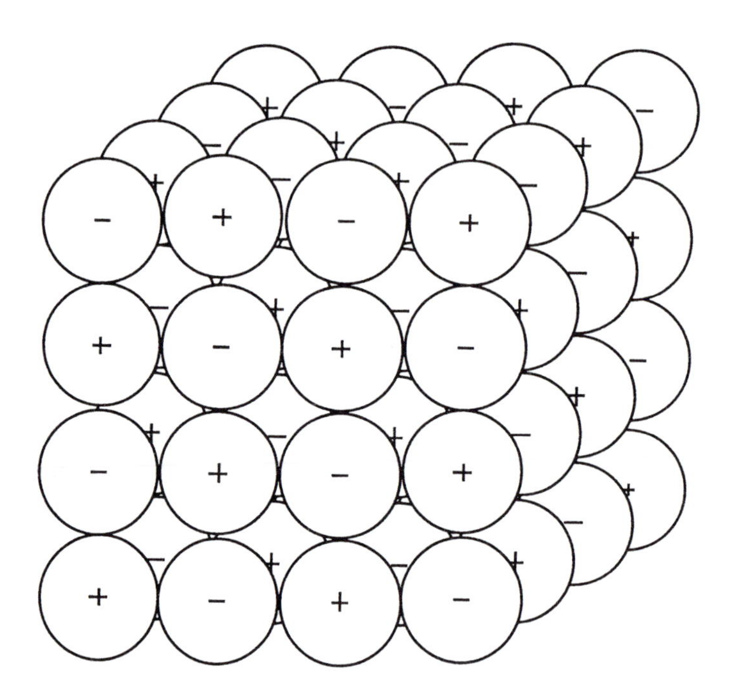

Fig. 1.8 The structure adopted by sodium chloride, NaCl. + = Na^+ and - = Cl^-. This is an ionic solid. I have shown the negative Cl^- anions as having the same size as the positive Na^+ cations. We shall find in Chapter 7 that this is not actually true.

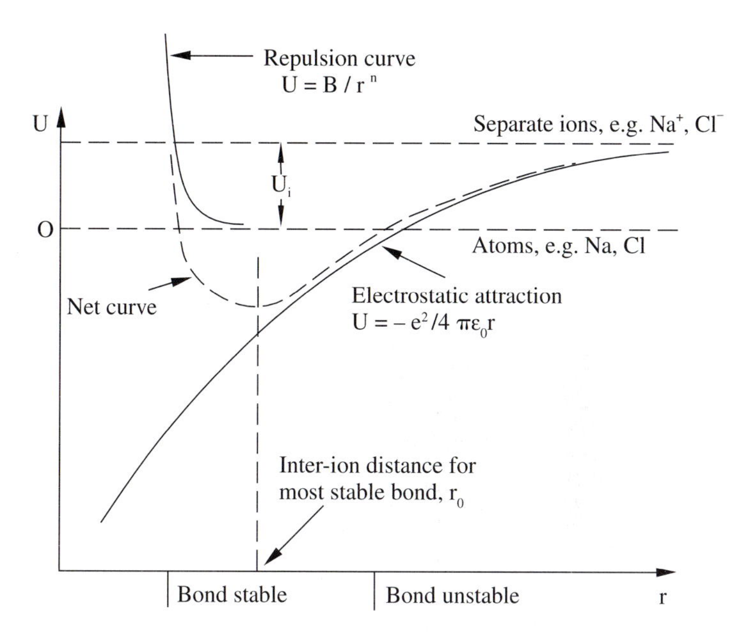

Fig. 1.9 The three energy contributions to a stable ionic solid: ionisation, electrostatic attraction and ion–ion repulsion. (B is a constant.)
(From *Engineering materials 1* by M.F.Ashby and D.R.H. Jones, courtesy of Elsevier Ltd.)

I should add that molecules of sodium chloride *do* exist—for example in the gaseous state. Molecular solids may not be very attractive for ionically bonded solids, but they are for the next type of bond we shall be dealing with—the covalent bond.

One point I have conveniently ignored is what stops the ions approaching closer and closer to each other. After all, the electrostatic attraction always pulls the ions together. The answer is, that when the electron shells about the nuclei start to impinge directly one on another, they repel very strongly. Thus we can plot the total energy of the system as a function of the ion separation, either for a complete solid or for two ions as shown in Fig. 1.9. The final, equilibrium, ion separation is where the total energy is at a minimum, or (the same thing) where the repulsive force due to the impinging electronic shells equals the attractive force due to the oppositely charged ions. This is shown in Figure 1.9 as r_0. From what has been said already, ionic solids will form from combinations of elements on the left hand side of the periodic table (see Fig. 1.6) with those from the right. So NaCl, LiF, KBr, etc. are typical ionic solids. It is also possible to transfer two electrons from one atom to another, and thus we can have calcium oxide, CaO, for example, where Ca loses its two 4s electrons, forming Ca^{2+}, and oxygen gains two electrons, forming O^{2-}. As the number of electrons needing to be transferred goes up, so the energy penalty increases and elements in the middle of the periodic table do not tend to form ionic bonds. Still, it is worth noting that Al_2O_3, alumina or aluminium oxide, where the aluminium atom has lost three electrons, is a good example of an ionic solid. Alumina is a common insulator and electronic substrate which will turn up again in Chapter 7.

To summarise the ionic bond, two elements achieve full s+p sub-shells by a *permanent transfer* of one or more electrons, the resulting electrostatic attraction between the ions holding them together. Ionically bonded solids tend not to be molecular, but to form 'supermolecules'—ionic solids—as shown in Fig. 1.7(a).

1.6 The covalent bond

We now come to the second type of strong bond, the *covalent* bond. In the case of the ionic bond, full s+p sub-shells were attained by a *permanent gift* of electrons. With the covalent bond, full s+p sub-shells are attained by *sharing* electrons.

Let us go back to the chlorine atoms which I used, along with sodium atoms, to illustrate the ionic bond. What if we have just chlorine atoms present? Do they bond together in any way? The electronic structure of chlorine is $1s^2\ 2s^2\ 2p^6\ 3s^2\ 3p^5$. The only way in which the ionic bond could help here would be if we were to strip seven electrons off one chlorine atom and donate them separately to other chlorine atoms, which might then be attracted electrostatically to the Cl^{7+} ion. To take 7 electrons from an atom requires a tremendous amount of energy, much more than is recouped via the electron affinities of the 7 surrounding chlorine atoms and the electrostatic attraction between the Cl^{7+} and Cl^- ions (not to mention the repulsion between the neighbouring Cl^- ions). An ionically bonded Cl_8 molecule is not feasible therefore. What actually happens is as follows.

As two chlorine atoms approach one another, the two incomplete 2p orbitals, one belonging to each atom and each containing one electron only, interact. Physically, the electrons in each orbital start to exchange with each other (very rapidly: $\sim 10^{18}$ times per second). The two separate 2p orbitals together form two combined orbitals which belong to *both* chlorine atoms at once. One of these combined orbitals has a lower energy than the separate 2p orbitals and one has a higher energy. The lower energy combined orbital is called the *bonding orbital* and the higher energy orbital is called the *antibonding orbital*. These are illustrated in Fig. 1.10. Notice that in the bonding orbital the electrons are concentrated between the atoms and act as a 'glue' holding the two ions† together.

Electrons in the antibonding orbital would lie mainly outside the two ions.

As I mentioned earlier, the Pauli exclusion principle applies under all circumstances. Thus, each of the separate unfilled atomic 2p orbitals can accommodate 2 electrons, of

† For such the atoms are if we notionally 'remove' two of the 2p electrons.

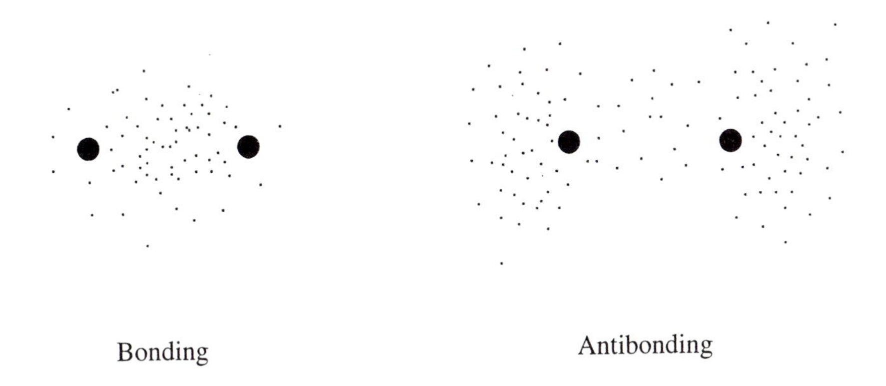

Fig. 1.10 Schematic view of bonding and antibonding orbitals. The density of dots indicates the probability of finding the electrons at any particular place.

opposite spin, and, similarly, each of the combined orbitals can accommodate 2 electrons each, also of opposite spin. The two bonding electrons belonging to the unfilled p orbitals of the isolated chlorine atoms both go into the lower energy bonding orbital. There is a drop in energy and a stable Cl_2 molecule is formed, held together by a covalent bond.

As with the ionic bond, it is the repulsion between the full inner shells of electrons belonging to the two atoms which sets the minimum distance of approach of the two chlorine atoms. In a similar way to chlorine, bromine atoms form Br_2 molecules and, for example, hydrogen atoms form H_2 molecules. In the latter case it is the $1s^2$ shell which is being satisfied—there is no 1p shell. When oxygen atoms form O_2 molecules, two covalent bonds are formed, and chemists write O_2 as O=O. The idea of two, half-filled orbitals combining to form a completely filled bonding orbital and an empty antibonding orbital, and thus a stable bond, leads naturally to the idea of 'sharing' to produce the inert gas structure. Chlorine has only one unfilled orbital and thus forms one covalent bond, which is 'saturated' (filled). We talk of two chlorine atoms 'sharing' two electrons, each thereby achieving a full inert gas structure. Oxygen atoms have two unfilled orbitals and form two covalent bonds: again we can say that the oxygen atoms are 'sharing' four electrons and achieving inert gas electronic structures. The situation for the metallic bond, which I will deal with next, is very different: there the bonds are *unsaturated*.

As we move away from the right-hand side of the periodic table towards its centre, the number of covalent bonds necessary to form a diatomic molecule rises and this situation becomes steadily less attractive. By the time we are across to carbon, where C_2 would require C≡C, there is a much easier, alternative solution. To see what this is, recall that when I was making a solid from ionically bonded atoms (see Fig. 1.7), there were two ways of doing it. Either a completely ionically bonded solid could be formed,

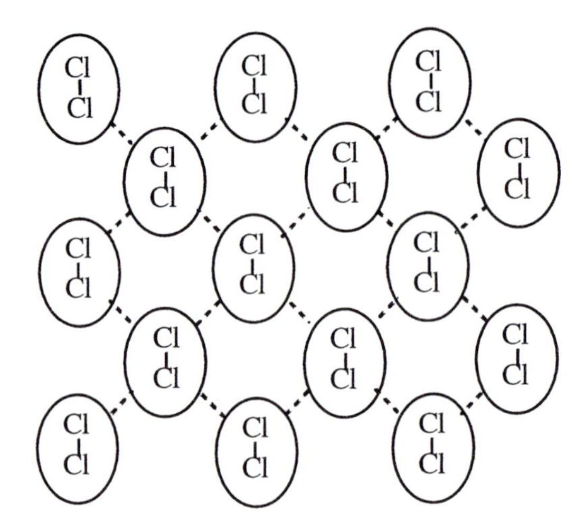

Fig. 1.11 Schematic diagram of solid chlorine (Cl_2). (Don't take the geometrical disposition of the atoms and bonds too seriously.)

where there are no distinct molecules, and the whole solid can be thought of as one 'supermolecule', or separate ionically bonded molecules could be assembled, held together by some as yet unspecified bond. For the ionic bond, it is the first of these two methods which is usually adopted. For the covalent bond, however, both types of solid are found. The second method is adopted by, for example, the diatomic molecules I have just mentioned: Cl_2, Br_2, H_2, O_2 etc.. They are called *molecular solids* (see Fig. 1.11).

The first method solves the problem of how to bond covalently carbon atoms (for example) together. Figure 1.12 shows the structure of one form of carbon: diamond. Each atom is connected by four covalent bonds to four other atoms. Thus every pair of atoms has one covalent bond between them. The covalent bonds are symmetrically disposed in space and they are as far from each other as they can be. If you connected up by lines the neighbours of one carbon atom you would form a tetrahedron and the atoms in diamond are said to be *tetrahedrally coordinated*. This structure is very important in electronic engineering because it is the structure adopted by *silicon* and many other semiconductors. Notice again that there are no *molecules* of carbon in diamond. The whole solid is one giant molecule. Solids which form like this are called *covalent solids*.

Figures 1.11 and 1.12 are both examples of *crystalline* solids in which a small group of atoms repeats itself regularly over all space. If the solid is formed by cooling a liquid rather quickly, then instead of a crystal, a *glass* may form. A glass is a less regular form of solid than is a crystal. Glasses will be discussed in Chapters 7 and 8. Ionically bonded

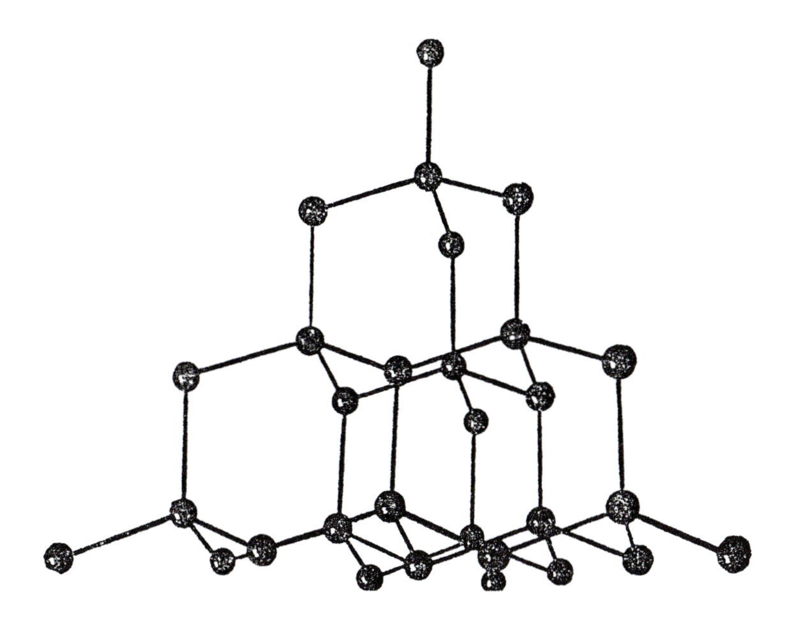

Fig. 1.12 The structure of diamond, a covalent solid. Each carbon atom is bonded to four other carbon atoms via four equally disposed covalent bonds. There are no molecules: the whole solid is a 'supermolecule'. (From *Atomic theory for students of metallurgy* by W. Hume-Rothery, courtesy of the Institute of Materials.)

solids may also form as glasses, although they are less likely to do so than covalently bonded solids. Metals (Section 1.7) rarely form glasses.

Although so far I have talked about covalent bonds between like atoms, as in an element, covalent bonds can also form between unlike atoms. For example, in water, H_2O, the bonds holding the hydrogen atoms to the oxygen atom are covalent bonds. When covalent bonds form between unlike atoms, however, the two atoms concerned will have different attractions for the electrons in the bonding orbital. They have different electron affinities, or, as we usually say in this situation, different *electronegativities*. Thus the electrons tend to move to one or other end of the bond, which is then said to show *ionic character*, or in other words the bond is a mixture of the two types of bond, ionic and covalent, which I have introduced so far. Such *polar* bonds will turn out to be quite important when we come on to using plastics in AC fields (Chapter 8).

Finally, I must end this section with a caveat. I have discussed covalent bonding in terms of completing s^2p^6 sub-shells. You will notice from Fig. 1.5 that as we go up the energy levels in the atom, there begins to be considerable overlap between the energy levels—e.g. 4s and 3d. Thus for heavier atoms, lower down the periodic table (Fig. 1.6), the details of covalent bonding—how many bonds are formed, and so on—can become quite complicated. The principle, however, remains just the same: two orbitals overlap, a bonding orbital is formed, and a shell or sub-shell is completed.

Now for the third and last type of strong bond, the *metallic* bond.

1.7 The metallic bond

So far we have considered the bonding between sodium and chlorine atoms, which gave us the ionic bond, and that between chlorine atoms, which introduced us to the covalent bond. Let us finish by taking sodium, an element at the left hand end of the periodic table, and considering what type of bond might hold *its* atoms together.

The sodium atom has one electron more than the stable $1s^2\ 2s^2\ 2p^6$ configuration and therefore forms ionic bonds very readily, as we have seen. Following an argument similar to that used in discussing chlorine, the only way one can imagine sodium atoms bonding to one another ionically is if one sodium atom were to receive one electron each from seven surrounding sodium atoms, forming $(Na^+)_7\ Na^{7-}$. As with chlorine, it would require a tremendous amount of energy to persuade a neutral sodium atom to accept seven extra electrons and so, what with this and the mutual repulsion of the seven Na^+ ions, sodium does not bond ionically with itself.

What about covalently bonded sodium? Two sodium atoms can bond covalently to form Na_2, which gives both atoms a 'full' 3s orbital (at least in the shared sense). This is good so far as it goes, but the Na atom really would like a full $s^2\ p^6$ outer shell. Bringing another sodium atom up provides one more electron, but the first two atoms have already used up their electrons on the first covalent bond. What if we form a sort of covalent bond that encompasses all three atoms? Again there are 'bonding' and 'antibonding' orbitals; the three electrons go into the lower energy 'bonding' type orbitals and are now free to visit any of the three atoms. As we add neighbouring atoms the 'bond' contains more and more electrons which are free to visit more and more atoms. We are moving, in fact, from unsaturated, covalent bonds towards a metallic bond. At some stage the *insulator → metal transition*, sometimes called (rather more excitingly) the *insulator → metal catastrophe*, takes place. We can look at this in two ways.

Each of the bonding electrons is relatively free and can move in such a way as to *shield* the other conduction electrons from the static ion cores. This concept of shielding is familiar to electrical engineers—think of using conductors to shield electromagnetic radiation. When the density of bonding electrons becomes high enough, the bonding electrons, because of their effect on each other, lose contact with the individual ion cores and are free to roam through the whole of the piece of metal. The approximate mathematical condition for this is

$$N_c^{\frac{1}{3}} r_H \approx 0.25 \tag{1.2}$$

where N_c is the number of bonding electrons per unit volume and r_H is the radius of the 'orbit' they would describe in an isolated atom.

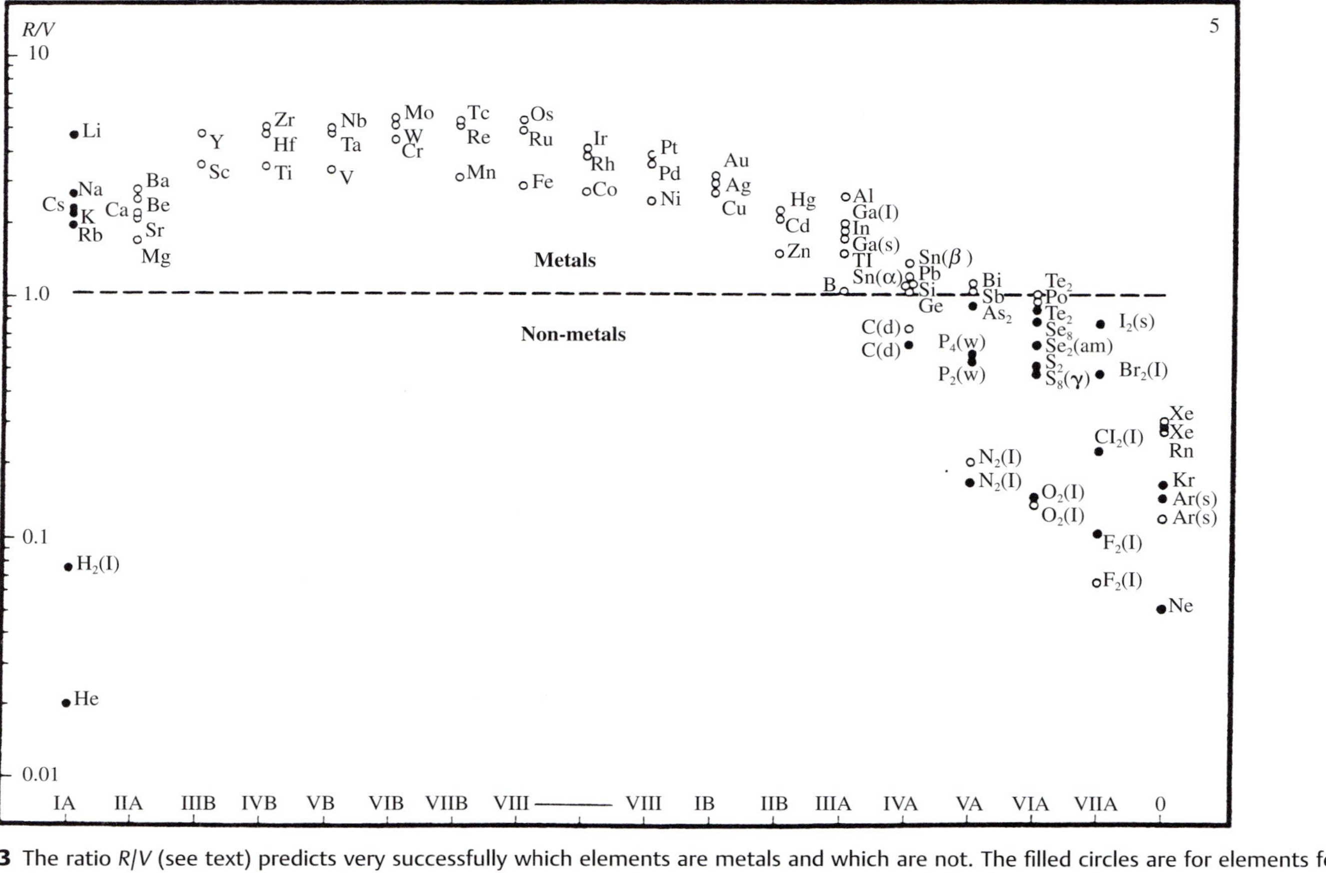

Fig. 1.13 The ratio R/V (see text) predicts very successfully which elements are metals and which are not. The filled circles are for elements for which R and V are both known. The open circles are for elements for which R has had to be calculated. (Taken from P.P. Edwards and M.J. Sienko (1982). *Acc. Chem. Res.* **15**, 87.)

A different, but equivalent, approach involves *polarisability*. Polarisability measures the extent to which electrons separate from their 'home base' positive ions when an electric field is applied, producing an electrical *dipole*. It is well known in the theory of insulators that the neighbouring dipoles amplify the effect of the externally applied electric field. (We shall come back to this sort of effect in Section III on insulators.) As the density of neighbouring dipoles increases, there comes a point where the 'springs' which hold the electrons to their ions are swamped by the effect of the neighbouring ion–electron pairs, and the electrons become free. This effect is described mathematically by

$$R > V \tag{1.3}$$

where R is the *molar refractivity* for the isolated atoms (i.e. their effect on light; molar means 'for a mole, or Avogadro's Number, of atoms'; Avogadro's Number is 6×10^{23}) and V is the *molar volume*. Equation (1.3) can be shown to be roughly equivalent to eqn (1.2). Figure 1.13 shows how well these simple criteria predict which elements are metals and which are not.

Whichever way we look at it, when the density of loosely bound electrons becomes high enough, the electrons become free and form a *metallic bond*. It is now known that *anything* can be made into a metal. All that needs to be done is to compress the material until N_c (eqn (1.2)) is large enough, or until V (eqn (1.3)) is small enough; the insulator–metal catastrophe will duly take place and a metal will form. Thus, for example, the hydrogen which forms the bulk of the large planet Jupiter is *metallic* hydrogen, because of the enormous pressure it is under.

As in the covalent bond, in the metallic bond the bonding electrons are shared and they glue together the positively charged ions. Unlike the covalent bond, however, in the metallic bond the ions are under no illusion that they have achieved the nirvana of a full s+p shell via the bonding electrons. On the contrary, the metal ions shed electrons until they reach the *preceding* inert gas structure, as they would do in an ionic bond. It is therefore the elements on the left hand side of the periodic table which will tend most easily to form metallic and ionic bonds. If you look at Fig. 1.6, where metals and non- metals are identified, it is also evident that *heavier* elements have a greater tendency to be metallic. This is because the greater number of inner electrons, which form complete shells and sub-shells, shield the outer electrons from the nucleus and make it easier for them to leave the atom and contribute to a metallic bond. Thus in Fig. 1.6 the line which separates metals from insulators slopes down and to the right.

Although I have discussed *elemental* metals, mixtures of elements also form metals. Sometimes the different atoms are all jumbled up, forming a *disordered metallic alloy*. Sometimes they form an ordered array and an *intermetallic compound* (which is still a metal) is formed.

Notice that I have not mentioned molecules at all. The metallic bond can hold liquids together as well as solids (not gases), but when a solid does form it is in the nature of the metallic bond that all the atoms contribute to the whole solid: there is no question of being able to identify a molecule, other than a supermolecule comprehending all the atoms in the solid, as in diamond.

The two important facets of the metallic bond are therefore

- free, or 'delocalised' electrons
- maximum number of atoms (or free electrons) per unit volume

These two simple facts control all the properties of metals with which we are so familiar, as we shall see in Chapter 2.

It is time now to summarise strong bonds.

1.8 Strong bonds: summary

Type of bond	Method of bonding		Type of solid
Ionic	Gift	→	Ionic solid
Covalent	Sharing	→	Covalent solid
		→	Molecular solid
Metallic	Sharing + delocalisation	→	Metal

1.9 Other bonds (weak bonds)

Apart from the three types of strong bond which I have just described, there are other, weaker, bonds. Two of these which are important for electrical engineering materials are the *hydrogen bond* and the *van der Waals bond*. Both are important in holding together the molecules in a molecular solid (see above). First of all, the hydrogen bond.

The hydrogen bond

I mentioned when talking about covalent bonds between unlike atoms that the electrons are always attracted more to one or the other of the atoms—the one which has the greater electronegativity. Thus the bond has some ionic character. Imagine molecules bonded together like this forming a molecular solid. The slightly positive end of one molecule will be attracted electrostatically to the other, slightly negative end of another molecule. This attraction, shown here by the dotted line, is the hydrogen bond. It is called the hydrogen bond because it occurs most strongly between covalently bonded molecules containing hydrogen bonded to nitrogen,

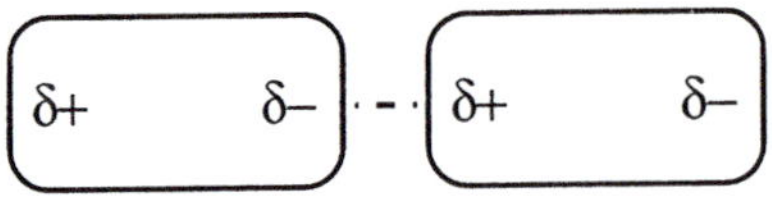

Fig. 1.14 The hydrogen bond: a weak electrostatic attraction.

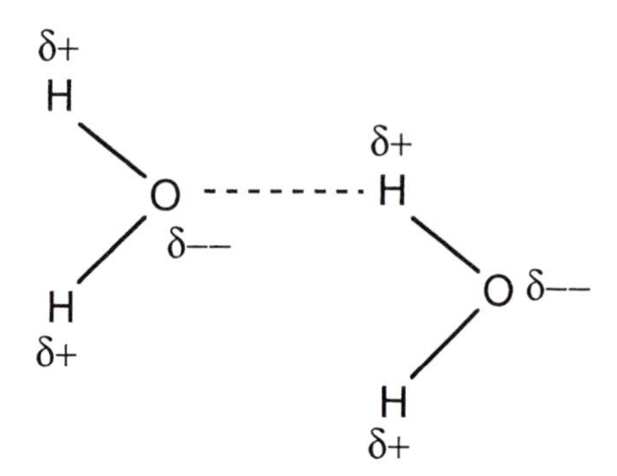

Fig. 1.15 Hydrogen bonding in ice and water. The extra stability conferred by the hydrogen bonding makes ice melt and water boil at relatively high temperatures and helps reduce the density of both: unusually, solid ice is less dense than liquid water, which is bad news for water pipes in cold climates!

oxygen or fluorine. For example in solid water (ice) the electronegative oxygen attracts the electrons of the covalent bond away from the hydrogen atoms, leaving an exposed positively charged nucleus. This 'hydrogen bonds' with a negatively charged oxygen atom belonging to another molecule. This type of bonding is why, for example, water freezes and, especially, boils at much higher temperatures than oxygen or hydrogen. Hydrogen bonds are important in polymers used as electrical insulators.

When I talk about 'weak' bonds, how weak is 'weak'? There is a tremendous range in strengths of strong bonds, but roughly speaking if we measure the strength of a bond by the energy needed to break it, a strong bond requires $\sim$1 to 10 eV, whereas a hydrogen bond requires 0.5 eV or less.

Van der Waals bonding

The other type of weak bond which is important to us as engineers is the van der Waals bond. This also relies on electrostatic attraction, but arises in a different way. If we imagine any isolated atom, *on average* the electron charge will be distributed evenly about the nucleus, giving a spherically symmetric cloud. At any *instant*, however, the

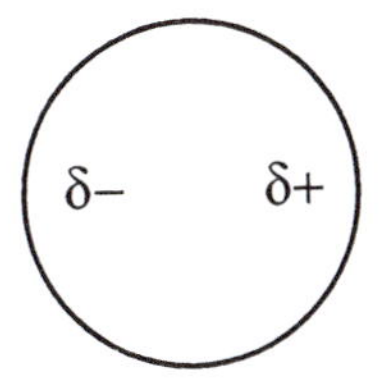

Fig. 1.16 A chance, momentary polarisation of a neutral atom.

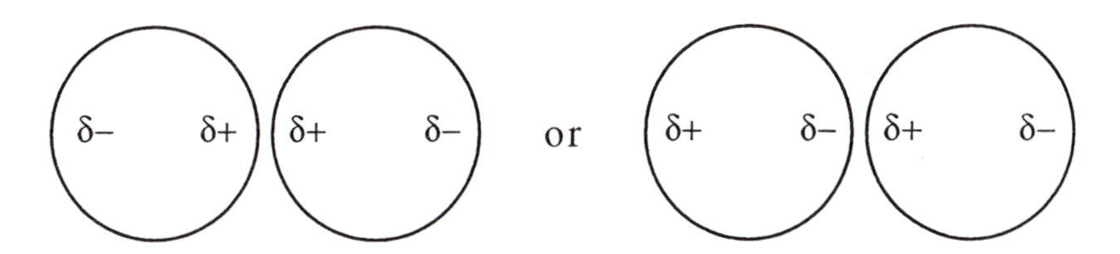

Fig. 1.17 Neutral molecules suffer momentary spontaneous polarisations. The arrangement on the right has lower energy than that on the left because unlike charges are next to each other.

electrons may by chance be at one side of the atom. Over time, of course, these variations average out.

Now imagine two atoms next to one another. The configuration on the right in Fig. 1.17 has a lower energy than that on the left. Thus the time variations in the electron charge distributions of two neighbouring atoms tend to correlate and, because the right-hand distribution has a lower energy, it occurs more frequently and there is thus a net attraction between the two (remember electrically neutral) atoms. This is the *van der Waals* bond. It always exists between any two atoms, but because it is so weak (typically 0.1 eV/bond) if any other form of bonding is present it is swamped. Van der Waals bonding is what holds molecular solids like oxygen, nitrogen and hydrogen together. Their very low melting and boiling points show how weak the bonding is.

Thus the molecules in a molecular solid are held together by weak bonds: the very weak van der Waals bond or the somewhat stronger hydrogen bond. This weak bonding is what confers some flexibility and resilience on plastics and is one of the reasons they are so popular as insulators (see Chapter 8) as compared with brittle ceramics.

1.10 Conductors, insulators and semiconductors

So far I have been describing the different sorts of bonds which hold solids together: now I would like to classify solids according to their electrical properties:

There are three types of strong bond:	**ionic** **covalent** **metallic**	and four types of solid:	**ionic** **molecular** **covalent** **metallic**

In a metallically bonded solid the bonding electrons are delocalised, that is each electron can (and does) move throughout the solid. What is more, if an electrical potential is applied across the solid, there is nothing to stop the bonding electrons moving in response to the electrical field. This is because the covalent bonds which went to make up the metallic bond were unsaturated. Metals are therefore good conductors of electricity. If we now move to a molecular solid, however, the situation is totally different. The electrons within each molecule fill completely a stable sub-shell (typically s^2 p^6) or at least the bonding orbitals derived therefrom. To rip electrons away from a stable situation like this and send them hurtling through what is effectively free space requires an enormous amount of energy and an enormous potential. Such conduction as does occur depends critically on impurities and defects in the solid. We shall return to the subject of *dielectric breakdown* in Section III (Chapters 7 and 8).

In ionic solids and covalent solids also there is no obvious easy mechanism for the long range movement of electrons. The sub-shells are full, by whatever method. Although we can imagine in a covalently bonded solid neighbouring bonds from the same atom exchanging electrons, this does not correspond to overall transport of electricity, as is required for electrical conduction (see Fig. 1.18).

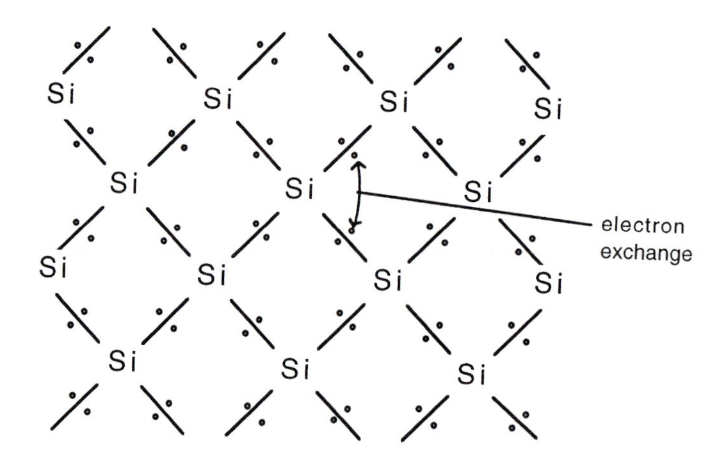

Fig. 1.18 In a covalent solid, although electrons can exchange between neighbouring bonds, this does not permit long range movement and electrical conduction. This is a schematic 2-D picture of the silicon structure. Silicon has the same structure as diamond (see Fig. 1.12).

Again, in real situations, conductivity depends very much on purity and perfection. If an atom is substituted by one with a different electronic structure, or is merely absent altogether, there is scope for limited movement of electrons through the structure. I shall return to this point very shortly.

So, metals are good conductors of electricity and everything else is an electrical insulator. The difference in the property of electrical conduction between the two types of solid is of cosmic proportions. A typical metal will have an electrical conductivity of the order of 10^7 $(\Omega m)^{-1}$ whereas a typical insulator will have a conductivity around 10^{-13} $(\Omega m)^{-1}$—a difference of 20 orders of magnitude.

There is a halfway house, though. In a covalent solid, the antibonding orbitals form an empty network stretching through the solid. If they have an energy not too far above that of the bonding orbitals, it will not cost very much energy to promote an electron into them. Once an electron gets into this network, it becomes free. (See Fig. 1.19.)

The many atomic orbitals contributing to the conducting orbitals mean that these consist of many individual levels (just as *two* overlapping orbitals yield one bonding and one antibonding orbital). These are very, very closely spaced and are referred to *en bloc* as the *conduction band*. The energy necessary to promote an electron from the bonding or valence level into the conduction band is called the *band gap*. How small the band gap is will dictate how many electrons chance, for example by thermal excitation, to find themselves in the conduction band rather than the *valence band* (the bound levels corresponding to the bonding orbitals). This will control the conductivity. (Note that this is different from breakdown: the electrons are not being torn from the atoms—those that happen to be free are being exploited.) A typical band gap for a

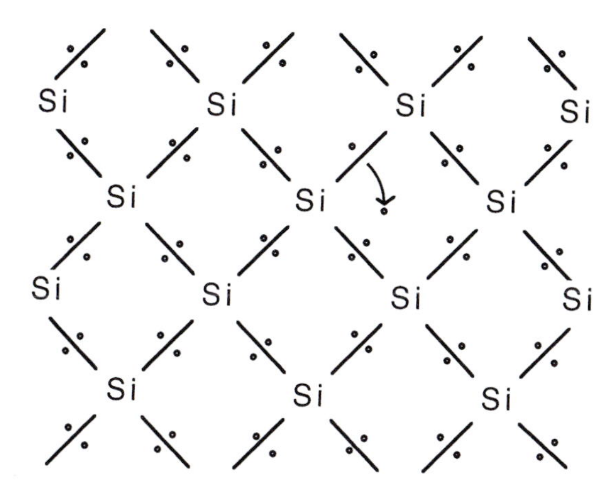

Fig. 1.19 An electron is freed from a covalent bond in silicon and can now move freely through the solid and provide electrical conductivity.

LIVERPOOL
JOHN MOORES UNIVERSITY
AVRIL ROBARTS LRC
TEL. 0151

semiconductor is about 1 eV[†]. The amount of thermal energy available is $\frac{3}{2}kT$, where k is Boltzmann's constant and T is the absolute temperature. kT is ~0.025 eV at room temperature and so ~0.04 eV is available (on average). Together, these two parameters (the band gap and kT) correspond to an intrinsic electrical conductivity for pure silicon with no impurities or structural defects of approximately 1 $(\Omega\text{m})^{-1}$. When electrons are promoted into the conduction band, this leaves *holes* in the valence band and makes possible another method of conduction, since now electrons can move from one bond to a neighbouring bond and thus provide conduction in an electric field. We can draw this as follows:

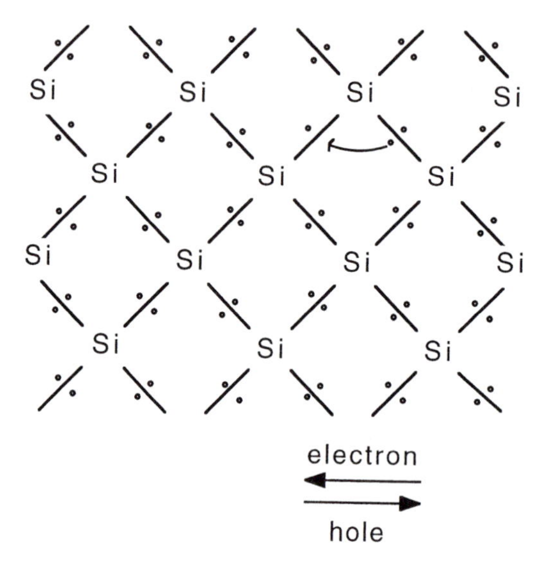

Fig. 1.20 Not only can the promoted electron provide electrical conductivity, so can the space, or 'hole', left behind in the covalent bond. An electron can now jump into this from a neighbouring bond. And so on. The hole migrates in the opposite direction to the flow of electrons. Although the electrons carry the electric charge, it is the hole which localises the conduction event — the jump. Thus we talk about 'conduction by holes'.

An electron jumps into a hole from a neighbouring bond. The electron moves one way and the hole moves the other. Although it is the electrons which physically carry the charge, it is the holes which localise the conduction event i.e. the jump. Thus we talk about 'conduction by holes'. The holes are effectively positively charged. This process is not necessarily so easy as electrons moving in the conduction band and we talk of the *mobilities* of electrons and holes being different.

[†] For comparison a good insulator will have a band gap of several eV.

This is *intrinsic* conduction. As implied above, conduction can be manipulated *extrinsically* by *doping* with impurities. Consider what happens if an atom possessing five outer electrons (such as arsenic, As) substitutes for a silicon atom:

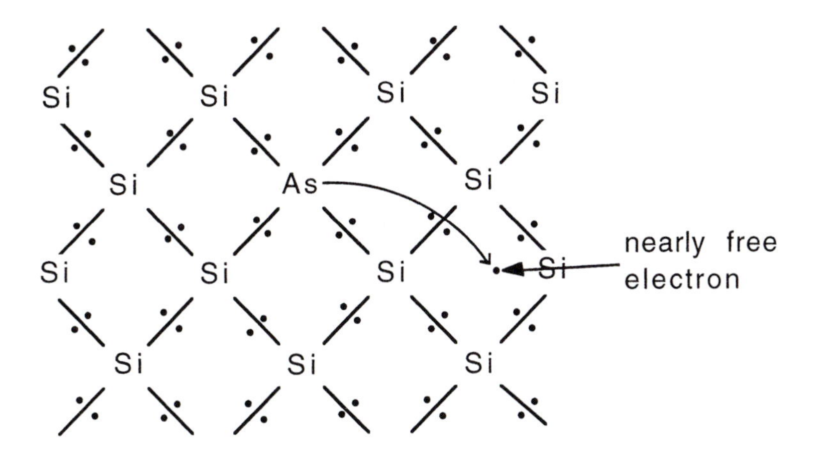

Fig. 1.21 An arsenic atom, possessing five outer electrons, substitutes for a silicon atom and as a result donates a nearly free electron to the silicon.

The arsenic atom forms the four tetrahedral covalent bonds, satisfying its immediate neighbours, but has one electron left over. This is only loosely bound to its quite well screened positively charged parent arsenic atom. The binding energy is ~0.05 eV. At sufficiently low temperatures this means the electron performs a large orbit about 2.5 nm in radius. At room temperature, the thermal energy available (average ~0.04 eV) is enough to detach frequently the electron and promote it to the conduction band. The arsenic is said to contribute a *shallow donor level*. A different way of presenting *some* of this information is the **band diagram** (see Fig. 1.22).

In view of the discussion above of the metallic bond it should come as no surprise that if the number of doping atoms is increased until the large nearly free electron orbits overlap, the silicon is turned extrinsically into a metal. Referring back to eqns (1.2) and (1.3), we have increased N_c by heavier doping, rather than by applying pressure, although here the free electrons are not the bonding electrons—the bonding is still provided by the saturated covalent bonds.

In terms of the periodic table (Fig. 1.6) and as far as elements are concerned, as we move from right to left in the table, or down the columns, we move from insulators to metals, for reasons explained earlier. In between the two regions there is a band of semiconducting elements which includes the most commercially important semiconductor, silicon. Some elements can take either form under quite normal

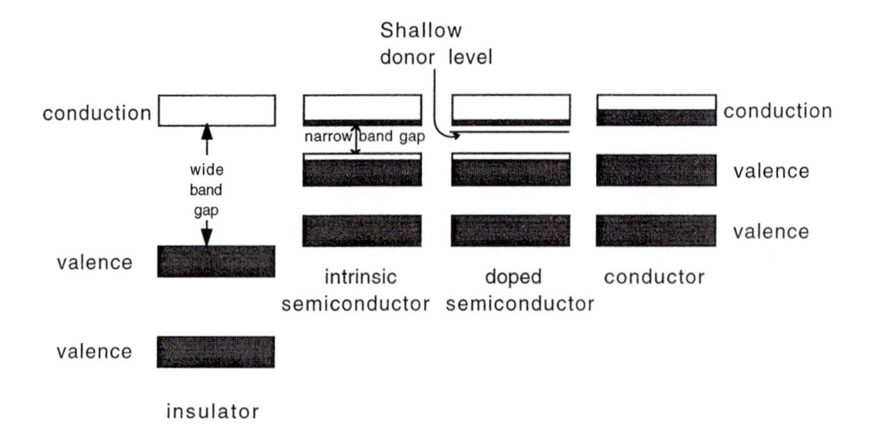

Fig. 1.22 The band diagram. Be aware that this diagram does NOT tell you some things — like where the electrons are and how many doping atoms there are. Nor does it tell you that a full band implies no electrical conduction: you have to *know* this, or work it out.

conditions—for example tin is a semiconductor at low temperature and a metal at normal temperaratures, as Napoleon's soldiers found in Russia.†

Chemical compounds as well as elements can be semiconductors, as we shall see in Chapter 9.

1.11 Summary

1. The atom consists of a small dense nucleus surrounded by a cloud of electrons.
2. The electrons move in a series of quantised orbitals with well-defined energies.
3. The periodic table presents the electronic structures of the elements in a form which illustrates their chemical and electrical properties.
4. There are three types of strong bond:
   - the ionic bond where a permanent gift of electrons enables the two participating atoms to achieve an inert gas structure;
   - the covalent bond where sharing of electrons achieves an inert gas structure;
   - the metallic bond, an unsaturated covalent bond where the bonding electrons are free and hold together the remaining ion cores, which have an inert gas structure.

† The tin buttons on their tunics changed from relatively ductile metal to brittle semiconducting ceramic and so turned to powder in the exceptionally cold Russian winter.

5. Ionic and covalent bonding make electrical insulators; metallic bonding makes conductors. If the band gap is small there is semiconduction, which may be intrinsic or extrinsic.

Recommended reading

The contents of this chapter generally appear as Chapter 1 or 2 of any general materials science text. I suggest one of the following:

Materials science and engineering by W.D. Callister (published by Wiley)

Engineering materials science by M. Ohring (published by Academic Press)

Introduction to materials science for engineers by J.L. Shackelford (published by Prentice Hall)

Principles of materials science and engineering by W.F. Smith (published by McGraw-Hill)

As the titles suggest, quite similar: lots of nice pictures.

Materials science (4th edn.) by J.C. Anderson, K.D. Leaver, R.D. Rawlings and J.M. Alexander (published by Chapman and Hall)

Popular with our students.

Modern physical metallurgy and materials engineering (6th edn.) by R.E. Smallman and R.J. Bishop (published by Butterworth)

The great virtue of this book is that everything is in there, somewhere.

An introduction to metallurgy by A.H. Cottrell (published by Institute of Materials)

If you really want to understand something.

Atomic theory for students of metallurgy by W. Hume-Rothery (published by Institute of Materials)

I used to hate this book as a student: now I love it. (Perhaps it's an old man's book.)

Questions (Answers on pp. 296–301)

1.1 Are the following solid materials electrical conductors, semiconductors or insulators? (a) wood, (b) steel, (c) stone, (d) silicon, (e) copper, (f) polythene, (g) gold, (h) porcelain, (i) salt, (j) aluminium, (k) glass, (l) silver, (m) polystyrene.

N.B. You will need to do some detective work for Questions 2, 3 and 4

1.2 Which is smaller, the ratio of the radius of the sun to that of the solar system, or the ratio of the nucleus of a hydrogen atom to its atomic radius?

1.3 Put the following elements in order of their constitution (by weight) of the earth's crust: copper; oxygen; aluminium; silver; hydrogen.

1.4 The isotopes of iron, their relative atomic masses and their relative abundances in the earth's crust are as follows:

	RAM	Relative abundance
Fe^{54}	53.94	5.80%
Fe^{56}	55.93	91.72%
Fe^{57}	56.94	2.20%
Fe^{58}	57.93	0.28%

What is the terrestrial relative atomic mass of iron? Would you expect this to be the same on Mars? In the Andromeda Galaxy?

1.5 Using the rules in Section 1.3, work out the maximum number of electrons which could occupy the third principal quantum shell in an element.

1.6 Using the rules in Section 1.3, derive the formula $2n^2$ for the maximum number of electrons in the nth principal shell.

1.7 Does the 3d level have a higher energy than the 4s level in sodium? In silver? Are the 3d and 4s levels occupied in these two cases?

1.8 Sometimes energies are written as eV atom^{-1} and sometimes as kJ mole^{-1}, where mole $\equiv$ Avogadro's number of atoms. Work out the relationship between eV atom^{-1} and kJ mole^{-1}. [The charge on an electron $e = 1.602 \times 10^{-19}$C; Avogadro's number, $N_A = 6.022 \times 10^{23}$ atoms mole^{-1}.]

1.9 Write down the electronic structures of potassium (K) and bromine (Br) in solid potassium bromide (KBr), of Al and O in Al_2O_3 (alumina) and of Ca and O in CaO (calcium oxide, or 'quicklime').

1.10 If the first ionisation energy of K is 4.34 eV and the electron affinity of Br is 3.36 eV, how much energy does it require to form a K^+Br^- ion pair? If the final equilibrium separation d_0 of the two ions is 0.334 nm, and the repulsive force between the two ions follows the law $\frac{B}{r^{10}}$, estimate the sublimation energy per ion pair. (The sublimation energy is the energy to convert something from solid to gas. The permittivity of free space, ε_0, is 8.854×10^{-12} Fm^{-1}.)

1.11 Work out the Madelung constant for the following linear chain of ions:

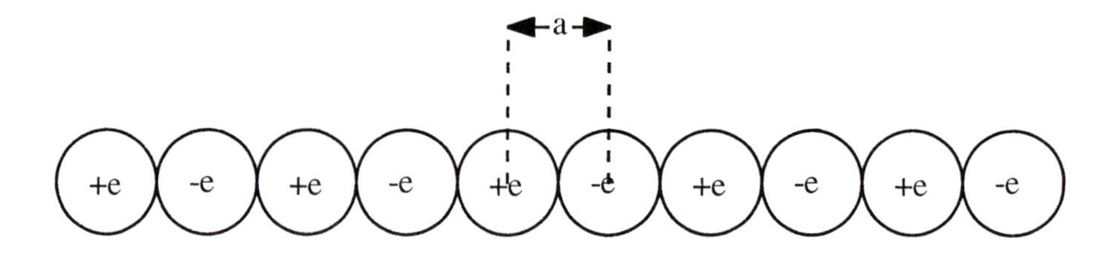

1.12 If the density of solid copper is 8960 kgm^{-3} and its relative atomic mass is 63.55, what are the weight and volume of a mole of copper atoms in the solid state?

1.13 The latent heat of copper (the energy required to make it melt) is 13 kJ $mole^{-1}$. How far would 13 kJ raise 1 mole of copper in the earth's gravitational field? (See also preceding question.) (The acceleration due to gravity, g, is 9.807 ms^{-2}.)

1.14 Nitrogen forms covalently bonded N_2 molecules in the gaseous, liquid and solid states. How many electrons are shared between each bound pair of nitrogen atoms?

1.15 What is the electronic structure of an isolated germanium atom? Solid germanium adopts the same structure as silicon. How many bonding electrons does each atom contribute? What is the electronic structure of a germanium atom (a) disregarding all the bonding electrons and (b) assuming all the bonding electrons seen by an atom are possessed entirely by that atom?

1.16 How many conduction electrons per atom would you expect the following metal elements to have?: (a) potassium (K), (b) magnesium (Mg), (c) copper (Cu), (d) aluminium (Al).

Part II **Electrical conductors**

Electrical and electronic engineering consists either in *encouraging* electrons to move, or in *preventing* them from doing so. Thus we have *electrical conductors* and *electrical insulators*. The semiconductors of microelectronics in fact achieve *both* functions in a very subtle way and on a very fine scale. In this section, Section II, we will be looking at electrical conductors, where we *want* the electrons to move. In Section III I will turn to *electrical insulators* and then in Section IV to *semiconductors*.

We are now, therefore, entering the world of *metals*...

2

An introduction to metals

Chapter objectives

- What makes a metal?
- The crystal structures of metals
- The microstructure of a metal: crystals, dislocations and vacancies
- The conduction of electricity through metals
- Which metals are used in Electrical Engineering?

2.1 What is a metal?

In Chapter 1 I concluded that the two important things about metals are that:

1. **The bonding electrons are free**: this is why metals conduct electricity
2. **The number of bonding electrons per unit volume must be high**: this is what makes *any* given material into a metal, although some find it easier than others—e.g. the elements on the left of the periodic table.

Now this is approaching metals from the scientific point of view. Let us switch to the other end of the scale and ask the question 'What is a metal?' from the common sense, everyday (engineering) point of view. Then, later in this chapter, I will tie the two approaches together.

If I hand to you a lump of something and ask you whether it is a metal, what do you do and think?

- Firstly, you will probably ask yourself: does this feel cold? Metals feel cold; insulators do not. This is because metals are *good conductors of heat*.

- Is the lump of material heavy? Metals tend to be *dense*.

- You flick the piece of material using your thumb and third finger. If it 'pings' when your fingernail strikes it, it is more likely to be a metal. If there is a dull thump, it is probably an insulator. Metals show *sonority*.

- If you are allowed the run of a very simple laboratory you may scratch or polish the material. If it is opaque and shiny, it is probably a metal. Metals show good *reflectivity*.

- You put the lump in a vice and squeeze. If it changes shape, it is a metal. If it shatters it is not. Metals show *plasticity*.

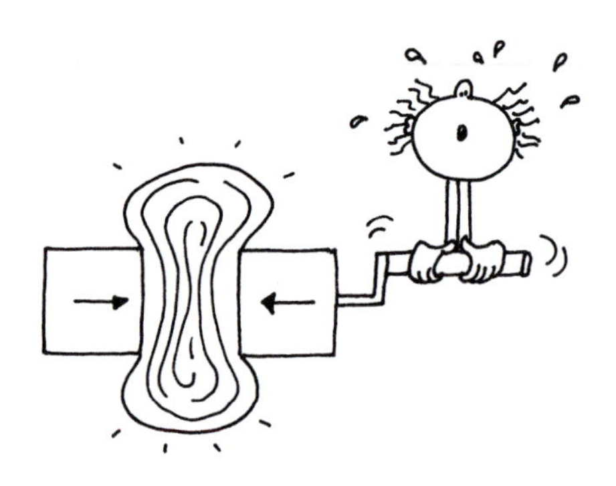

- Finally, but most importantly for us, you attach a battery and an ammeter. The rest, I think, you know! Metals conduct electricity.

Let me collect all this together. In general:

Metals:	Insulators:
conduct electricity well	do not
conduct heat well	do not
absorb and reflect light	do not
show sonority	do not
are heavy	are not
are plastic	are not

Now why, why, why? Why do the two properties at the beginning of this chapter—free electrons and lots of them per unit volume—result in all these differences? I will answer these questions very shortly, but in order to do so, I need to explain a little bit about the *structure* of metals. In fact, I need to explain about the *microstructure* of metals.

2.2 How metals are constructed: microstructure and crystals

Most solids, when they are cut up, look fairly *homogeneous*: that is, you can't see anything very interesting. It is a fact that what controls most properties of a solid takes place on a level which we cannot see with the naked eye. When you think that an atom is only 2×10^{-10} m across, maybe this is not too surprising.

The structure of a solid from the point at which our eyes give out—say 0.1 mm—down to the atomic level, is called its *microstructure*. The microstructure therefore extends from 10^{-10} m to 10^{-4} m.

The easiest way to tell you about microstructure is by way of an example. Copper is the most common electrical conductor we shall come across. Let us take a piece of copper, cut it in half and then look at its structure at increasing magnifications. We are going on a voyage of discovery, which will take us from the everyday to the world of the atom.

Stage 1

We slice a piece of copper, so that we know the surface is clean and truly representative of the inside of the copper. We polish the copper to form a flat, smooth surface and, as I said above, we can see absolutely nothing (except a smooth, reddish reflection)! So,

Stage 2

We put our piece of copper under an optical microscope (Fig. 2.1). Because copper is opaque to light, we look *at* the copper, rather than through it (as a biologist would do). Still we can see nothing! Is the magnification sufficient? We turn up the magnification until we see detail of the order of the wavelength of the light—$\sim \frac{1}{2} \mu\text{m}$ (0.5×10^{-6} m)—this is the best we can do with the light microscope—still ... nothing. Is the copper really homogeneous on this level? The answer is 'no, it certainly is not'. In polishing the specimen to make it flat and suitable for a microscope, we have removed all the interest. So now:

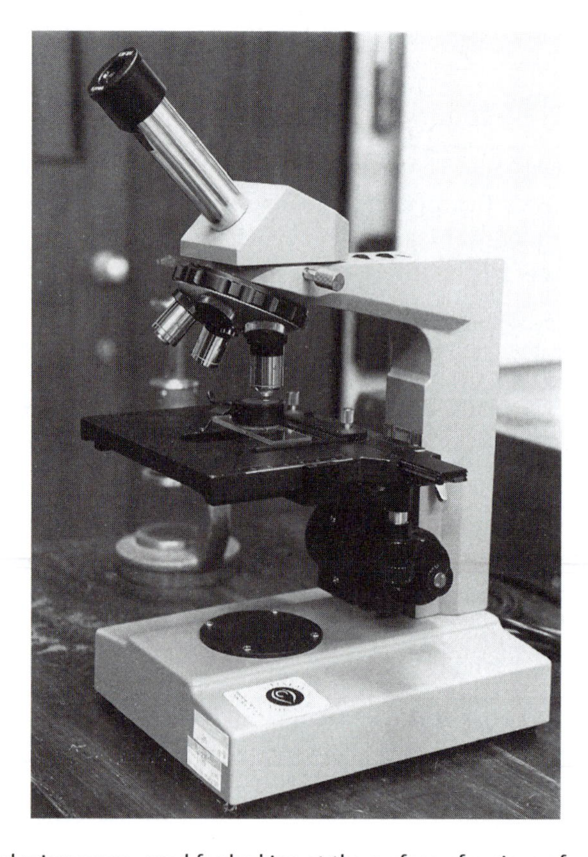

Fig. 2.1 An optical microscope, used for looking at the surface of a piece of metal.

Stage 3

we *etch* the specimen—that is we dip it in a chemical solution which attacks the microstructure of the copper and allows it to be seen. Suddenly our specimen comes to life and we can see the following:

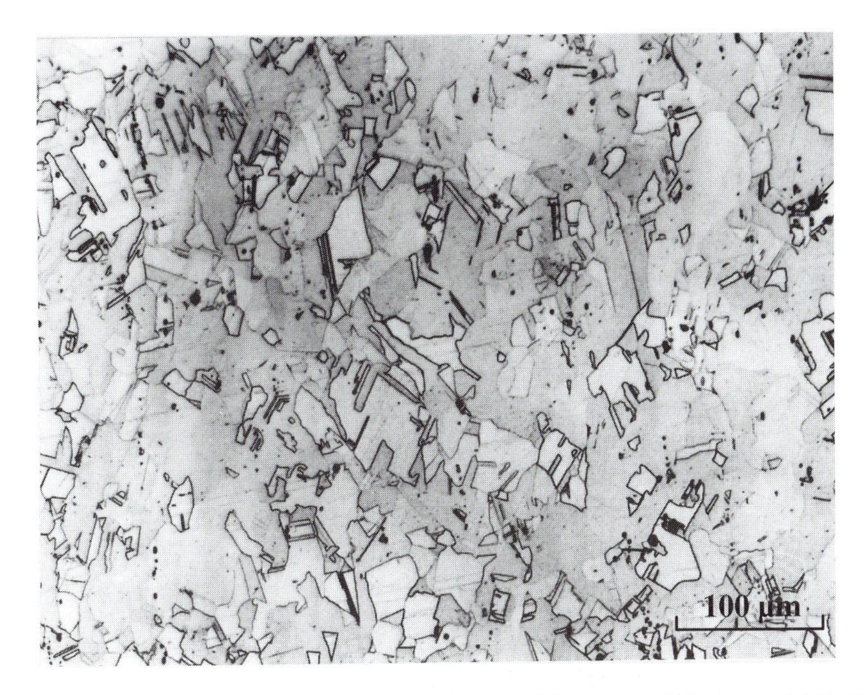

Fig. 2.2 Copper photographed in an optical microscope. (This is an *optical micrograph*.) The line on the photograph has a true length of 100 μm, or $\frac{1}{10}$th of a mm. This is about the smallest size your eye can resolve. The little shapes, surrounded by lines, are crystals, or *grains*, of copper. The lines are *grain boundaries*. These etch more quickly in the chemical solution and that is why we can see them. (The original was rather a voluptuous reddish-orange colour, but unfortunately colour costs money and even the splendid Oxford University Press has to make a profit!)

The copper consists of a network of lines, a little like looking down on England from an aeroplane—but not so regular as English fields and hedgerows. Each of the enclosed irregular shapes is a crystal of copper—very small, only 0.1 mm across in this case, but nevertheless a crystal, just as you have seen large crystals in geological museums, or perhaps as grains of salt, or sugar, or diamonds ... In metal science—metallurgy—these crystals are called *grains* and the boundaries between them are called *grain boundaries*. I shall come back to *what sort* of crystals copper forms later on. All metals† are crystalline.

Sometimes the crystals are large, like those in this piece of copper; sometimes much

† *Nearly* all metals—see Chapter 11.

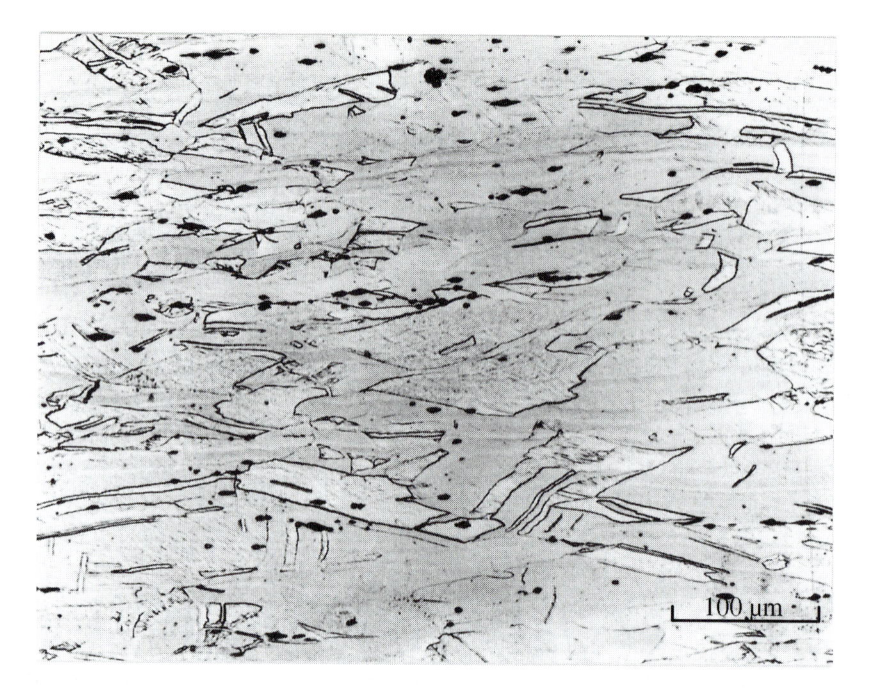

Fig. 2.3 An optical micrograph of a piece of copper which has been rolled. The thickness of the piece was reduced by one half in the rolling process. Notice how the grains of copper have been squashed out.

smaller. It all depends on how the metal is treated, or *processed*: grain size—and shape—are something an engineer can control. For example, Fig. 2.3 shows a photograph of a piece of copper which has been *rolled*. You can see that the grains have been squashed out.

In copper, there is little more to be seen by looking at the *surface* of the metal, even if we change to another type of microscope which will allow us to see down to the scale of atoms. (Later on we will have reason to do just this on a different type of specimen.) Further on in this chapter we will resume our voyage of discovery by going *inside* the piece of copper. For the moment, however, I would like to look in a little more detail at the *crystals* which make up the copper and other metals.

2.3 What are metal crystals like?

A crystal is something completely regular. The same small group of atoms repeats itself throughout space with the same spacing and with the same orientation. What sort of crystals do metals form? To answer this I would like to consider a metal *element*—where all the atoms are of the same type. Each atom is therefore *spherical* (averaged over time) and of the same size.

Recall that the essence of metallic bonding is

- delocalised bonding electrons,
- maximum number of bonding electrons per unit volume.

Every metal atom has one, two or three bonding electrons to contribute, depending on its position in the periodic table (see Chapter 1, Fig. 1.6). For example, copper has one. To maximise the number of bonding electrons per unit volume we therefore need to maximise the number of atoms per unit volume. How, then, can we pack the maximum number of equally sized spheres into the smallest volume?

Let us attack this problem in two stages. First of all, in two dimensions. The best *packing density* we can achieve in two dimensions is a hexagonal array, as shown in Fig. 2.4(a). This will come as no surprise to the snooker players amongst you. We must now stack these rafts of atoms together to make a three dimensional crystal. Remembering that we want to maximise the number of atoms per unit volume, each atom will fit into the dip formed by the three atoms below it. The important thing to notice is that there are twice as many dips or *interstices* in each raft as there are atoms in each raft itself. Thus if I label one set of interstices by the letter 'B' and the other set 'C' (see Fig. 2.5), *either* the 'B' positions can be filled, or the 'C', but not both. (I have reserved the letter 'A' for the centres of the original atoms.) Now it doesn't matter which set we choose: either set gives the best packing of atoms which we can achieve. For the sake of argument, let's say that we fill the 'B' interstices, so that so far our packing sequence reads

AB...

This is shown in Fig. 2.6(a). What happens next? There are two sets of interstices in the second raft of atoms, which must be filled by the third layer. One set is the 'C' interstices which were not filled when I chose to fill the 'B' interstices. The other set is what we have labelled as 'A'—directly over the original atoms. To convince youself of this, look at the schematic diagram in Fig. 2.6(b). And so the same atom positions keep recurring. An 'A' layer is followed by a 'B' or 'C' layer. A 'B' layer is followed by an 'A' or 'C' layer and it is easy to convince yourself, either with a model or a drawing, that a 'C' layer is followed by an 'A' or a 'B'.

The rule is that we can have any sequence of A, B and C, *provided we never repeat a letter*, because this would spoil the efficiency of our sphere packing. All sequences which obey this rule are equally efficient as far as achieving the maximum number of atoms per unit volume is concerned. So which sequence is actually adopted? This is where *second order* effects come in. The free electrons which glue the atoms together in the metallic bond travel in all directions through the structure. These electrons are

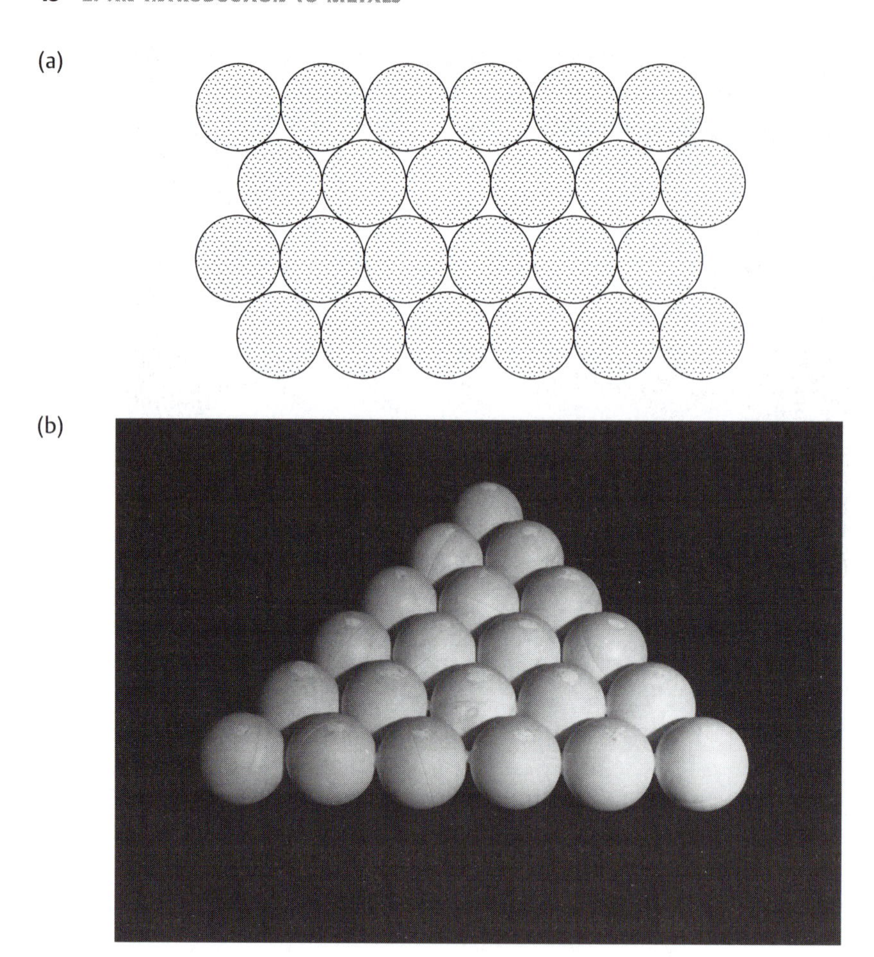

Fig. 2.4 Close-packed spheres (atoms) forming a hexagonal array: (a) schematic diagram, (b) photograph of model using table tennis balls for atoms.

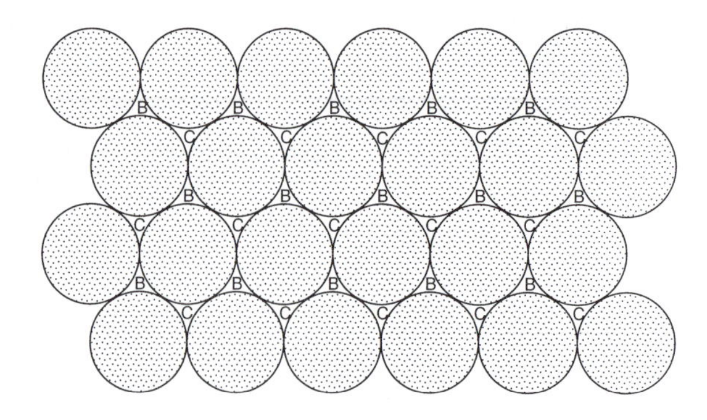

Fig. 2.5 Each layer of spheres has *two* possible positions for the next layer to rest in.

(a)

(b)

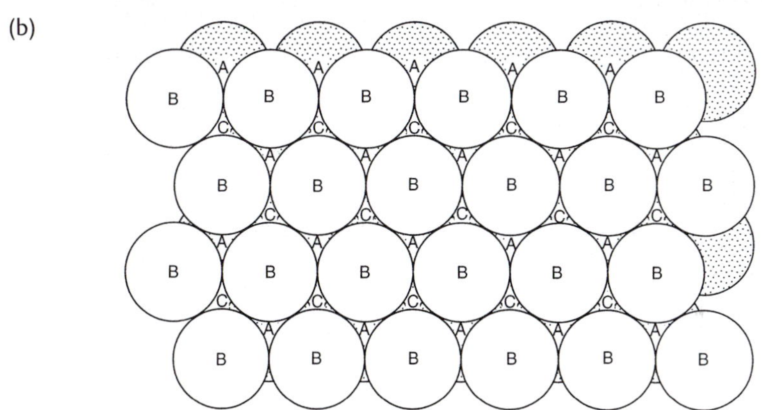

Fig. 2.6 (a) Two planes of 'atoms' fitted together to form a close packed array. (b) Schematic of two layers of atoms stacked one above the other. Note that the same interstices and thus atom positions repeat again and again.

scattered by the atoms in the structure as they pass. They much prefer a simple array of atoms to a complicated one and as a consequence have a lower energy in a simple structure. Thus the two sequences which are actually found in nature are:

ABABAB...
and
ABCABC...

The first sequence gives a hexagonal structured crystal which reflects the symmetry of the individual layers of atoms. Because of the high packing density of atoms it is called *close packed hexagonal*, c.p.h. for short. Many common metals, such as zinc (Zn), cadmium (Cd), magnesium (Mg) and titanium (Ti) adopt this structure.

The other structure, ABCABC . . ., actually gives a *cubic* structure, that is, one based on a cubic arrangement of atoms. Now, this is not at all obvious, but if you look at a model of the structure (Fig. 2.7), you can see the square array of atoms typical of a cubic arrangement turning up on the faces.

Fig. 2.7 When the hexagonal layers stack together in the ABCABC . . . sequence they form a *cubic* arrangement. Can you see the square on the exposed facet?

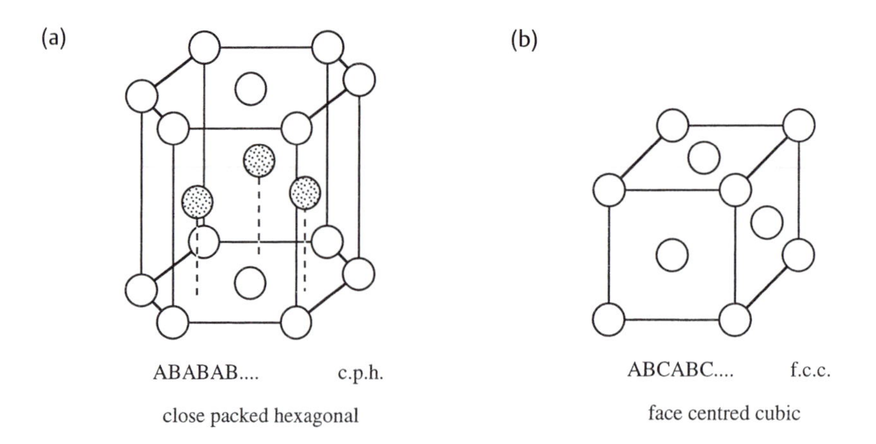

Fig. 2.8 The *unit cells* for (a) c.p.h. and (b) f.c.c.

If we break a crystal down and down and down into its smallest components, we arrive at the smallest group of atoms which has the symmetry of the crystal itself. This is called the *unit cell*. The unit cells for our two structures are shown in Fig. 2.8.

The ABCABC ... structure has a cubic unit cell with atoms at the centres of the faces: it is therefore called *face centred cubic* (f.c.c.). The connection between the c.p.h. unit cell and the ABABAB ... sequence is obvious; less so for the f.c.c. unit cell. An f.c.c. cell with the ABC ... layers highlighted is illustrated schematically in Fig. 2.9.

Like h.c.p., f.c.c. is also a popular choice of structure for metals. All the commonly used electrical conductors are f.c.c. elements, for reasons which I shall explain shortly: copper (Cu), gold (Au), silver (Ag), aluminium (Al) and many more without significant electrical applications such as nickel (Ni), platinum (Pt), rhodium (Rh) etc.

Incidentally, fruiterers obviously would make good materials scientists (see Fig. 2.10).

There is a third common metallic crystal structure which is not quite so closely packed as the two examples I have just given you. It is called *body centred cubic* (b.c.c.) and is adopted by most of the remaining common metals: iron (Fe), chromium (Cr), tungsten (W), molybdenum (Mo) and sodium (Na). The unit cell is cubic, like f.c.c., but the extra atom is at the centre of the cube, hence 'body centred cubic'. It is illustrated in Fig. 2.11.

Table 2.1 gives examples of the three common metal element crystal structures (at room temperature) plus a few exceptions.

In c.p.h. crystals the bonding is not *perfectly* metallic and a little covalent bonding can cause the structure to expand or contract perpendicular to the A and B planes of atoms. Thus in real life c.p.h. crystals are not perfectly close packed. This cannot happen to f.c.c. crystals, because the symmetry would be destroyed (see Fig. 2.8). So f.c.c. metals are the only perfectly close packed ones. In fact f.c.c. metals are the perfect metals. All the properties I listed at the beginning of this chapter as characterising a metal, f.c.c. metals have to the greatest degree. They are ductile

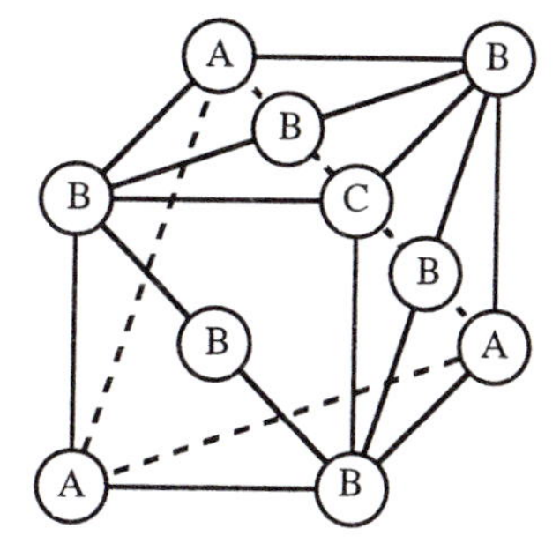

Fig. 2.9 How the face centred *cubic* (f.c.c.) structure arises from *hexagonal* arrays of atoms. Each letter refers to a separate close packed plane. The magical thing is that we can do this starting from *any* of the eight corners. The f.c.c. structure is very symmetrical and beautiful!

Fig. 2.10 Some close packed apples. What I have deduced heuristically here has just been proved mathematically in a mere 250 pages by Professor Thomas Hales. 'Kepler's stacking problem', as it was known, remained unsolved for 300 years since it was first posed by Sir Walter Raleigh in connection with the stacking of cannon balls (of course). (Mathematician Thomas Harriot passed it on to Johannes Kepler—hence the name.) I hope you will be satisfied with my rather more innocent approach!

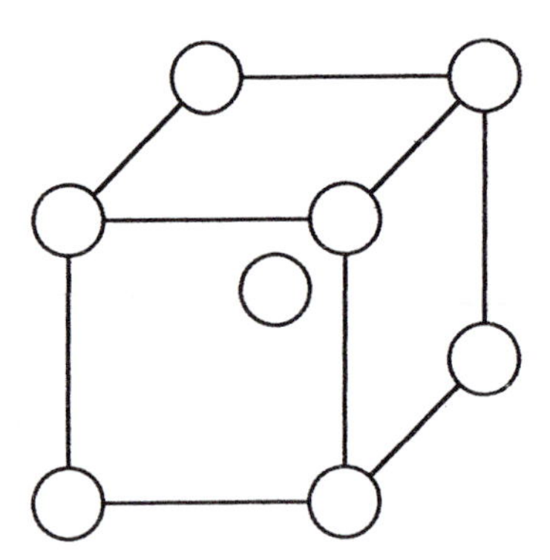

Fig. 2.11 The body-centred cubic (b.c.c.) unit cell. The extra atom is in the *centre* of the cube (compare with Fig 2.8(b)).

Table 2.1 Crystal structures at room temperature of some common metal elements. (See Table 1.1 if you don't know some of the chemical symbols.)

c.p.h.	f.c.c.	b.c.c.	others
Mg	Cu	Fe	Hg*
Zn	Ag	Cr	Mn
Cd	Au	W	U
Ti	Al	Mo	Bi
Zr	Ni	Na	Gd
Be	Pt	Ta	Nd

*At low temperature: Hg is liquid at room temperature.

and malleable, they have the best thermal conductivity and most importantly of all for us, they are the best electrical conductors. Thus the conductors we come across most in electrical engineering are f.c.c. metals: copper, silver, gold and aluminium.

2.4 Defects in crystals

So far we know that metal elements consist of small crystals, or grains, joined together at grain boundaries. Each little grain has a crystal structure which is f.c.c., c.p.h. or b.c.c.. These crystals are not perfect, however. Just as it is the defects in each of us which make us interesting as people, so it is the defects in crystals which control many of their properties!

Do you remember the piece of copper we were examining? We got as far as looking at the surface. Now, what about the *inside* of the copper?

Stage 4

Metals are opaque to light (all those free electrons) and to look inside a piece of metal we have to use *electrons*. These are the negatively charged particles which are a fundamental part of all matter, as I described in Chapter 1, and which carry the electrical current when a voltage is applied to a metal. To free electrons from a metal and use them independently, the metal is heated to a high temperature and a strong electric field applied to it. The electrons can then be used in a *transmission electron microscope* which is illustrated in Fig. 2.12.

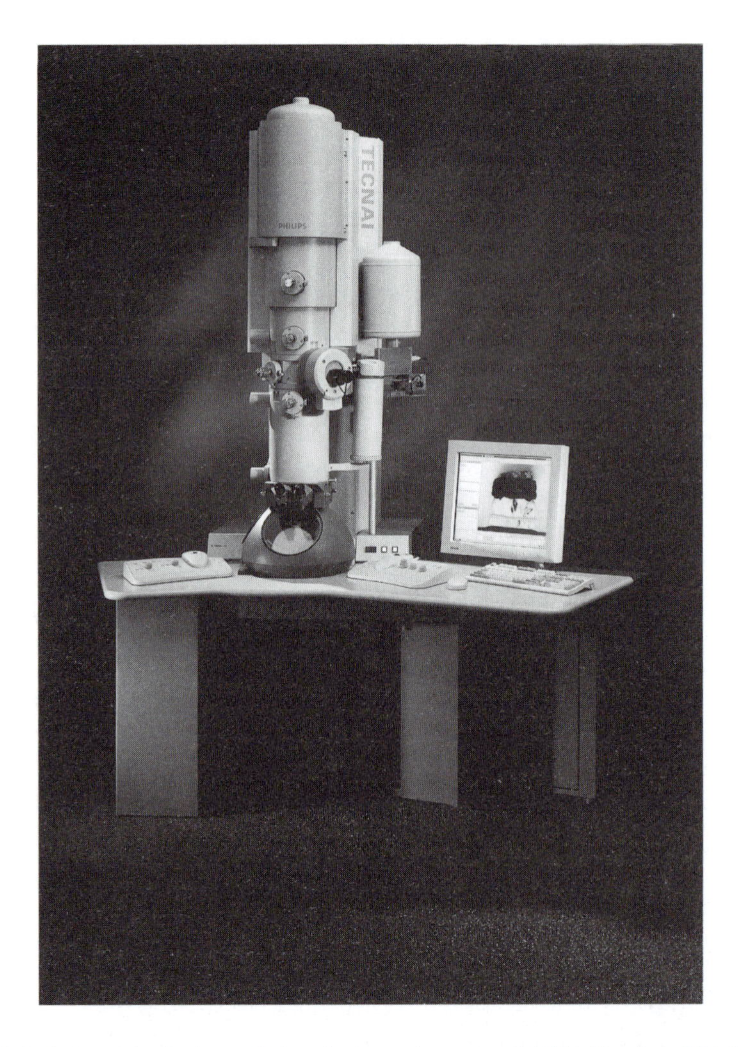

Fig. 2.12 A photograph of a transmission electron microscope (T.E.M.) (Courtesy F.E.I. Inc.) The source of electrons—the *gun*—is at the top of the microscope. The thin specimen goes on the end of a rod whose black outer end appears halfway down the microscope. The picture may be viewed through the glass at the bottom of the microscope or on the big screen, as here. The 'lenses' are actually circular magnetic coils and the whole microscope has to operate under vacuum. The little Dewar halfway down the column, on the right, is to do with analysing the X-rays excited from the specimen by the electron beam. This allows chemical analysis nearly on an atomic scale.

The electrons have a wavelength λ associated with them,[†] where

$$\lambda = \frac{h}{mv}$$

[†] This is a quantum mechanical property of *all* particles, not just electrons.

h is Planck's constant ($= 6.6 \times 10^{-27}$ J s), m is the mass of the electrons ($= 9.1 \times 10^{-31}$ kg) and v is the electrons' velocity. $\frac{1}{2}mv^2 = eE$ where E is the electric field applied, in volts. A typical value of E is 200 kV, so $\lambda = 2.7$ pm.† Remembering that resolution is limited by wavelength, we clearly have a very powerful microscope indeed!‡ Electrons penetrate all forms of matter (to a small extent), and we can therefore use our electron microscope to look *through* the copper.

What do we see?

At low magnification we see the same pattern of grains and grain boundaries as we have already seen in the light microscope (Fig. 2.13). Because electrons do not penetrate matter very easily, we use a very thin specimen—typically 10^{-7} m in a transmission electron microscope. Since this is much less than the grain size, the pattern is essentially two dimensional, like the optical pictures, although with the electrons we are looking *through* the specimen.

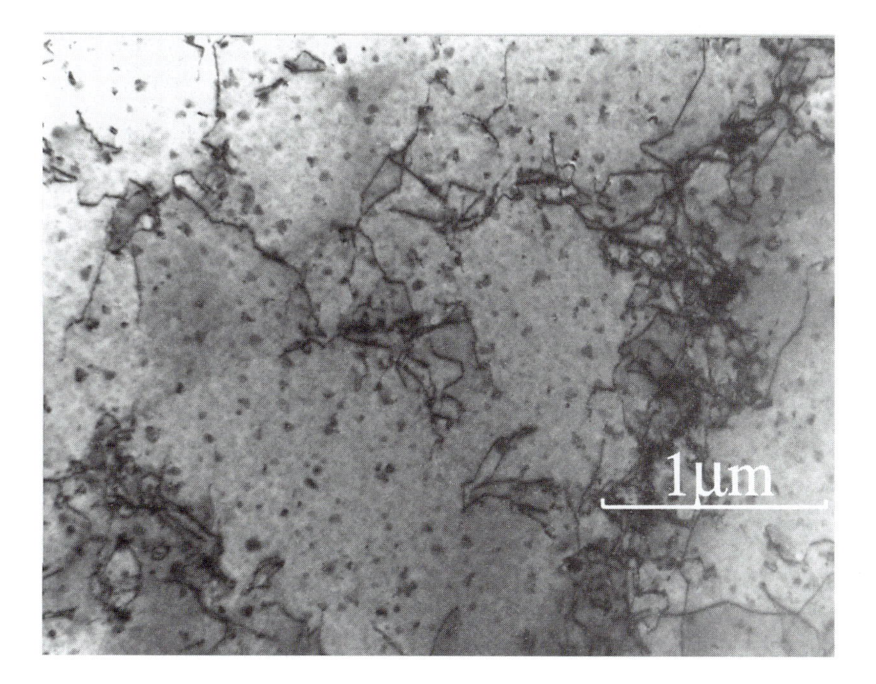

Fig. 2.13 How copper looks in a *transmission electron microscope*. The copper has been *annealed*—heated to remove damage from the lattice. The wiggly black lines are called *dislocations*. (Courtesy of Mr A.J. Burbery.)

† A more accurate *relativistic* calculation gives λ as 2.5 pm.

‡ In fact electron lenses are nothing like as efficient as light lenses, so the resolution is not quite so wonderful as λ suggests.

Fig. 2.14 Rolled copper in the T.E.M. Now the grains are squashed flat and there are masses of dislocations. The dislocations are what accomplish the plastic deformation (see text). (Courtesy of Mr A.J. Burbery)

Let us raise the magnification by another hundred times. What we see is illustrated in Figures 2.13 and 2.14.

In both pictures you can see a series of fine black lines. These are called *dislocations*. They are *line defects* in the copper crystals that make up our piece of copper. *Line* defect means one dimensional, like a piece of string, to distinguish them from *area* or surface defects or *volume* defects. Notice that there are far fewer dislocations in Fig. 2.13, which is of soft 'annealed' copper (annealing is heating gently), than in Fig. 2.14, which is of rolled copper. Dislocations are very much connected with the deformation of metals. They are VERY important for a metal's *mechanical* properties, but not very important for its *electrical* properties. I shall explain the structure of dislocations in Chapter 3.

Although we can use an electron microscope to resolve down to the level of atoms in a specimen, we wouldn't actually see any more than we have already in this type of specimen. Sometimes, like for semiconductors, it *is* useful to image directly the columns of atoms.

2.5 Stage 5: point defects

There is, finally, one type of defect which is very important to electrical properties but which we cannot routinely image with any type of microscope. This is called a *point defect*, because it exists at one atomic site (i.e. at a point). There is only one type of point defect intrinsic to metals: it is called a vacancy and is where an atom is missing from its site in the crystal.

The number of vacancies is very sensitive to temperature. If the energy to form one vacancy is E_f^v then the concentration of vacancies c_v (i.e. the fraction of atoms missing from their sites in the crystal) is

$$c_v = e^{-\frac{E_f^v}{kT}} \tag{2.1}$$

where k is Boltzmann's constant and T is the temperature measured in Kelvin. The vacancy concentration is said to obey Maxwell-Boltzmann statistics. There is another type of statistics called Fermi–Dirac statistics which apply to electrons rather than to atoms.

A typical value for E_f^v is 1 eV[†]. Since k is 8.6×10^{-5} eV K^{-1} it is easy to see that the concentration of vacancies at room temperature is very small. For example, c_v for copper is 10^{-17} at room temperature. Only at very high temperatures does c_v become appreciable ($\sim 10^{-4}$ near the melting point).

There is a second *extrinsic* type of point defect, which is an impurity atom. Up to a certain limit one solid element will *dissolve* in another, just as salt dissolves in water. The original element is called the *solvent* and the element which dissolves into it is

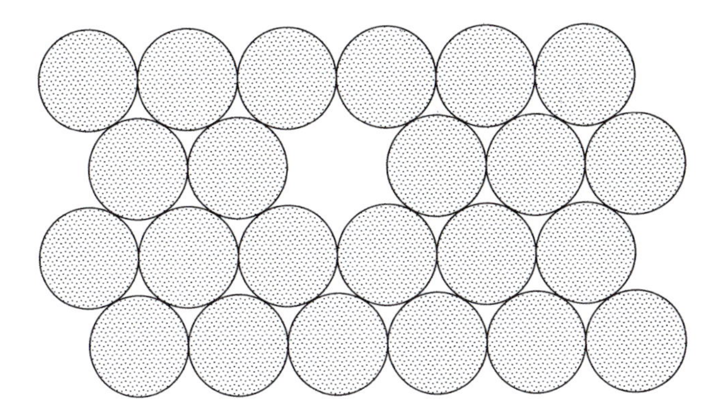

Fig. 2.15 When an atom is missing the vacant site is called a *vacancy*. A vacancy is a 'point' defect.

[†] See Chapter 1, Section 1.5 for definition of eV.

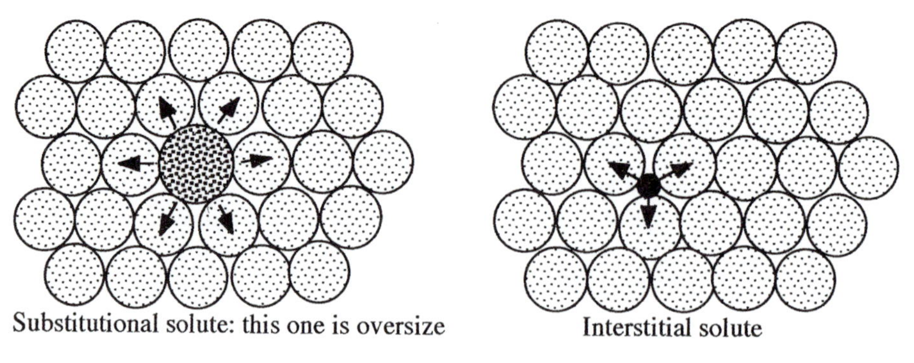

Fig. 2.16 Solids form solutions as well as liquids. Here are the two types: *substitutional* and *interstitial*.

called the *solute*. When the solvent is solid, we talk of a *solid solution*. The limits of solid solubility are usually quite small, of the order of a percent of the solute. Solid solutions can be *substitutional*, when an atom of the solute *replaces* one of the solvent, or they may be *interstitial*, when an atom of the solute squeezes in between those of the solvent (see Fig. 2.16). When the solvent is a metal, the atoms are very closely packed together, and so substitutional solutions are the rule. Only the very small atoms in the first row of the periodic table will dissolve interstitially—i.e. C, N, O etc..

Although their concentration is so small, vacancies are very important to metals because they enable *solid state diffusion* of the matrix (solvent) atoms or of substitutional solutes to take place. Look at Fig. 2.17. A neighbouring atom to the vacancy can, with the help of thermal vibrations, jump into the vacancy. It has diffused, i.e. moved through the solid; equally the vacancy has diffused in the other direction. Diffusion as a process is very sensitive to temperature. Not only is it proportional to the concentration of vacancies, it is also proportional to the rate at which the atoms jump into them. The concentration of vacancies obeys eqn (2.1) above; it turns out that the rate of jumping obeys a similar expression. Thus the *diffusion coefficient*, D, a measure of how quickly diffusion occurs, is proportional to $e^{-\frac{2E_f^v}{kT}}$. Diffusion is very important to how metal alloys behave and will turn up again in Section 3.9 on creep, Section 4.5 on solder and Section 4.6 on precipitation hardening. Diffusion is equally important in non-metallic solids, for example DC conductivity in ceramics and polymers (Chapters 7 and 8) and for doping of semiconductors (Chapter 9).

2.6 Metal properties revisited

Early in this chapter I listed a series of characteristics which would enable you or me to tell whether or not something is a metal. Using the two properties or prerequisites of the metallic bond

- free bonding electrons

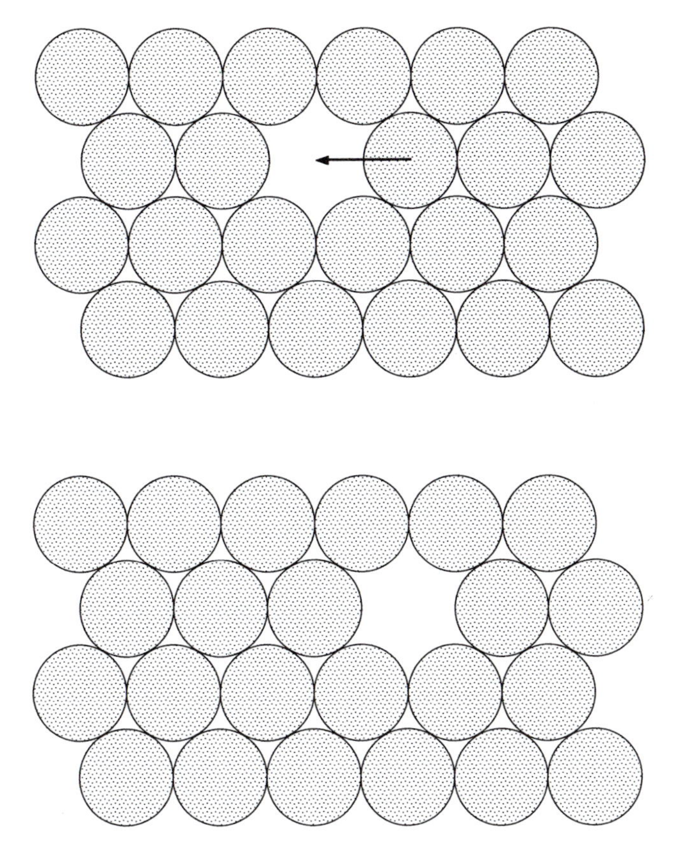

Fig. 2.17 How vacancies enable solid state diffusion to occur. An atom next to the vacancy jumps into the vacancy: it has *diffused*. Note that the vacancy has *diffused* the other way.

- high number of bonding electrons (and therefore atoms) per unit volume

and what we have learnt of the internal structure of metals, let us go back and understand where these properties come from. As enumerated in Section 2.1 metals conduct electricity and heat well, absorb and reflect light, are sonorous, dense and deform plastically.

- The good conduction of electricity and heat both come directly from the free electrons. I shall be looking shortly in more detail at how the electrons conduct electricity. Heat is the internal energy of a solid which is stored as atom vibrations. As the temperature of a body increases, so the vibrations grow larger and quicker. The free electrons are in thermal equilibrium with the vibrating atoms and act as inefficient but speedy transporters of heat.

- The absorption and reflection of light also come directly from the 'delocalisation' or freedom of the electrons. A light ray consists of oscillating electrical and magnetic fields. It is the electrical component of light rays which is important in their interaction with matter. The electric field makes the free electrons vibrate with the same frequency. The vibrating electrons may act as an aerial and emit a light photon of the same frequency (i.e. reflect the light) or they may transform the light energy into heat in the same way as I have just been describing.

- The property of sonority is a question of the transmission of sound waves. Sound, like heat, consists of atom vibrations, but the sound vibrations are of much lower frequency and are artificially superimposed on the atoms of the solid, unlike heat vibrations which are natural internal vibrations of the solid. (I would like to say that sound vibrations are *extrinsic*, but I have already just used this word in a different sense.) Non-metals, for reasons which will become clearer in Chapters 7 and 8, contain many small holes, called pores. These scatter and absorb the sound, resulting in the 'deadness' associated with non-metals. Reasonable sonority is taken as a good indicator of a well made (i.e. dense) ceramic, for example, and so this distinction between metals and non-metals is sometimes a rather indirect one. In polymers the pores are actually holes between the large, ill-fitting molecules.

- Metals tend to be dense for two reasons: firstly because of the close atomic packing and secondly because heavy atoms become metals more easily (see Chapter 1, Section 1.7).

- The plasticity of metals is because of the delocalised nature of the bonding electrons. The atoms in a metal can move past each other without any very great disruption of the metallic bonding—all the metallic bond cares about is preserving the atom packing at a high density. This is not true of the two other types of bonding, ionic and covalent. In ionic structures moving atoms past one another forces like charged ions to be next to each other and with two covalently bonded atoms the bond likes to be in a certain direction and of a certain length. Trying to move two atoms past one another outrages the bond. A much more detailed description of plastic deformation, how it takes place and why metals find it so easy will be found in Chapter 3 (Section 3.4).

Table 2.2 summarises these points.

Table 2.2 Metal properties explained via the metallic bond.

Metal characteristic	Electron delocalisation or close packing?
Conduction of electricity	Delocalisation
Conduction of heat	Delocalisation
Absorption and reflection of light	Delocalisation
Sonority	Close packing (and more complicated reasons)
Density	Both
Plasticity	Delocalisation

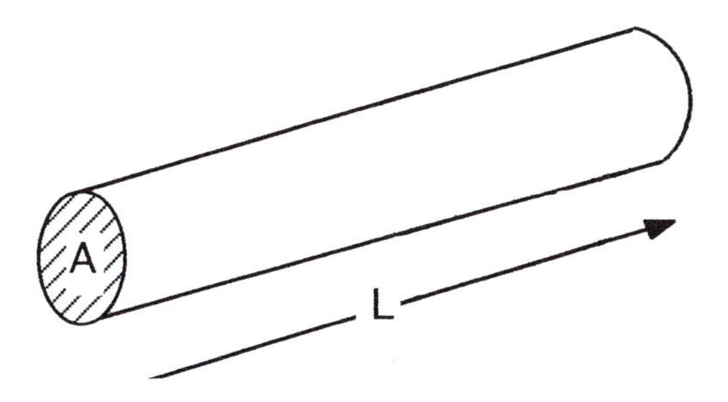

Fig. 2.18 Resistivity and resistance. There is a potential drop, *V*, across *L* and as a result a current *I* flows through *A*, with an associated current density *I*/*A*.

2.7 The conduction of electricity through metals

Ohm's law states that

$$V = IR \tag{2.2}$$

where V is the potential difference across a piece of metal, I is the current and R the resistance. R depends on the size and shape of the specimen; *resistivity*, ρ, is a more intrinsic material property. If we imagine a volume of material where the electric field, E, is constant—say a cylinder (see Fig. 2.18), then $V = EL$ and $I = JA$ (J is the current density: amps per unit area). $V = IR$ thus becomes $EL = JAR$ or

$$E = \frac{RA}{L} \times J = \rho J \tag{2.3}$$

where ρ is the resistivity of the material. Equation (2.3) is equally a statement of Ohm's Law. σ, the *electrical conductivity*, is $\frac{1}{\rho}$. ρ is measured in ohm-metres (Ωm) and σ in $(\Omega \mathrm{m})^{-1}$.

Where does the resistivity come from? In a metal, when an electric field is applied, the conduction electrons[†] are accelerated:

$$ma = eE$$

where m and e are the mass and charge of the electron and a the resulting acceleration.

In a perfect crystal lattice, with all the atoms in position, the electrons would accelerate away until the field was cancelled in some way. In a real situation, however, the atoms never are exactly in the correct crystalline position. Thermal vibrations cause them to be out of position; the defects in the lattice—the dislocations, vacancies and grain boundaries—do so also and impurity atoms, whether substitutional or interstitial, have the same effect. Thus the conduction electrons are scattered. Each time they are scattered their velocities are randomised. Then they are accelerated by the field again, then scattered again and so on. The scattering events constitute the electrical resistivity.

[†] In fact, quantum mechanics shows that it is only the fastest of the conduction electrons which are affected.

Imagine there is a scattering event, on average, every τ seconds. Then the maximum extra velocity achieved by the electrons in the direction of the field will be $v_{\max} = a\tau = \frac{eE}{m}\tau$. The *average* velocity, which we call the *drift* velocity v_D of the conduction electrons in the electric field, is then $\frac{1}{2}\frac{eE}{m}\tau$. The drift velocity is a lot smaller than the random velocity of the electrons as they buzz around the lattice. If there are n conduction electrons per unit volume, then the current density $J = nev_D = \frac{n}{2}\frac{e^2E}{m}\tau$. Looking back to eqn (2.3), we can identify the electrical resistivity ρ as:

$$\rho = \frac{2m}{ne^2\tau} \tag{2.4}$$

Now *Matthiessen's Rule* states that

$$\rho_{\text{total}} = \rho_{\text{defects}} + \rho_{\text{impurities}} + \rho_{\text{thermal}} \tag{2.5}$$

that is, the total resistivity is made up of *independent* contributions from the defects, impurities and thermal vibrations in the metal.

In a pure element at room temperature the impurity contribution will be by definition zero. Under these circumstances the thermal contribution dominates. For example, for a well annealed† piece of copper at room temperature

$$\frac{\rho_{\text{defect}}}{\rho_{\text{thermal}}} = 2 \times 10^{-3}$$

For silver and gold the same ratio is 5×10^{-4}, for aluminium it is 3×10^{-5} and for iron 10^{-3}. In a heavily worked‡ piece of metal the defect contribution at room temperature may rise to 5%.

At normal temperatures ρ_{thermal} is proportional to temperature. At low temperatures this becomes a dependence on T^3 or T^5. When impurities are present their effect on resistivity is comparable with, or dominates, ρ_{thermal} and is much greater than those of other defects, like vacancies or dislocations. This is not so much because an impurity atom has more effect than a vacancy or a dislocation, but because there are usually so many more of them—for example, a concentration of 10^{-9} vacancies is typical of copper at 300 °C (see eqn (2.1)), but it is easy to have several % of a solute element.

The room temperature resistivities of some common metal elements are shown in Table 2.3. The small defect contribution has been subtracted off.

The low resistivity group of metals: silver, copper, gold and aluminium, are the most widely used conductors in electrical engineering and I will be saying quite a lot about them over the next few pages. They all have the quintessentially metallic f.c.c. crystal

† The defect content of a metal is sensitive to how it is handled. 'Annealing' is heating to a moderate temperature—about a third of the melting point, expressed in K—to remove most of the defects—see Section 2.4 above.

‡ *heavily worked* means 'strongly deformed plastically'—e.g. by extrusion or rolling (see Chapter 4).

Table 2.3 Electrical resistivities of some common metal elements. The defect contribution has been removed.

Range	Metal element	Resistivity (10^{-8} Ωm)	
Low	Ag	1.61	Used in electrical engineering
	Cu	1.70	
	Au	2.20	
	Al	2.74	
Medium	W	5.3	
	Zn	5.92	
	Ni	7.0	
	Cd	7.27	
	Fe	9.8	
High	Pt	10.4	
	Cr	12.9	
	Ti	43.1	
	Nd	59.0	
	Mn	139.0	

structure and the other metallic properties in the highest degree—for example they are very plastic: fabricability is not a problem with electrical conductors—unfortunately the same cannot be said of price.

Notice that the rare earth elements (represented here by neodymium, Nd), manganese (Mn), chromium (Cr) and so on have very high electrical resistivities. This is tied up with their unpaired d and f electrons (see Chapter 1) and their magnetic properties.

I will return to the effect of impurities or solutes in Chapter 4, Section 4.6, which is about strong conductors.

2.8 Which conductors are used in electrical engineering?

What are electrical conductors used for in electrical engineering? Wires, cables and conducting bars account for a large proportion. The other two main applications are motors/generators and transformers. Rough percentages are shown in Table 2.4. Note

Table 2.4 How electrical conductors are used in Electrical Engineering.

Wires, cables etc	55%
Rotors and generators	15%
Transformers	10%
Other	20%

that wires form part of motors, generators and transformers and that most of the shapes involved are fairly simple: e.g. a wire. Usually the electrical resistivity is required to be low to minimise I^2R power losses. (Sometimes this is not the case—e.g. heating elements or light bulb filaments.) In the vast majority of cases, where we want to minimise resistivity, what material do we choose?

The scientific answer to this question is fairly straightforward. We already know that we need to use a pure metal element, preferably in a well annealed condition. Assuming that usage will be at or near room temperature, reference to Table 2.3 shows that the best conductor is silver.

But we are not scientists, we are engineers. We are not just interested in understanding *how* things work (fascinating as this is): we are interested in seeing them used in the real world. This means that we have to worry (amongst other things) about cost, as well as physics and chemistry. This doesn't make life less interesting—just a little more challenging!

The cost of an article can be split up as follows:

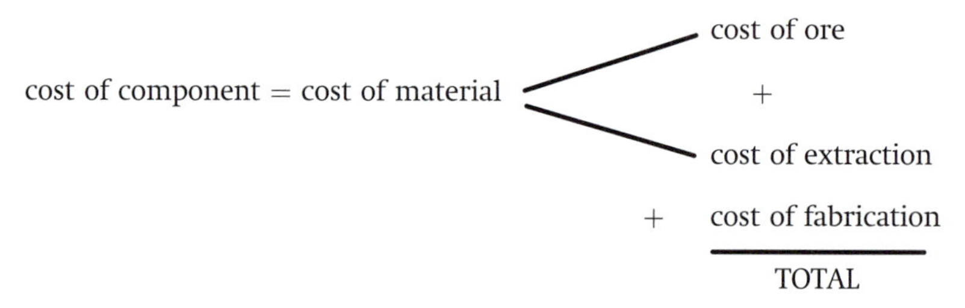

The four candidate metals are silver, copper, gold and aluminium. Gold and silver ores are expensive, copper ores moderately so and aluminium ore is relatively cheap. Gold and silver are called 'noble' metals because they are loath to combine with non-metals. *Pari passu* this means that gold and silver are easy to extract from their ores. (Ores are usually chemical compounds of the metal with a non-metal—normally oxygen (→ oxide) or sulphur (→ sulphide). Gold is so noble that it occurs as the native metal—this is very unusual.) Gold and silver are very cheap to extract, therefore, and copper fairly so. Aluminium is decidedly ignoble (= base) and it costs a lot of money to separate

Table 2.5 Which metal do we use for electrical conductors?

Element	Resistivity ($\times 10^{-8}$ Ωm)	Cost of ore	+	Cost of extraction	=	Cost of raw material ($/tonne)	Cost of fabrication
Ag	1.61	high		low		100 000	low
Cu	1.70	medium		medium		2700	low
Au	2.20	high		low		12 400 000	low
Al	2.74	low		high		1850	low

aluminium from the oxygen with which it is usually combined. (Aluminium oxide occurs as *bauxite*, named after Les Baux† in Provence, France.)

All four metals are cheap to fabricate, for the most part (because they are so ductile, as explained above).

Therefore, for the majority of uses, **copper** is the preferred material:

- it is moderately priced
- it has excellent electrical conductivity (almost the best)

and • it has good corrosion resistance.

Copper is *by far* the most common material used as a conductor in electrical engineering.

Silver and gold are too expensive for everyday bulk use. Silver is used as an alloying element and both silver and gold are used to *plate* other components. I shall deal with both alloying and plating later on.

Aluminium, because of its low cost, does find use in applications where electrical conductivity is non-critical. Because of its low density ($2.7\,\text{Mg m}^{-3}$ c.f. $8.9\,\text{Mg m}^{-3}$ for copper) it is used in suspended power transmission cables.

2.9 Summary

1. Metals tend to conduct electricity and heat well, absorb and reflect light, show sonority and to be dense and plastic.
2. Metals are (poly-)crystalline.
3. Metal crystal structures are close packed. Most metal elements adopt face centred or body centred cubic crystal structures or close packed hexagonal.
4. Metal crystals contain line defects—dislocations—and point defects—vacancies.
5. Electrical resistance is caused by scattering of the conduction electrons.
6. In an element, most of the scattering is caused by thermal vibrations of the atoms.
7. In a solid solution, scattering by the solute atoms can be equally important.
8. The most common metals used in Electrical Engineering are the elements copper, aluminium, silver and gold.

† Well worth a visit for (1) its spectacular cave dwellings which make an other-worldly horizon against the setting sun and (2) its 3 Michelin rosette restaurant.

Recommended reading

All the general books recommended at the end of Chapter 1 contain appropriate material to back up this chapter. For some background information about the electrical resistivity of metals, I suggest

Electrical resistance of metals by G.T. Meaden (published by Heywood Books) and ***The electrical properties of metals and alloys*** by J.S. Dugdale (published by Edward Arnold) which are rather old books (1966 and 1977 respectively) and out of print—you will have to obtain them via your library.

I found ***The electrical resistivity of metals and alloys*** by P.L. Rossiter (published by Cambridge University Press) far too difficult for me.

Questions (Answers on pp. 301–305)

2.1 If f.c.c., c.p.h. and b.c.c. crystals are made up from solid spheres which touch, what fraction of space is occupied by matter in each case?

2.2 How many atoms are there in the f.c.c., c.p.h. and b.c.c. unit cells?

2.3 Is there any distinguishable difference between the surroundings of the four atoms in the f.c.c. unit cell?

2.4 An f.c.c. crystal is made up from solid spheres which touch. Where in the unit cell is the largest interstice and what is the size of the largest sphere which could be inserted into it?

2.5 Answer Question 2.4 for c.p.h. and b.c.c. structures.

2.6 How many nearest neighbours does an atom in an f.c.c. crystal possess?

2.7 Dislocation densities are measured in lines per unit area (like lines of magnetic flux cutting a surface). A well annealed metal sample will contain about 10^8 dislocation lines per m^2 whereas a heavily cold worked sample may contain 10^{16} lines per m^2. What are the average separations of the dislocations in the two cases?

2.8 A good rule-of-thumb states that the concentration of vacancies in a metal is about 10^{-4} at the melting point T_m. Express the energy of formation of a vacancy, E_f^v, in terms of T_m. What is the concentration of vacancies at $\frac{T_m}{2}$? $\frac{T_m}{4}$? If the melting point of copper is 1085 °C, what is E_f^v? (Boltzmann's constant $k = 8.6 \times 10^{-5}$ eV K^{-1}.)

2.9 The electrical resistivity of copper at 160 K is 0.83×10^{-8} Ωm and that at 280 K is 1.69×10^{-8}Ωm. The resistivity of copper–1% nickel at 280 K is 3.12×10^{-8} Ωm. What is the resistivity of copper–3% nickel at 220 K?

2.10 If the diameter of a copper wire is 1 mm, what would be the diameter of an aluminium wire with the same resistance per unit length? What would be the ratio of the two masses per unit length? (The density of copper is 8.9 Mg m^{-3} and that of aluminium is 2.7 Mg m^{-3}.) (Use Table 2.3.)

2.11 What would be the ratios of costs of silver, copper, gold and aluminium wires of the same resistance per unit length? (The densities are 10.5, 8.9, 19.3 and 2.7 Mg m^{-3} respectively.) (Use Table 2.5.)

3

Mechanical properties

Chapter objectives

To understand

- stress and strain
- strength and ductility
- hardness, wear and impact resistance
- fracture toughness
- fatigue
- creep

and how to measure them.

3.1 Why should an electrical engineer know about *mechanical* properties?

We have already seen that the ideal conductors are elements. Unfortunately, metal elements are very weak. Sometimes this doesn't matter too much, but sometimes it does. In this chapter I am going to talk about *mechanical properties*. In the next I will explain how we can make electrical conductors strong without ruining their electrical properties. Why should an *electrical* (or electronic) engineer need to worry about *mechanical* properties? After all, that is what mechanical engineers are for! No man is an island, however, and no engineer can afford to isolate himself from other engineering disciplines. Any electrical device, if it is to be of any practical use, must satisfy certain mechanical criteria. ANY AND EVERY DEVICE. It may need a certain *strength*, but mechanical properties do not begin and end with strength. What about the propensity to crack? This is a mechanical property. For example, most microelectronic circuits are made on a *substrate* of silicon. If the silicon is so brittle that it cannot be handled without cracking into pieces, it will not be of much use to an electronic engineer. So

LIVERPOOL JOHN MOORES UNIVERSITY
LEARNING SERVICES

what I shall have to say in this chapter will be important not just for the metals of Part II, but for the materials of Parts III and IV also.

3.2 Mechanical properties

Mechanical properties generally describe a *shape change* as a result of some mechanical *stress* applied to a specimen or component. The shape change, or *deformation* of the body is usually described as a *proportional* deformation, which is called *strain*. So we have

stress $\overset{\text{causing}}{\longrightarrow}$ strain

The strain may be

temporary or permanent

Permanent strains may be

gradual or catastrophic (i.e. the specimen cracks in two (or three ...))

The stress may be constant, or increasing, or it may oscillate. So we have some important concepts here. The common sense terms and their scientific equivalents are shown in Table 3.1.

Table 3.1 Everyday terms for mechanical properties and their scientific equivalents.

Everyday language	Scientific terminology
stress	stress
constant stress	creep
oscillating stress	fatigue
thermal stress	stress produced by change of temperature
change of shape	deformation
proportional change of shape	strain
temporary strain	elastic strain
permanent strain	plastic strain
gradual strain	creep strain

The mechanical properties I shall discuss are as follows:

strength
ductility
hardness
abrasion/wear resistance
toughness
creep resistance
fatigue resistance

First I shall define a little more exactly what I mean by *stress* and *strain*. Then I shall explain how these are generated in real materials. And only then will I explain for each property:

1. what it is;
2. how it is assessed;

and give 3. examples of where it is important.

3.3 What are stress and strain?

I used the words *stress* and *strain* above. *Stress* involves the force applied to a body (specimen, component etc.) and *strain* describes the resultant change in shape. What is the difference between force and stress?

stress = force / unit area

It has units of Pa(scal) i.e. N m^{-2}.

Imagine a little cube, such as you might cut out of any piece of material:

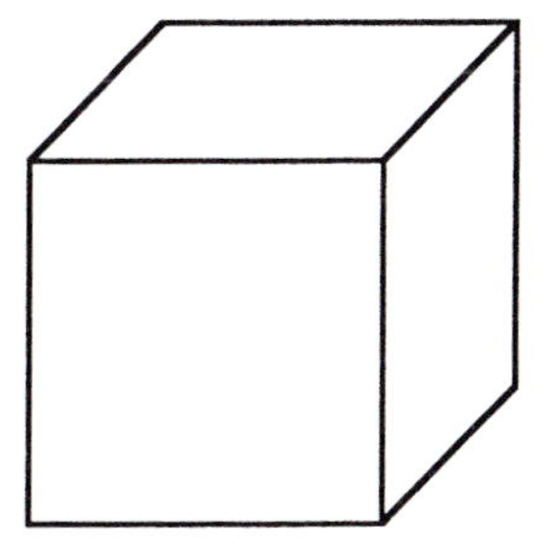

Each face can have three components of force on it:

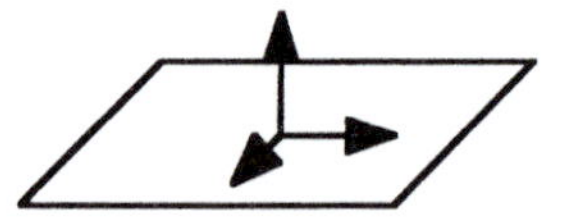

The two *in* the face are called *shear* forces and the other, normal to the face, is called a *tensile* or *compressive* force (depending on its sense). Thus we have shear stresses, tensile stresses and compressive stresses. Three forces, or stresses, times three pairs of faces make nine different stresses.

Similarly for strains: to describe a general change in shape of our small cube, each face can move in three different directions, so we also have nine shear strains and tensile/compressive strains. What is the connection between the movement of a face and strain? Strain is *proportional* shape change, so we divide the movement by the original dimension. For tensile/compressive strains,

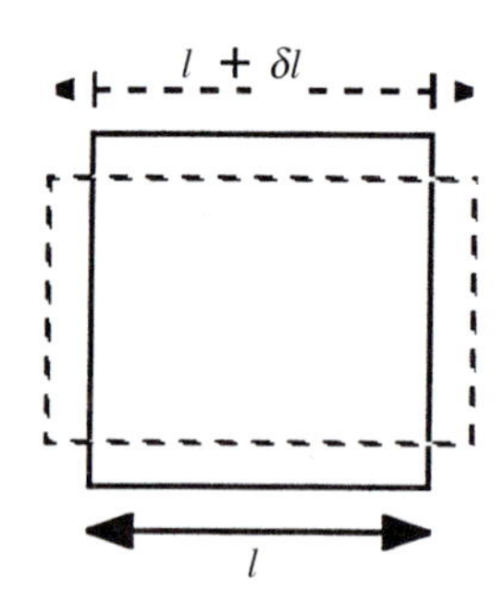

$$\text{Tensile strain} = \frac{\delta l}{l}$$

Compressive strain is when δl is negative

For shear strain

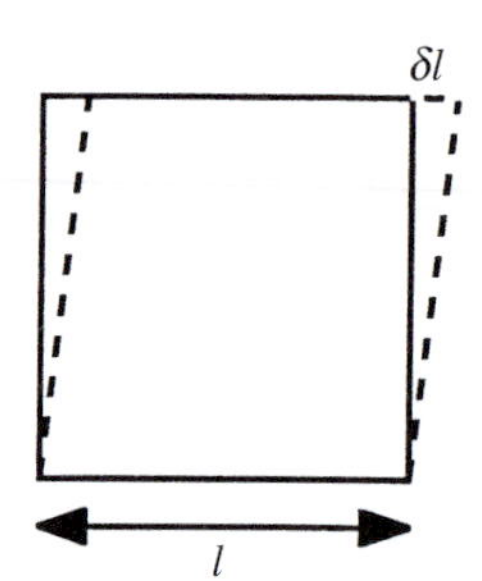

$$\text{Shear strain} = \frac{\delta l}{l}$$

but this time δl and l are perpendicular

For obvious reasons, strain has no units.

We therefore have a total of nine stresses and nine strains to describe completely a body's state of stress and the resultant strain. This is very complicated! Luckily, in Electrical and Electronic Engineering we can usually simplify radically this situation and, certainly for this book, I shall use only the following concepts:

stress	force/unit area
tensile/compressive stress	force perpendicular to surface
shear stress	force parallel to surface
strain	displacement/length
tensile/compressive/shear strain	analogous to stress (see above)

3.4 Temporary and permanent strain (elastic and plastic)

When a stress is applied to a body it produces a strain. When it is removed, the strain disappears. This temporary strain is called *elastic* strain. If the stress is increased, at some point part of the strain becomes permanent. This is called *plastic* strain.

How does strain happen?

As stress increases, the *bonds* between the atoms stretch:

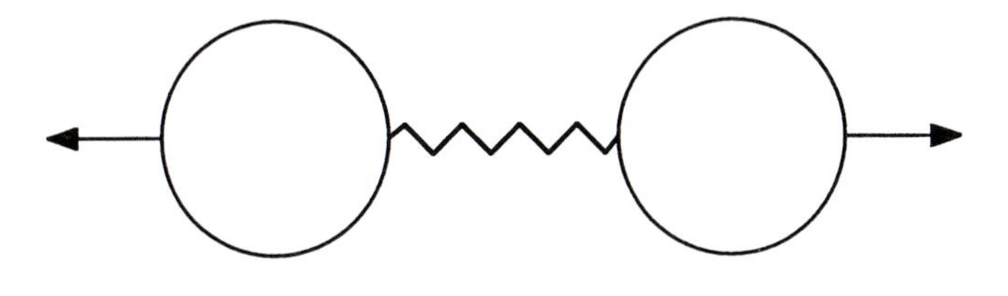

or

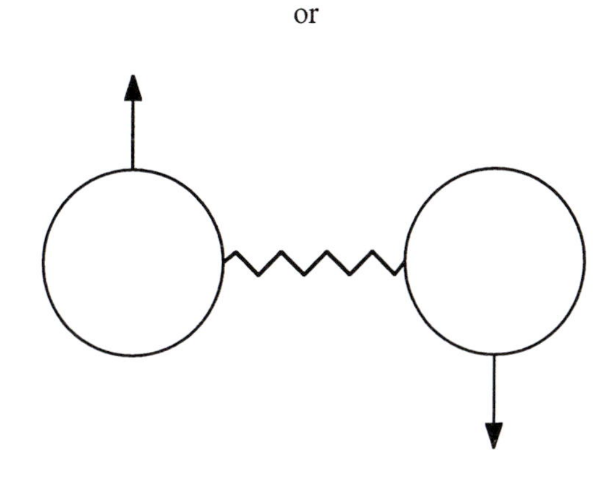

If the stress is removed, the atoms return to their equilibrium separation. This is *elastic strain*. The ratio of stress to strain is called the *elastic modulus*. Tensile (compressive) stress/tensile (compressive) strain is Young's modulus. Shear stress/shear strain is called the shear modulus.

There comes a point, however, when the bond snaps. There are two possible results:

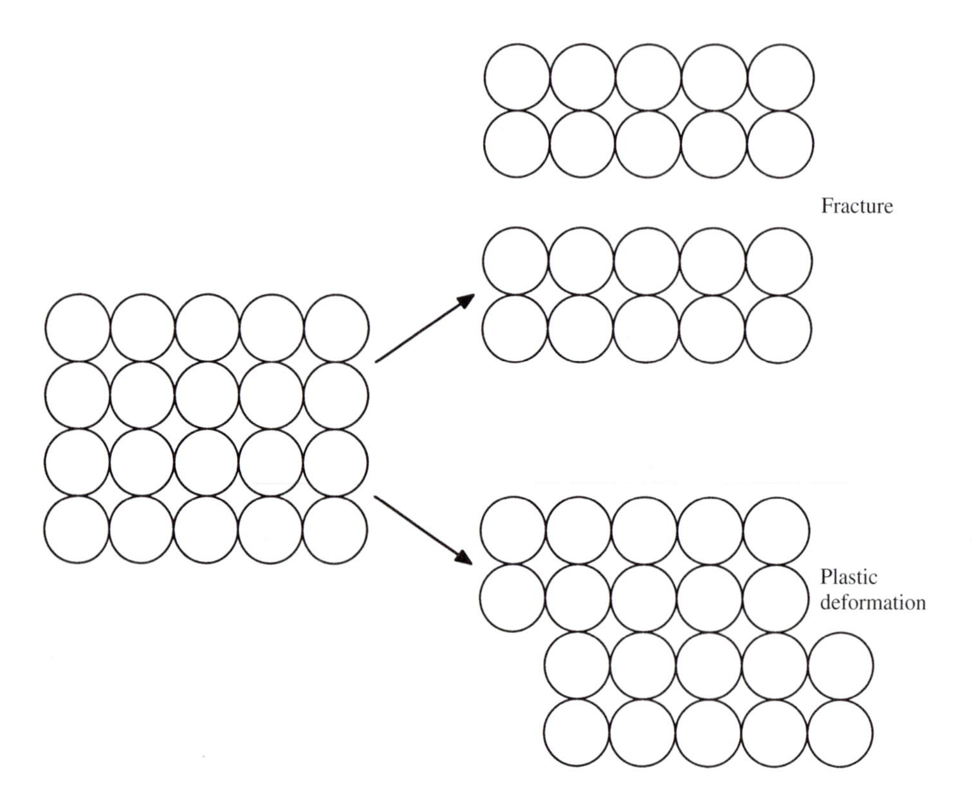

Either the body splits in two (fractures) or it changes shape *permanently* (plastic strain). You can see that plastic strain is essentially of the *shear* type.

How does plastic strain happen? This may seem obvious, but in fact it isn't. Let us estimate the shear stress necessary to produce plastic deformation in a crystal.

As the shear stress increases, so does the shear strain. There will come a point, however, when one atom will change its allegiance from the atom below it to the next one along—for example, in Fig. 3.1, when atom no. 1 reaches halfway between atoms 2 and 3. Once this point is reached, the atom will click into the next position of its own volition—it would actually need to be held back—and the stress *required* becomes negative. Cooperative movement across a plane will give the result shown on the right hand side: the crystal has sheared permanently by one atom spacing.

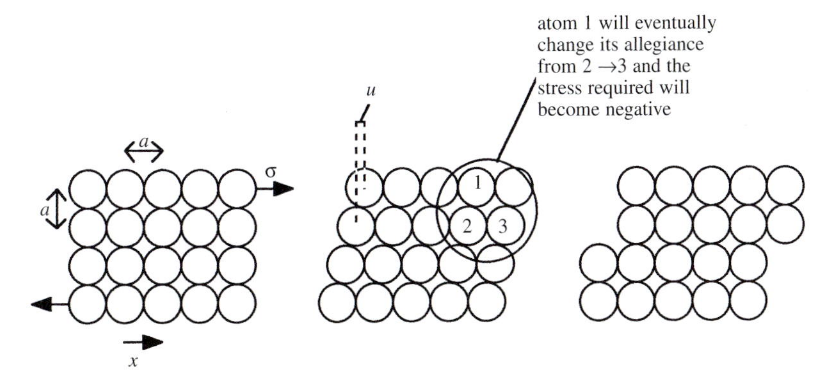

Fig. 3.1 Estimating the stress necessary for plastic deformation.

Roughly, then, we can say that the requisite stress

$$\sigma = \sigma_{\max} \sin \frac{2\pi u}{a}$$

What we want to know here is $\sigma_{\max}$. At small strains (the elastic regime)

$$\frac{\text{stress}}{\text{strain}} = \text{elastic modulus} = \textit{shear} \text{ modulus}, \mu$$

$$\therefore \frac{\sigma}{u/a} = \mu$$

$$\text{and } \sigma_{\max} = \frac{\frac{\mu u}{a}}{\frac{2\pi u}{a}} \quad \text{(remember } \sin x \cong x \text{ for small } x)$$

$$\text{Thus } \sigma_{\max} \cong \frac{\mu}{6}$$

Measured values of $\sigma_{\max}$ for metals are lower than this by factors of 10^4 and more! We cannot have got the answer so wrong! *Plastic deformation does not happen as I have shown it.* What actually happens involves movement of the *dislocations* introduced in the previous chapter (Fig. 3.2). The deformation is concentrated at the dislocation, which moves across the crystal. Thus the deformation is *sequential*—it doesn't happen all at once. The dislocation mechanism operates at much lower stresses than *G*/6. The change of allegiance of atoms in one plane to atoms in the plane below, and across which the shear will occur (referred to above, using atoms 1, 2 and 3) is accomplished very easily at the centre of a dislocation, because the crystal is so deformed there (Fig. 3.3). This is why the stresses involved are so low.

All plastic deformation happens via dislocations

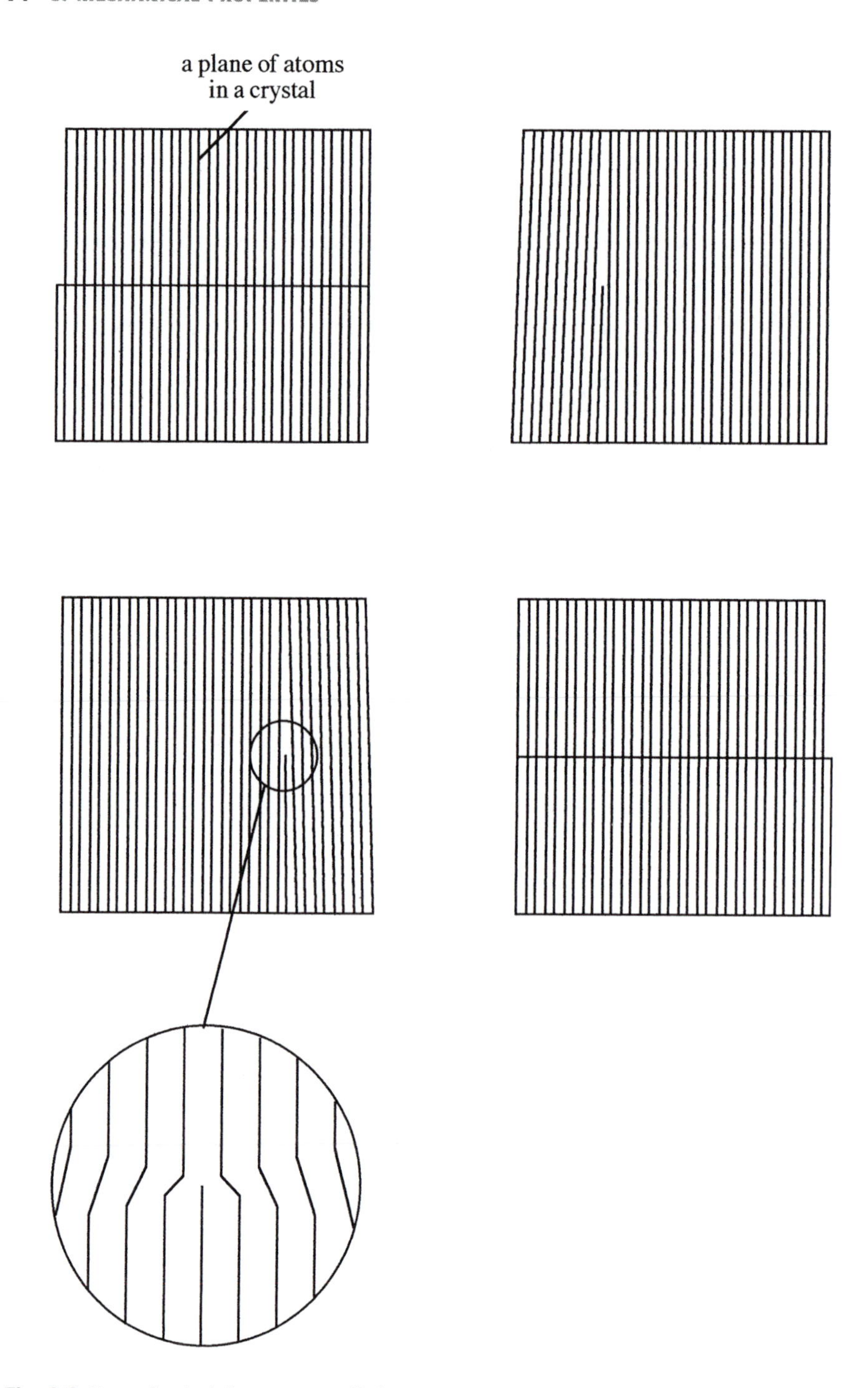

Fig. 3.2 How plastic deformation really happens. The dislocation moves from left to right and the top of the crystal moves from right to left (relative to the bottom).

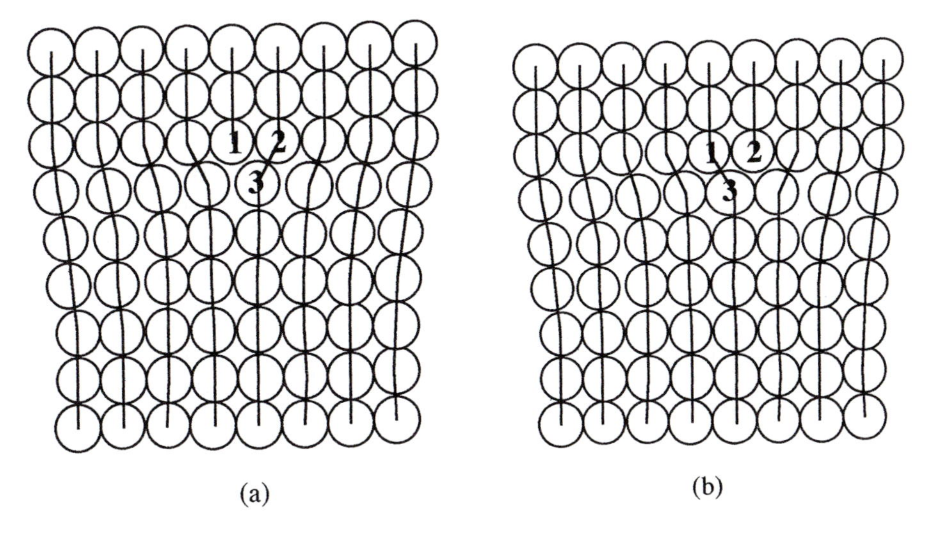

Fig. 3.3 Plastic deformation occurring via dislocation movement. In (b) the dislocation has moved one plane to the right relative to where it was in (a). Notice how miniscule are the necessary movements of the atoms. This is why plastic deformation occurs so much more easily by dislocation movement.

A nice analogy which helps to explain why a dislocation accomplishes plastic deformation at a lower stress than that corresponding to slip across the whole plane at once is illustrated in Fig. 3.4. As you may already have experienced, trying to pull a whole carpet across a floor at once is not a trivial endeavour, but putting a ruck (≡ dislocation) into the carpet and moving the ruck instead is considerably easier!

You can see that a dislocation is like a wave spreading across the crystal. More generally (Fig. 3.5) the dislocation is the boundary between the sheared and the unsheared crystal.

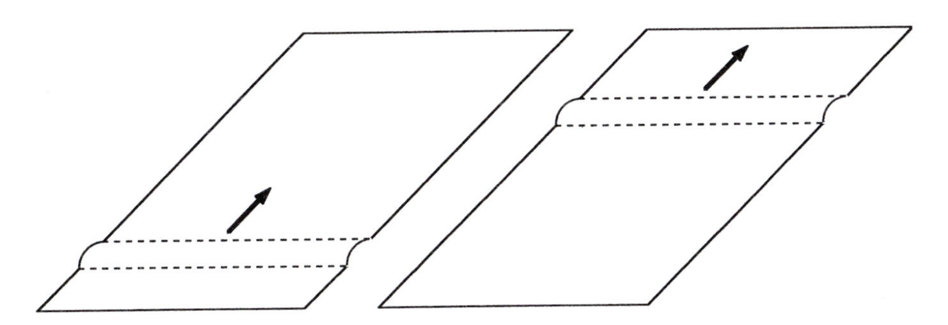

Fig. 3.4 An analogy showing why shear via dislocation motion is so much easier. A carpet takes less effort to move if a ruck is introduced and moved, instead of the whole carpet at once.

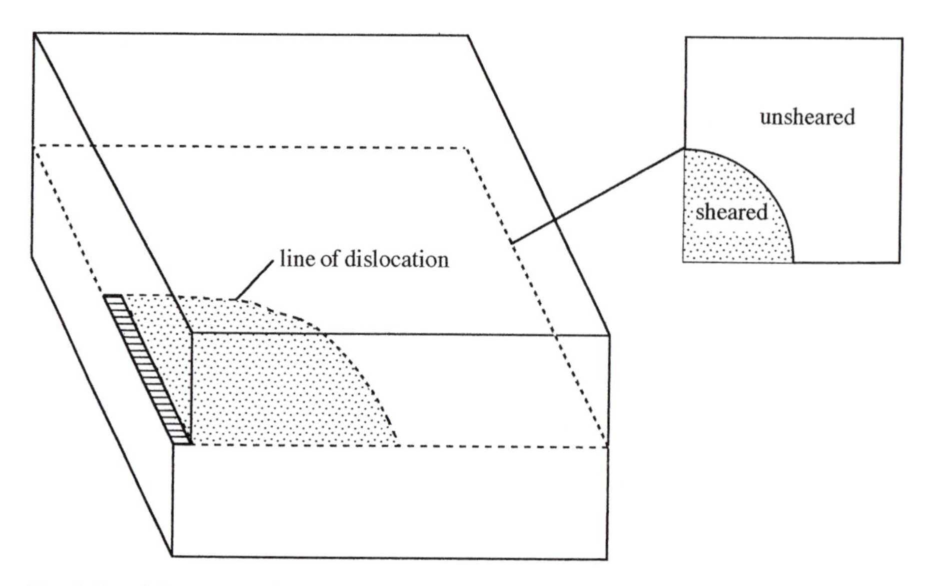

Fig. 3.5 A dislocation is the boundary between the sheared crystal and the unsheared. The diagram in Fig. 3.2 corresponds to where the dislocation emerges from the front of the block in this picture.

Shear happens via dislocations in metals only
∴ only metals show plastic deformation

This isn't quite true! All materials will deform plastically close to their melting points. It is true, however, that the only solids which show much inclination to plastic deformation well below their melting points are metals.

This is VERY important. It means that when a conducting component is made, it can be *shaped* plastically. Insulators cannot. Hence we can draw wire, make thin sheet, stamp out tabs and connectors and crimp them round the wire, just to name a few examples. Why do dislocations only exist in any quantity in metal crystals? It is because of the nature of the bonding. If you look at the dislocation moving in Figs 3.2 and 3.3, the metallic bond, which only depends on the number of atoms per unit volume, not on their exact spatial relationship with one another, is not really bothered by the dislocation. An ionically or covalently bonded crystal would, however, be VERY UPSET by a dislocation. This is why metals can be shaped and insulators cannot.

The bonding controls both the electrical and mechanical properties

(This is probably the most important sentence in this book!)

Mechanical properties and how they are measured

I will now begin to work through the list of mechanical properties given in Section 3.2, explaining what they are, how they are measured and where in Electrical Engineering they are important.

3.5 Strength and ductility: the tensile test

In a tensile test a plate-shaped

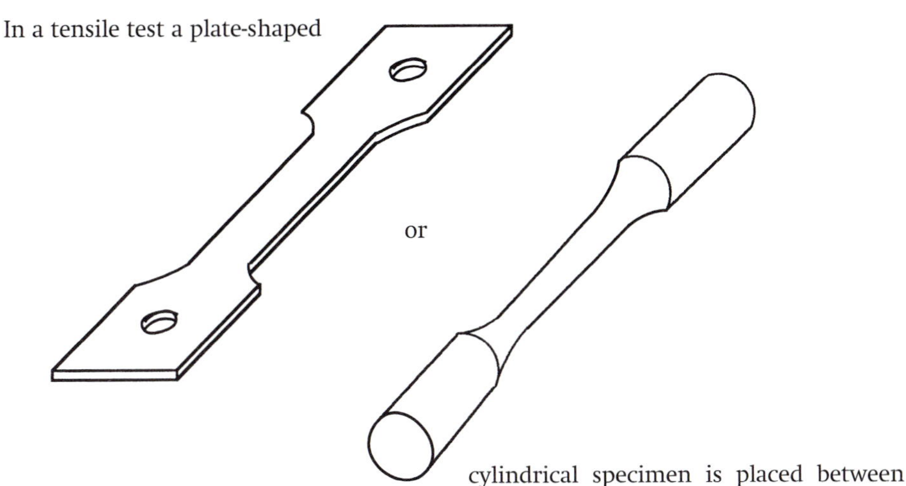

cylindrical specimen is placed between the two jaws of a tensile testing machine (Fig. 3.6). The jaws move apart at constant speed, imposing a constant *strain rate* on the specimen and the force necessary is measured as a function of displacement. The specimens, whether plate or cylinder, have a narrow bit in the middle to ensure that all the deformation takes place there. (Because the cross-sectional area is lower, the stress is higher than elsewhere.) The load or force applied is divided by the cross-sectional area of the gauge (the narrow part) to convert it into stress and the relative displacement of the jaws is likewise converted into tensile strain by dividing by the gauge length. The resulting graph is shown in Fig. 3.7. The first, straight line, part of the curve is the elastic region. This extends as far as the stress σ_y, which is the *yield stress*. If the stress were to be relaxed along this part of the curve the specimen would return to its original length, i.e. to the origin of the graph. The gradient of the graph in the elastic region is called *Young's modulus*, *E*, for the material, although the tensile test is not a very good way of measuring Young's modulus. At the yield point ($\sigma = \sigma_y$) dislocations start to move and the graph starts to curve over (why isn't it horizontal?—see below). If the stress is removed along this part of the curve, the specimen follows a line such as AA′ parallel to the original elastic part of the curve. The residual strain OA′ is called the *plastic* strain: it is a *permanent* strain.

Fig. 3.6 A tensile testing machine and some tensile specimens. The specimen shown being tested (between the two black jaws in the centre of the photograph) is of the plate shaped variety (see text).

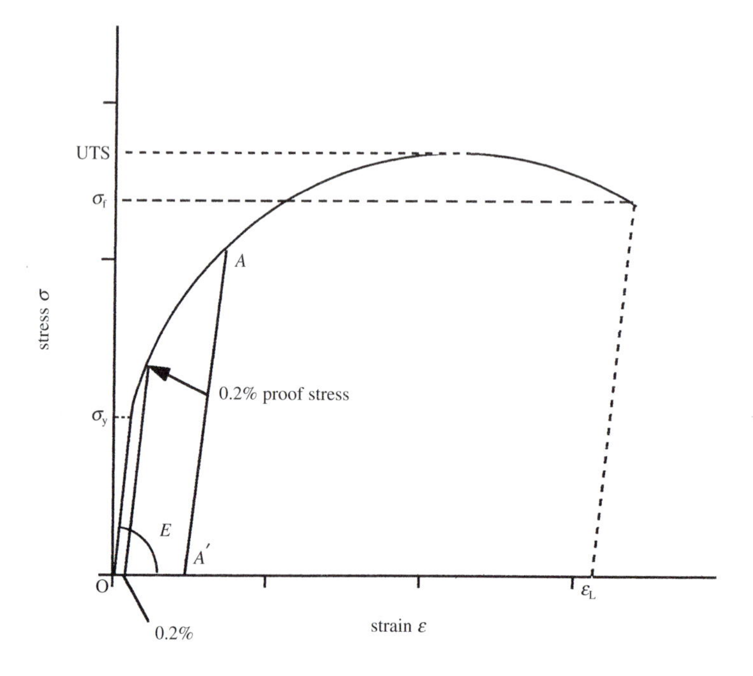

Fig. 3.7 The output from a tensile test is a graph of stress vs strain. Several important parameters can be measured (see text).

The yield stress is a very important and useful parameter and tells us a lot about the material. Sometimes it is difficult to see where the curve starts to bend over. A more useful engineering parameter is the *proof*, or *offset*, stress. This is the stress where a particular value of plastic strain, e.g. 0.2%, has been achieved. Thus: $\sigma_{0.2}$.

The greatest stress which the material can survive is called the *Ultimate Tensile Strength* (often just *Tensile Strength*). This is a much used parameter without much fundamental significance, because it depends on the details of the specimen geometry. Finally the specimen *fractures* at σ_f, having achieved a *tensile ductility* (total plastic strain) of ϵ_L. At this point try Questions 3.3 and 3.4 at the end of this chapter. These will teach you more than my general description above.

Notes on the tensile test

1. Why isn't the curve horizontal after σ_y?

Why isn't the tensile stress–strain curve as shown in Fig. 3.8? This is because dislocations become tangled up and get in each other's way. As a tensile test proceeds, the density of dislocations goes up and it becomes more and more difficult for them to force their way through. This is not specific in any way to tensile tests and is called *work hardening*. The yield stress is the stress to make the dislocations move. When the applied yield stress is resolved onto the plane on which the dislocations move (the 'slip plane') and in the direction of slip, the resulting 'critical resolved slip stress', which is the stress needed to move a dislocation through a relatively uncluttered and dislocation-free crystal, is a very fundamental parameter which tells us a lot about the material (in contrast to the tensile strength, for example).

2. Why does the curve go over a maximum and start to come down?

This is for two reasons. When I worked out the stress and strain I used the *original* cross-section and length throughout. This gives the *nominal* or *engineering* stress and strain. As

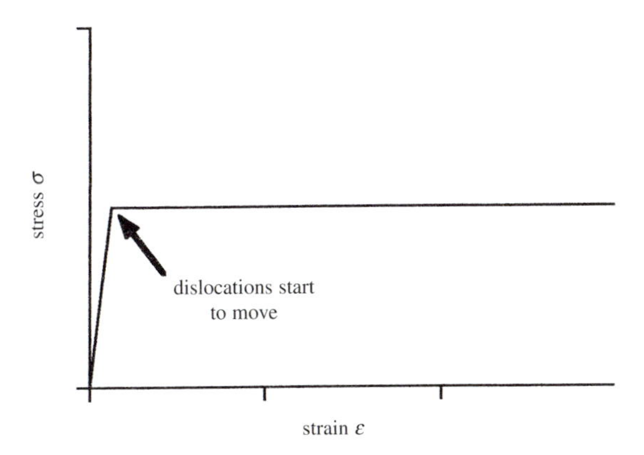

Fig. 3.8 How we might expect a tensile stress–strain graph to look.

the test progresses, however, the gauge length increases and the gauge area decreases. It would be more logical to use the length and area appropriate to every point of the test. This gives the *true* stress and strain. The other and more important reason why the curve starts to come down is connected with the failure of the specimen, which I shall deal with now in more detail.

3. Why does the specimen fail suddenly?

A naïve view of materials would suggest that the specimen should just get longer and longer:

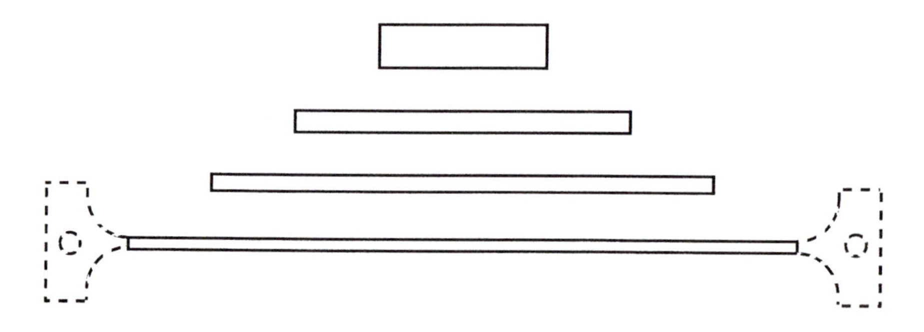

etc, etc (I've run out of page width!)

Now although chewing gum might behave roughly like this, or hot glass fibres when they are drawn, it is a fact that metals, at any rate, don't.

The first reason is called 'necking'. What happens is this:

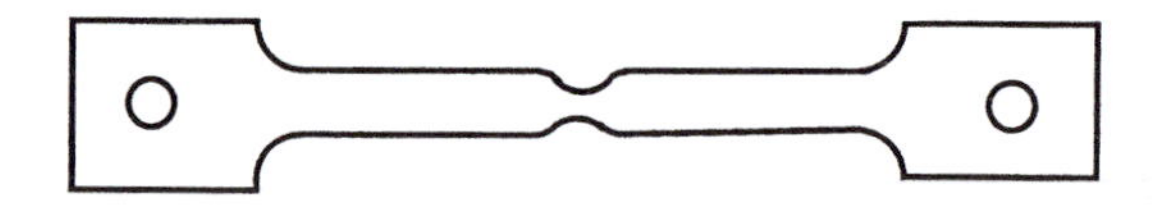

A part of the gauge length thins in a very localised way. The material prefers that a small part of it deforms a lot, rather than all of it a little. This happens when the gradient of the tensile test curve (the work hardening rate) falls (see Question 3.7 at the end of the chapter). Necking, or localised thinning, also makes the tensile curve lower than the true materials properties would dictate and removes the puzzling maximum in the stress–strain curve, which is thus seen to be an artefact (Fig. 3.9).

This isn't the whole of the story, however. Failure by necking would lead to what is called 'ductile rupture':

Now this is very rare—one of the few examples amongst the materials which you will come across is very pure copper. What is far more common, and what I showed in the example at the beginning of this section, is that a crack forms and terminates abruptly the test. Why is this?

The reason is sufficiently important for us to start a new section. We are therefore leaving the tensile test and moving on to ask

3.6 Why do cracks form? Fracture toughness

A crack is a geometrical instability which changes the shape of a component in an abrupt and usually catastrophic way. Cracks are bad news to engineers. Engineers don't mind so much a component slowly changing its shape in a civilised way—often there is

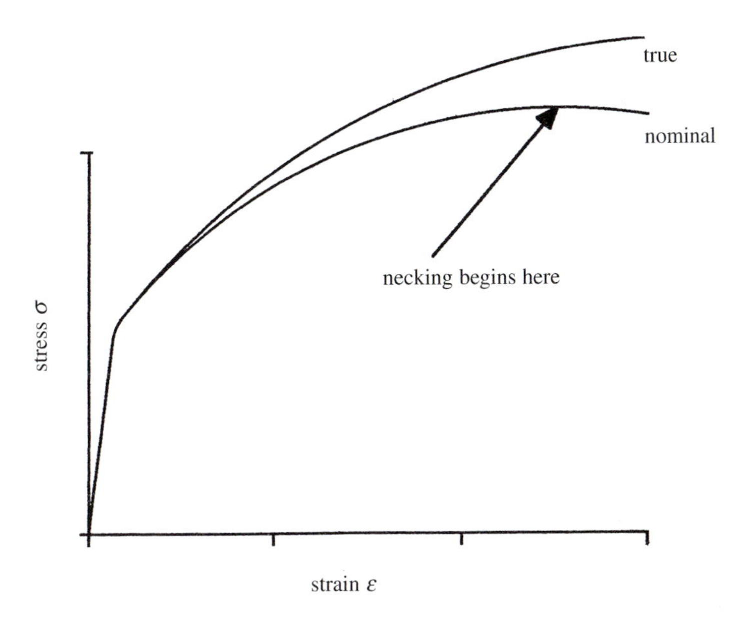

Fig. 3.9 Nominal and true stress and strain.

plenty of time to do something about the situation. But a crack is something else. You don't know when it's going to happen and when it does, it happens like lightning and all you are left with is pieces and a disaster. Cracks make engineers tremble to the ends of their boots.

We must have some way of predicting whether a material is likely to crack

This is the subject of *fracture mechanics.*

Let us start by examining an incipient crack in a plate and asking the question 'when will it spread catastrophically?'. Imagine (Fig. 3.10) a plate containing a crack subject to a tensile stress σ perpendicular to the crack. If the crack of length 'a' grows by δa, we both do work and receive work. The work done is to increase the crack area. If this has energy G_c/unit area we have to supply

$$t\,\delta a\,G_c \qquad (3.1)$$

The energy we get back is the released strain energy corresponding to a region connected to the tip of the crack. Try this experiment: tear a piece of aluminium cooking foil a third of the way across and then pull from either end of the sheet (see

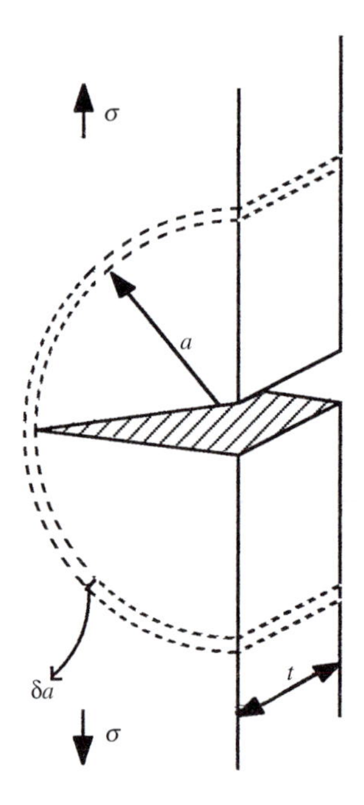

Fig. 3.10 A schematic diagram of a crack in a plate. We are asking the question 'when will it run out of control?'.

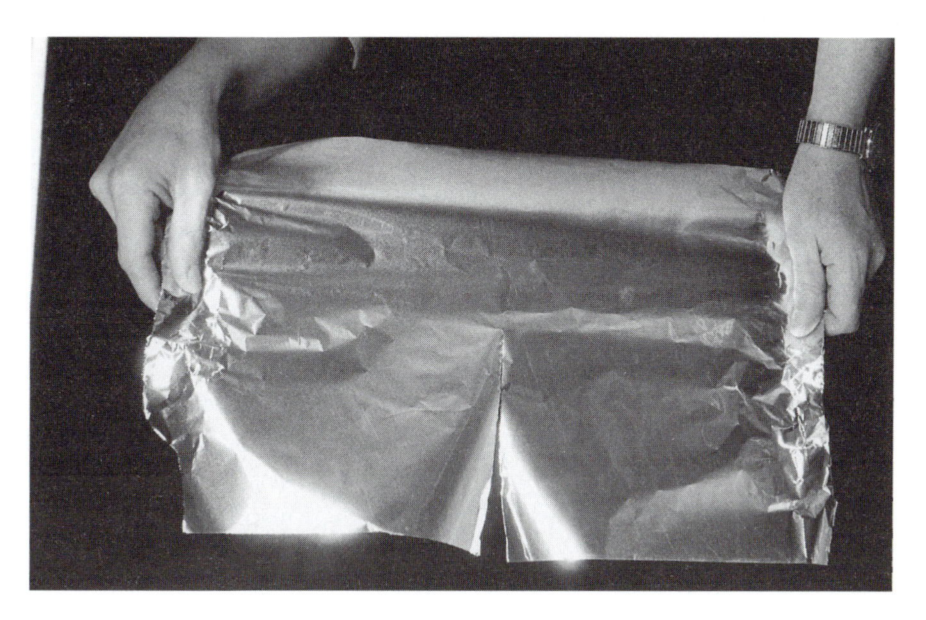

Fig. 3.11 A piece of cooking foil with a tear in it is subjected to a tensile stress. Notice how the foil near the mouth of the crack is flapping loose. The strain energy is stored in the remainder of the sheet. For how and where, see the main text.

Fig. 3.11). Notice how the foil near the beginning of the crack flaps loose. The strain energy is all in the untorn (≡ uncracked) part. The originator of this approach, a Dr Griffith, cleverly realised that the advance of the crack by δa releases strain in a semicircular annulus passing through the tip of the crack. How much strain energy is there per unit volume? Thinking of the elastic part of the tensile test:

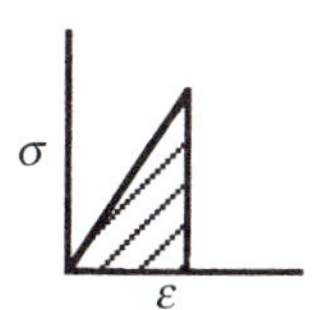

the work done is $\frac{1}{2}\sigma\epsilon$ per unit volume $= \frac{\sigma^2}{2E}$. Thus the energy released by the crack is

$$2\pi a \times t \times \delta a \times \frac{\sigma^2}{2E}$$

$$= \pi a t \delta a \frac{\sigma^2}{2E} \quad (3.2)$$

If expression (3.2) is greater than expression (3.1) the crack will spread catastrophically. The critical point, where $\sigma = \sigma_f$ the fracture stress, is obtained by setting (3.1) = (3.2):

$$t\,\delta a\,G_c = \pi a t \delta a \frac{\sigma^2}{2E}$$

$$\text{or} \qquad \sigma_f = \sqrt{\frac{2\,G_c E}{\pi a}} \tag{3.3}$$

Noting that $\sigma_f \propto \dfrac{1}{\sqrt{a}}$,

Big cracks are more dangerous than small cracks

Rearranging this equation by putting all the material parameters on one side and all the externally imposed parameters on the other:

$$\sigma_f \sqrt{\frac{a\pi}{2}} = \sqrt{G_c E} \tag{3.4}$$

$\sigma\sqrt{\dfrac{a\pi}{2}}$ is called the *stress intensity* at the end of the crack. When $\sigma = \sigma_f$, the stress intensity becomes the *critical stress intensity* which eqn (3.4) tells us is equal to $\sqrt{G_c E}$. This is a material parameter which is more usually called the *fracture toughness* K_c. It tells us how resistant a material is to cracks. It is a VERY important parameter. Table 3.2 shows some typical fracture toughnesses. The units of fracture toughness are stress $\sqrt{\text{distance}}$—usually MPa $m^{0.5}$. If you see K_{Ic} referred to, this merely means 'K_c for mode I testing'—i.e. with a tensile stress across the crack, as in Fig. 3.10.

Measuring fracture toughness

A reasonably thick plate has a notch machined in it (Fig. 3.12). The notch is extended by a very fine crack (with a sharp end), produced by *fatigue* (I shall deal with fatigue later on). The specimen is loaded incrementally in a tensile testing machine and the displacement measured. At the point where the curve passes over a maximum, and thus the crack begins to spread catastrophically, the fracture toughness can be measured from the load and displacement. The actual expressions used are more exact and complicated than that I gave above, but this need not concern us here. Fracture mechanics—the study and prediction of cracks—is a commercially very important activity nowadays, as you might guess.

Real engineering materials

Most metallic materials used in engineering lie between the two extremes which I have described: ductile rupture and perfectly brittle fracture.

Table 3.2 Some typical fracture toughnesses (from *Engineering materials I* by M.F. Ashby and D.R.H. Jones (by kind permission of Elsevier)).

Material	G_c(kJ m^{-2})	K_c(MPa m$^{0.5}$)
Pure ductile metals (e.g. Cu, Ni, Ag, Al)	100–1000	100–350
Rotor steels (A533: Discallov)	220–240	204–214
Pressure-vessel steels (HY 130)	150	170
High-strength steels (HSS)	15–118	50–154
Mild steel	100	140
Titanium allows (Ti 6A1 4V)	26–114	55–115
GFRPs	10–100	20–60
Fibregiass (glasstibre epoxy)	40–100	42–60
Aluminium alloys (high strength–low strength)	8–30	23–45
CFRPs	5–30	32–45
Common woods, crack ⊥ to grain	8–20	11–13
Boron-fibre expoxy	17	46
Medium carbon steel	13	51
Polypropylene	8	3
Polyethylene (low density)	6–7	1
Polyethylene (high density)	6–7	2
ABS polystyrene	5	4
Nylon	2-4	3
Steel-reinforced cement	0.2–4	10–15
Cast iron	0.2–3	6–20
Polystyrene	2	2
Common woods, crack ∥ to grain	0.5–2	0.5–1
Polycarbonate	0.4–1	1.0–2.6
Cobalt/tungsten carbide cermets	0.3–0.5	14–16
PMMA	0.3–0.4	0.9–1.4
Epoxy	0.1–0.3	0.3–0.5
Granite (Westerly Graniter)	0.1	3
Polyester	0.1	0.5
Silicon nitride, Si_3N_4	0.1	4-5
Beryllium	0.08	4
Silicon carbide SiC	0.05	3
Magnesia, MgO	0.04	3
Cement concrete, unreinforced	0.03	0.2
Calcite (marble, limestone)	0.02	0.9
Alumina, Al_2O_3	0.02	3-5
Shale (oilshale)	0.02	0.6
Soda glass	0.01	0.7-0.8
Electrical porcelain	0.01	1
Ice	0.003	0.2*

*Values at room temperature unless starred.

Recall that the fracture toughness $K_{Ic} = \sqrt{G_c E}$. Surface energies of solids are $\sim 1\,\text{Jm}^{-2}$, so we might expect G_c to be $\sim 2\,\text{Jm}^{-2}$ (a crack has two surfaces). A typical value for E is $\sim 10^{11}$ MPa. Putting these in the equation for K_{Ic} suggests that it should be of the order

(a)

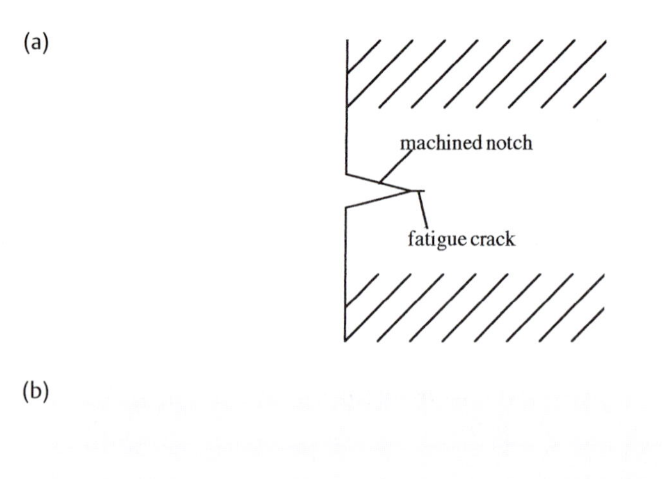

(b)

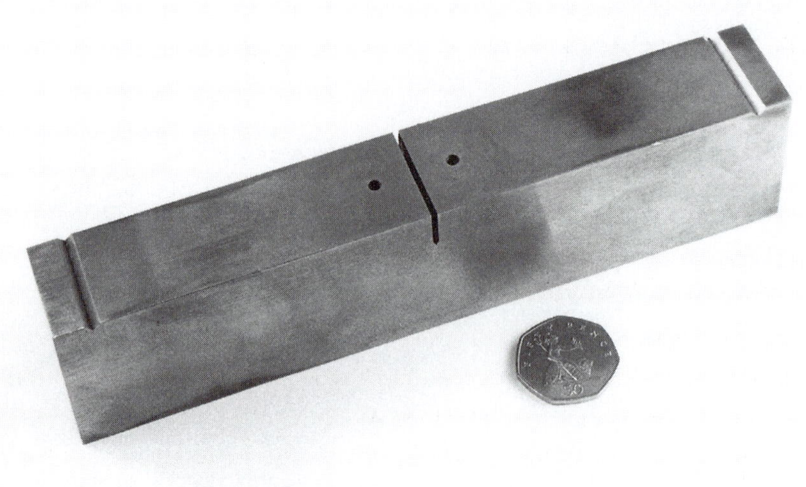

Fig. 3.12 Making a fracture toughness specimen (a) schematic (b) a real specimen. (See also Fig. 3.19.)

of, or less than, 1 MPa $m^{0.5}$. Table 3.2 shows that most metals used in engineering have a fracture toughness considerably greater than this. In fact, a reasonable minimum for K_{Ic} for a metallic alloy which is to be used in a traditional way is $\sim$ 20 MPa $m^{0.5}$. Where does the difference come from? In fact it comes from an underestimate of G_c. When a crack forms in a metal with some plastic ductility, the material at the end of the crack, where the applied stress is intensified (this is the stress intensity factor), yields (Fig. 3.13). The crack energy G_c is augmented by all the plastic work done ahead of the crack, resulting in the much higher fracture toughnesses of Table 3.2. For any traditionally useful metal, by far the majority of the fracture toughness comes from

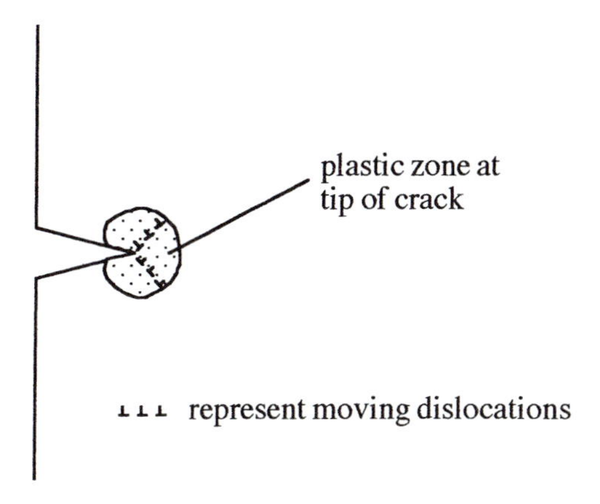

Fig. 3.13 At the root of a crack in a piece of metal there is a zone where plastic deformation occurs.

plastic work done at the crack tip, i.e. by dislocations moving. Now the *strength* of a material lies in *preventing* the dislocations from moving. Thus, all other things being equal,

making materials strong makes them brittle

This is the saddest fact in materials science!

At the other end of the brittle-ductile scale was ductile rupture, which I described above, where the material necked and drew down to a point. In between this rare extreme and materials which undergo relatively brittle fracture, there is a whole class of materials which undergo what is called *ductile fracture*. At the myriad of little included foreign particles which most real metal alloys contain, the material pulls down locally to a point. Macroscopically this leads to a fracture surface with a fibrous appearance, whose overall shape is not flat, as with more brittle materials, but conical (Fig. 3.14), resulting in a so-called 'cup and cone' fracture. Fracture toughnesses *can* be measured for such fractures, but with increasing difficulty as ductility increases. Generally, the more useful a material is, the more difficult it is to measure its fracture toughness. Thus there is a whole spectrum of ductility (or brittleness):

totally brittle →	brittle appearance →	ductile fracture →	ductile rupture
$K_{Ic} \sim 1$ MPa m$^{0.5}$	but with plastic deformation		
	$K_{Ic} \sim 10$ MPa m$^{0.5}$	$K_{Ic} \sim 100$ MPa m$^{0.5}$	$K_{Ic} \sim 500$ MPa m$^{0.5}$

The very ductile metals which are used as electrical conductors are right over on the right of this table. Most steels (see Chapter 5) are more in the centre. The ceramics

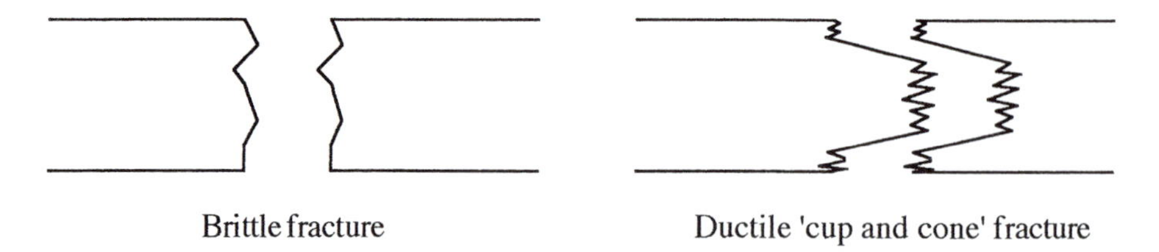

Fig. 3.14 Two types of fracture.

which form an important class of electrical insulators are right over on the left. Remember that all this comes from the **bonding**: electrical and mechanical properties alike.

3.7 Cheap and cheerful mechanical tests

The two mechanical tests I have described so far take a certain amount of trouble on the part of the engineer: making the specimen, using quite expensive equipment, interpreting the results etc. I am now going to describe two mechanical tests which tell us a lot quite quickly (and easily and simply and cheaply!). These are the **hardness test** and the **impact test**.

The hardness test

In the hardness test a hard object is pressed into the specimen and the size of the indent measured. The harder the specimen, the smaller the indent. There is a variety of hardness tests employing different indenters. I shall describe just one as an example: the Vickers Hardness Test. Here a square pyramidal diamond is pressed into the specimen and the lateral size of the indent measured (Fig. 3.15).

$$\text{The Vickers Hardness Number (VHN)} = \frac{1.854P}{L^2}$$

(P in kg; L in mm)

Notice that the VHN has units of stress. The different hardness tests correlate with different combinations of yield strength and tensile strength, as measured in the tensile test. The VHN, described above, is closely related to the yield stress. (VHN $\sim 3\sigma$. As the load is increased, σ moves from yield stress to UTS.)

The hardness test is easy to perform and requires little in the way of specimen preparation—indeed, a finished component can be hardness tested as a quality control check (provided a part of it can be found that is not sensitive to the small indentation).

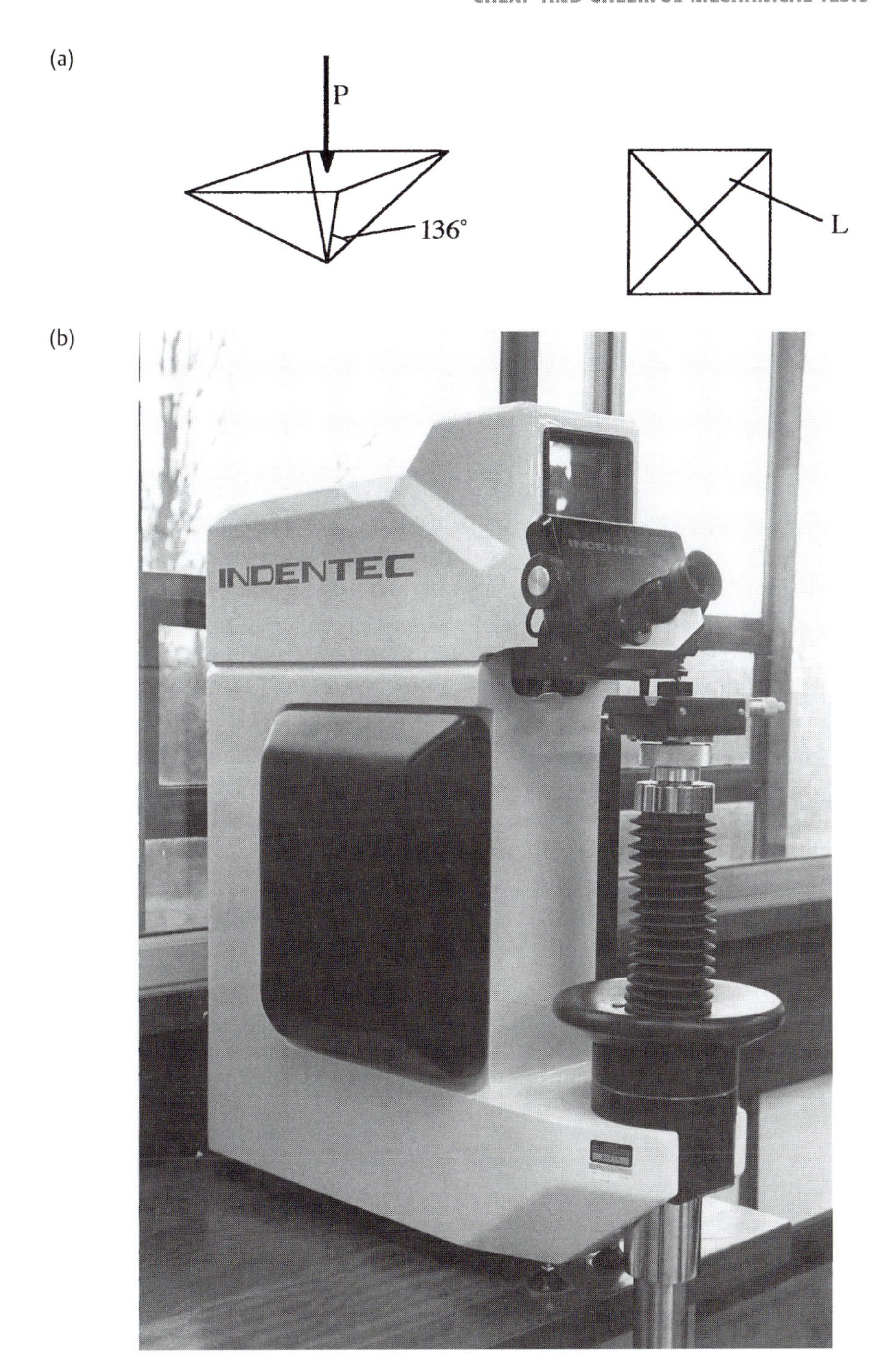

Fig. 3.15 The hardness test (a) schematic (b) a hardness tester. (Note the diamond being pushed into the specimen (near the top of the picture)).

The information obtained, of course, is neither as sophisticated nor as full as that obtained from the tensile test.

Abrasion or *wear* is something which electrical components frequently suffer from—for example overhead power lines for an electric train. A hardness test tells us something about abrasion resistance, although there are special tests specifically designed to measure abrasion properties. Think of mechanical tests as constituting a spectrum right from the very general and basic tests, like the tensile and hardness tests, down to very specific tests linked to very specific applications. Ultimately, each application is its own most appropriate mechanical test. The other side of the coin, though, is that the information obtained is of very limited relevance.

The impact test

Just as the hardness test can be thought of as a simple version of the tensile test, so the impact test is a simple version of the fracture toughness measurement. A notched bar is broken by a pendulum and the energy absorbed in the fracture is measured (Fig. 3.16). Two common specimen geometries are the *Charpy* and the *Izod*. The energy absorbed is measured in Joules. The Charpy V-notch energy C_V is $\sim$ 5 J for a brittle specimen and $\sim$ 100 J for a ductile or tough one. Impact tests find their greatest application with steels (see Chapter 4), which go through a *ductile-brittle transition* with temperature (Fig. 3.17). It is dangerous to use a steel below or near its DBTT.

3.8 Fatigue and fatigue tests

Fatigue is the response of a material to an *oscillating* stress. The important word here is *oscillating*. If I give you a paper-clip to break, you will support it with your two pairs of forefingers and thumbs and bend it to and fro[†] until it snaps. You are instinctively using the fact that *fracture stresses are lower when the stress is oscillated*. Figure 3.18 shows this a little more scientifically as a graph. The fracture stress of any material will come down and down as the number of stress cycles to which it is subject goes up:

Nearly all failures in engineering are due to fatigue

In some materials (e.g. steel) the fracture stress bottoms out at $\sim \frac{1}{3}$ or $\frac{1}{2}$ of the UTS. In most electrical conductors, however, the fracture stress under fatigue conditions continues to drop as N increases. Fatigue failure is always preceded by plastic deformation and the movement of dislocations. It is useful to distinguish two types of fatigue: *low cycle fatigue* and *high cycle fatigue*. In low cycle fatigue (LCF), the applied stress

[†] Unless you are very strong. Or stupid. (Or both.)

(a)

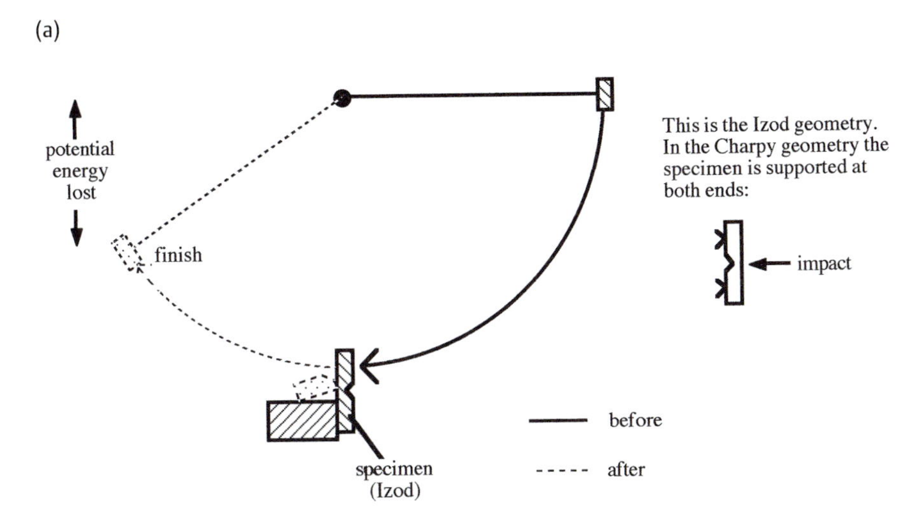

(b)

Fig. 3.16 The impact test: (a) schematic with Izod specimen in place, (b) impact tester with Charpy specimen about to be fractured. Left: the pendulum is swinging from left to right towards the specimen. Right: close-up of the pendulum approaching the specimen (the small rectangular block bridging the gap).

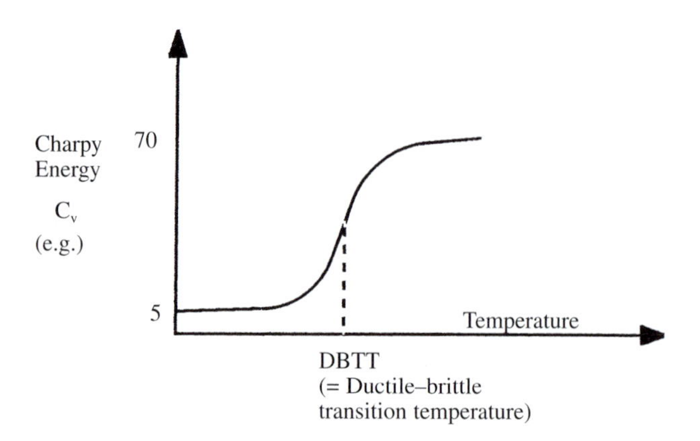

Fig. 3.17 The ductile-brittle transition in steels.

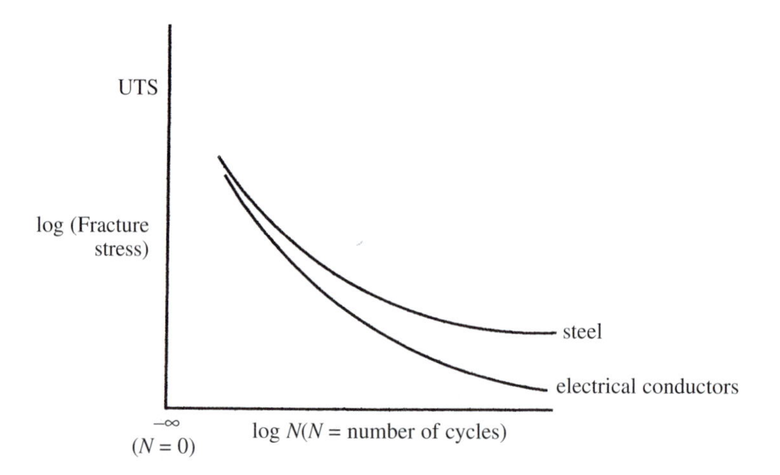

Fig. 3.18 For an oscillating stress, the greater the number of cycles, the lower the stress at which something breaks. The graphs here show two different types of behaviour.

goes above the yield stress and failure occurs in $\sim 10^4$ cycles or less. In HCF the applied stress stays always below the yield stress and failure occurs (if at all) for $N > 10^4$ cycles. Because the applied stress is below the yield stress, there needs to be some *stress raiser* in the component to cause fatigue: for example, a notch ★< or a change in cross section ↳★ where the *local* stress (*) goes above σ_y, even if the averaged out macroscopic stress does not.

Now you may object that you do not intend to subject any of your electrical or electronic apparatus or machinery to oscillating stresses. In fact, these are almost universal, as the following examples will show.

Take motors and generators. The power shaft will be subject to periodic stresses with a frequency equal to that of rotation. There are also low frequency stresses connected with starting and stopping. Most large machinery is subject to mains hum. Any projecting component is liable to oscillate at its natural frequency (classic examples include aeroplane wings and ship bodies). Even electronic components are subject to fatigue—frequently of thermal origin. For example, in a computer chip there is the clock frequency and many other lower frequencies with lots of potential notches—soldered wires are a good example.† The filament in an electric light bulb suffers from *thermal fatigue* (as does the winding of an electric fire) complicated by oxidation. For the light bulb, the design lifetime is very dependent on oxygen pressure in the light bulb.

How do we avoid fatigue? Common sense in design, first and foremost. Avoid sudden changes in cross-section and thus potential stress raisers:

Secondly, it is a fact that most fatigue cracks start from a surface. Putting the surface of a component into compression by some suitable treatment (e.g. carburising a steel) prevents fatigue cracks from opening up. For the same reason in the opposite sense, corrosion can interact with fatigue in a particularly horrible way.

How do we test for fatigue? Fatigue tests are designed to provide two types of information: the number of cycles to failure as a function of stress (an example was shown above: these are called *S–N* curves) and the rate at which fatigue cracks spread. The first type of test is usually made on a polished specimen, the second on a notched specimen, so that the engineer knows where to look for the crack. There are four stages in the life of a fatigue crack:

- initiation
- Stage I crack growth is along the slip planes on which the dislocations glide
- Stage II crack growth is normal to the applied stress. The crack surface shows characteristic ripples (corresponding to the oscillations)
- catastrophic failure

† There is an annual conference on Fatigue in Microelectronic Circuits whose popularity may owe something to the fact that it takes place in San Diego!

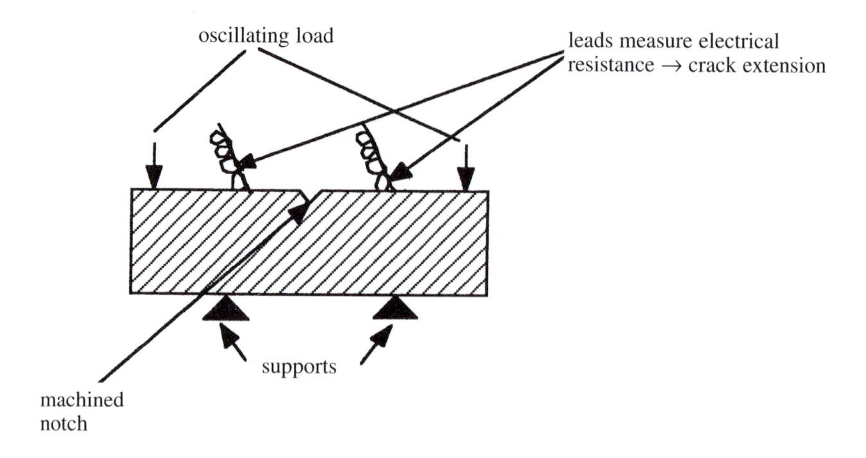

Fig. 3.19 A specimen for a fatigue test (see also Fig. 3.12).

A typical fatigue test specimen is illustrated schematically in Fig. 3.19. Because of their tremendous engineering significance, fatigue and fracture prediction are now an enormous industry in their own right, bristling with mathematical equations and 'lifing strategies'. You, as an electrical or electronic engineer, need not concern yourself with this—but you do need to be aware of the tell-tale signs of fatigue and have some idea how to avoid it.

3.9 Creep and creep tests

Now we come to the last of our mechanical properties and tests. Recall that fatigue involved *oscillating* stresses. Creep, the last of our topics, involves *constant* stresses. So we have

Tensile test	:	Constant strain rate (increasing stress)
Fatigue	:	Oscillating stress
Creep	:	Constant stress

Now you may object that a constant stress is not very significant—either it *is* above the fracture stress, or it *isn't*. Waiting is not likely to change anything. This is true at normal temperatures, but at high temperatures a slow viscous deformation takes place in materials which ultimately ends in fracture. This is *creep*.

One high temperature application important in electrical engineering and where creep matters is in a turbine for generating electricity. The hotter the turbine, the more efficient it is. The very hot blades spinning round are subject to considerable centrifugal forces. If they were to grow in length (i.e. to creep) they would eventually catch on the casing. Ordinary plastic deformation takes place via dislocation movement, as we have seen. Creep deformation requires *diffusion* as well. Diffusion in solids requires high temperatures, and this is why creep only occurs in hot materials. What is 'hot' for some materials, however, is not for others. The important parameter here is 'homologous temperature'. This is the temperature in Kelvin (i.e. degrees absolute) divided by the melting point of the material (also in Kelvin). Thus for a steel or a nickel alloy, which melts between 1500 °C and 2000 °C, 'hot' means several hundred °C, whereas for electrical solder, say, which melts at 180 °C, room temperature might be considered 'hot'.

Why does diffusion matter? Diffusion, in creep, acts either as an *enabler* of dislocation motion, or at very high temperatures it provides the creep deformation directly. Diffusion enables dislocation motion by allowing dislocations to negotiate past barriers—formed for example by other dislocations as in work hardening. If you want to understand more about this topic, consult one of the references given at the end of the chapter.

A typical creep curve is shown in Fig. 3.20. A fast spell at the beginning (I) is followed by a long straight part (II) which gradually accelerates (III) towards fracture (*). Creep data are obtained from a tensile testing machine which has a furnace fitted round the specimen (Fig. 3.21). Temperature control is quite important—a few °C difference can make a great difference to the creep rate. Creep tests often go on for a very long time—months or years—so there must be some back-up in case of power failure.

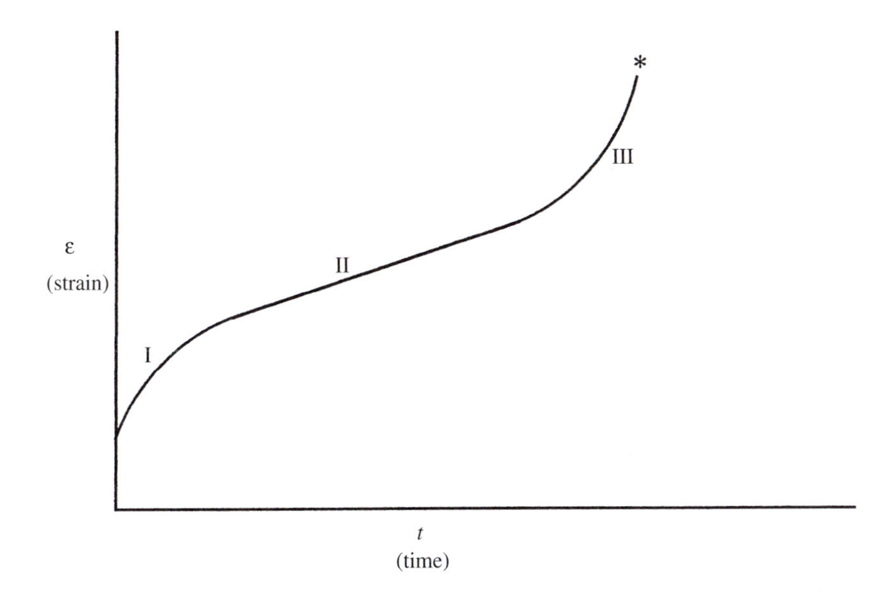

Fig. 3.20 A typical creep curve showing strain as a function of time.

Fig. 3.21 Some creep testing machines.

Creep data are not usually presented as in the graph above, because it doesn't contain enough information. Two typical ways of presenting creep data to engineers are illustrated in Fig. 3.22.

How do we reduce creep? As with fatigue, avoiding high stresses and using strong materials both help. One difference between creep and fatigue, however, relates to grain size (size of the little crystals which make up the metal). Generally speaking a small grain size in a metal is a good thing (this isn't true just of fatigue—it is quite general). Not for creep, though. It turns out that a quite important part of creep deformation consists of grains sliding past one another. The smaller the grain size, the worse it becomes. High tech turbine blades (more appropriate for aero gas turbines than land based power generators, in fact) consist of very large grains or even of single crystals! Now this is very bad for fatigue, strength, ductility and just about everything else. As so often in engineering, we have to come to the best compromise between several competing and mutually incompatible requirements.

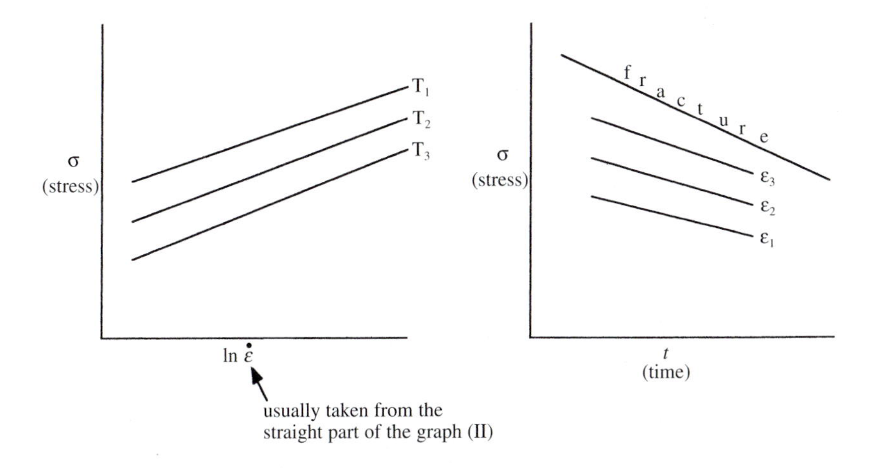

Fig. 3.22 Two common ways of presenting creep data.

3.10 Summary

1. Stress is force/area.
2. Strain is displacement/distance.
3. Tensile and compressive stresses have the force perpendicular to the area they act upon.
4. For a shear stress the force is parallel to the area it acts upon.
5. A tensile or compressive strain has the displacement and reference length perpendicular to a surface.
6. A shear strain has displacement parallel to a surface, but the reference length perpendicular to it.
7. Elastic deformation is temporary, plastic deformation permanent.
8. Plastic deformation occurs via dislocations.
9. A tensile test measures yield strength, proof strength, (ultimate) tensile strength, tensile ductility and (not very well) Young's modulus.
10. Hardness is measured by forcing a hard object into the surface. Vickers Hardness Number is $3 \times \sigma_{\mathrm{y}}$.
11. Charpy and Izod impact tests involve measuring the energy absorbed in fracture from a quickly moving pendulum by a notched specimen.
12. The propensity of a material to crack is measured by its fracture toughness.
13. Fatigue is exposure of an object to an oscillating stress.

14. Fatigue failure stresses are lower than tensile ones: most industrial failures are fatigue failures.
15. Creep is deformation under a constant stress. It is usually a high temperature phenomenon.

Recommended reading

Mechanical metallurgy by G.E. Dieter (published by McGraw-Hill)

Engineering materials I by M.F. Ashby and D.R.H. Jones (published by Elsevier)

Questions (Answers on pp. 305–315)

(The acceleration due to gravity, g, is 9.81 ms^{-2}.)

3.1 A metal cube with an edge length of 2 cm is placed on a smooth flat table. A mass of 100 kg is balanced on top of it. What stress acts on the cube? If Young's modulus for the cube is 10^{11} Pa, what is the resulting decrease in height of the cube?

3.2 The cube of Question 3.1 is glued to the table and a rod glued to the top side:

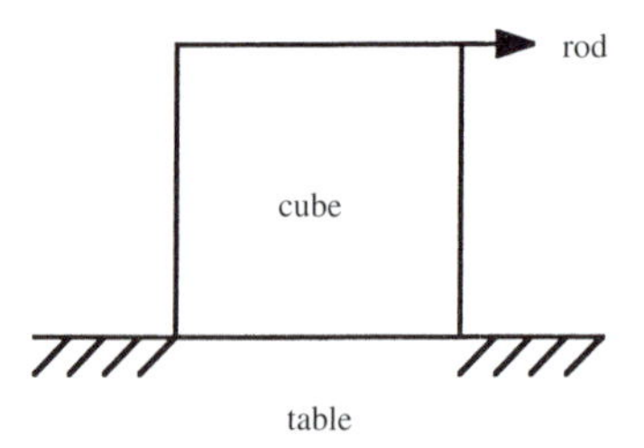

Via the rod a force of 3 kN is applied. If the shear modulus is 5×10^{10} Pa, describe the final shape of the cube and the strain within it.

3.3 In a tensile test, the gauge length of the cylindrical mild steel specimen was 5 cm and its diameter 13 mm. Yielding occurred at a load of 36.6 kN when the extension of the gauge length was 0.067 mm. The maximum load was 60.4 kN at an extension of 15.4 mm and the specimen finally fractured at an extension of 17.0 mm. Using these figures, work out

(a) the yield stress,

(b) Young's modulus,

(c) the (ultimate) tensile stress,

(d) the true strain at the onset of necking

and (e) the tensile ductility.

3.4 A piece of machinery is suspended using two vertical solid cylindrical bars. The bars are 1 cm in diameter and 30 cm long. The material from which the bars are made had previously been assessed using a tensile test, from which the following parameters had been derived:

Yield stress	=	300 MPa
Young's modulus	=	200 GPa
U.T.S.	=	450 MPa
Tensile ductility	=	20%

(a) What weight of machinery would cause plastic yielding of the supporting bars?

(b) What would be the length of the bars when plastic yielding commenced?

(c) What weight of machinery would break the bars?

(d) What would be the length of the bars just before they broke?

3.5 The following data are taken from a Copper Development Association brochure:

Typical mechanical properties of wrought coppers and high conductivity copper alloys

		Designations - ISO and BS									
		Wrought coppers and copper alloys					*Heat treatable alloys*				
Typical mechanical properties	**Condition**	**Cu-ETP Cu-FRHC Cu-FRTP CuAg C101, C102, C104**	**Cu-DHP Cu-OF Cu-OFE C106 C103, C110**	**Cu S CuTe C111, C109**	**CuCd C108**	**Condition**	**CuBe2 CB101**	**CuCo2Be A3/1**	**CuCr1 CuCrlZr CC101 CC102**	**CuNi2Si A3/2**	**CuNip A4/1**
Tensile strength (N/mm^2)	a	220	220	230	280	b	500	310	230	310	230
						c	1160	740	400	635	450
	h	385	385	310	700	d	1400	850	510	740	495
0.1% Proof stress (N/mm^2)	a	60	60	60	60	b	185		45	80	60
						c	930	650	265	480	340
	h	325	325	265	460	d	1080	770	430	650	420
Elongation (%)	a	55	60	40	45	b	50	32	50	50	45
						c	5	15	22	15	25
	h	4	4	8	4	d	2	10	20	10	20
Hardness (HV)	a	45	45	50	60	b	100	80	65	60	60
						c	370	220	125	170	140
	h	115	115	100	140	d	400	240	160	210	175
Softening temperature (°C)		150†			250	c	300	500	500††	500	475

Condition

a, annealed; h, hard; b, solution heat treated; c, solution heat treated and aged; d, solution heat treated, cold worked and aged.

Softening temperature—the lowest temperature that, if maintained for 2 hours, will give a reduction in hardness of 20% of the difference between the hardest as received condition and the softest possible condition of that material.

† for Cu-ETP

†† CuCrlZr 52 °C

(Taken with permission from *High conductivity coppers—properties and applications*, CDA Technical Note 29, Table 7 (replaced by Publication 122).)

Using the data in the table, work out at what load a tensile specimen of solution heat treated, cold worked and aged CuCr1 would break. The gauge of the testpiece is circular with a radius of 2 mm.

3.6 Again using the table above, estimate the flow stress of aged CuBe2 from its hardness and compare with the yield stress and UTS.

3.7 The plastic part of the tensile true stress–true strain curve of copper is well approximated by the expression $\sigma = 320\sqrt{\varepsilon}$ MPa. Assuming that the volume of a copper tensile specimen does not change during plastic deformation, at what strain does necking begin?

3.8 In an impact test a pendulum of mass 9.84 kg swings down and strikes a Charpy specimen, fracturing it. Its final height is 40.2 cm less than its starting height. What is the impact energy of the specimen?

3.9 The surface of an aluminium alloy tensile specimen is intersected by a 1 mm crack. At what stress would the crack cause fracture? (The fracture toughness K_{Ic} of this particular aluminium alloy was 30 MPa $m^{0.5}$.)

3.10 Steady state creep rates obey the following relation:

$$\dot{\varepsilon} = A\sigma^n e^{\dfrac{-Q}{RT}}$$

where $\dot{\varepsilon}$ is the strain rate, A and n are constants, Q is an 'activation' energy, R is the Gas constant and T the temperature in K.

Use the following data, derived from creep tests on solder specimens, to determine A, n and Q and thus predict the steady state creep rate at a stress of 6 MPa and temperature of 25 °C.

T (°C)	σ (MPa)	$\dot{\varepsilon}$ (s^{-1})
20	5	3.54×10^{-6}
20	10	6.11×10^{-5}
40	5	2.97×10^{-5}

(The Gas constant $R = 8.31$ J mol^{-1}.)

We have now reached p. 101 of this book and I think it is time to do something useful! Like making things . . .

Manufacturing conductors

Chapter objectives

- Where metals come from
- Making simple shapes
- Soldering
- Strengthening metals

4.1 Producing elemental copper

This chapter is dominated by copper, which, for reasons discussed at the end of Chapter 2, is by far the most popular choice of material for electrical conductors. Have you ever wondered where all the metal used nowadays and which is all around us, comes from? In this section I would like to give you a brief account of how copper arrives at a factory ready to be made, or *formed*, into useful articles like wire etc. (This is because I think you will find it interesting! This is not a book on metallurgy and you are not supposed to be metallurgists.)

All metals are dug up out of the ground, mainly in the form of *ore*. An ore consists of a *mineral* mixed with a lot of earth and rock. The mineral is a compound of a metal with a non-metal:† usually oxygen or sulphur. The earth and rock have to be separated off first, by mechanical means. The mineral is more stable than the metal and so we have to supply energy to convert it into the metal. This energy is supplied as heat or electricity and often other chemicals are involved. This whole process is gradually reversed later on, when the metal has been made into a component, by *corrosion* (see Fig. 4.1 and Chapter 6).

† Notice that minerals are *ceramics* (see Chapter 7). It is because ceramics have a far lower energy than metals that lavatories, washbasins etc are so (relatively) cheap.

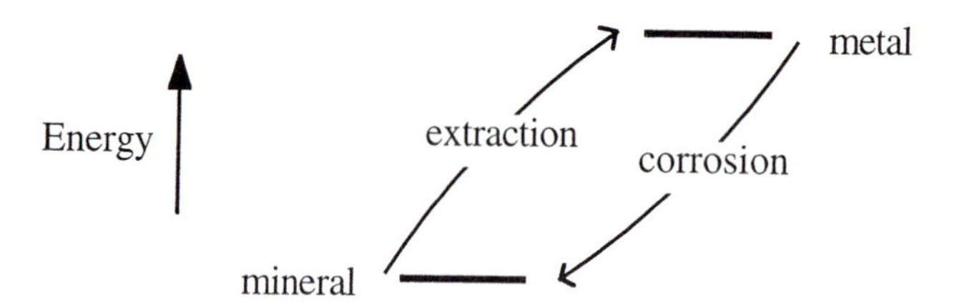

Fig. 4.1 Minerals are more stable than metals. We have to supply energy to extract the metal from the mineral. Corrosion then gradually returns the metal to its native state.

Copper minerals are generally copper sulphides. They are quite rare—for example there is 10000 times less copper in the earth's crust than aluminium or iron. They are converted to copper by oxidising the sulphur to sulphur dioxide in a furnace. As the copper solidifies, the escaping bubbles of sulphur dioxide mark the surface and the product is thus called 'blister copper'. It is about 99% pure, containing small amounts of iron, nickel, sulphur and oxygen. The blister copper is cast in the form of plates (1 m × 1 m × 40 mm thick) which then are made the anode of an electrolytic cell (see Chapter 6, Section 6.2). The electrolyte is copper sulphate and the voltage applied is about $\frac{1}{4}$ V. The copper dissolves off the anode into the electrolyte and then at the other end of the cell it plates onto the cathode. The impurities fall to the bottom of the cell to form a sludge. 'Copper cathodes' (also shaped like plates) are about 99.99% pure, but do contain enough hydrogen to make them brittle. They are remelted to remove the hydrogen and cast into shapes appropriate to their final fate (e.g. wire, sheet etc). This type of copper is called *electrolytic tough pitch* or ETP and is the most common grade used in the electrical industry. A little oxygen (~0.02%) is left in the copper to remove harmful impurities which degrade the electrical conductivity. Conductivity (and therefore quality) is measured in terms of %IACS (International Annealed Copper Standard). This was set up in 1913. The purity of copper has improved since then to the extent that a good quality ETP copper now has a conductivity equivalent to 101.5%IACS.

Copper is a commodity metal. This means that sufficient of it is used—it is sufficiently economically important—for it to be traded on the London Metal Exchange (LME) (sometimes Comex for Americans). The price of copper can vary quite violently and rapidly. This is because there isn't much of it in the Earth's crust (see Section 5.1 and Table 5.1) and is also due in part to the political instability of some of the countries where copper mines are found.

For a manufacturer buying copper its price is of course of crucial importance. Some of the variations shown in Fig. 4.2 can be ironed out by trading in copper 'futures'—promises to buy copper at a given time in the future at a given cost. A spectrum of times is used and this is what averages out the price. Copper nowadays is usually transported from the mine as cathodes. Typically, the copper on a ship (say) will be bought and sold several times while it is on its way from its country of origin to the country where it will be used. Convention dictates that whoever owns it when it hits the quay must actually

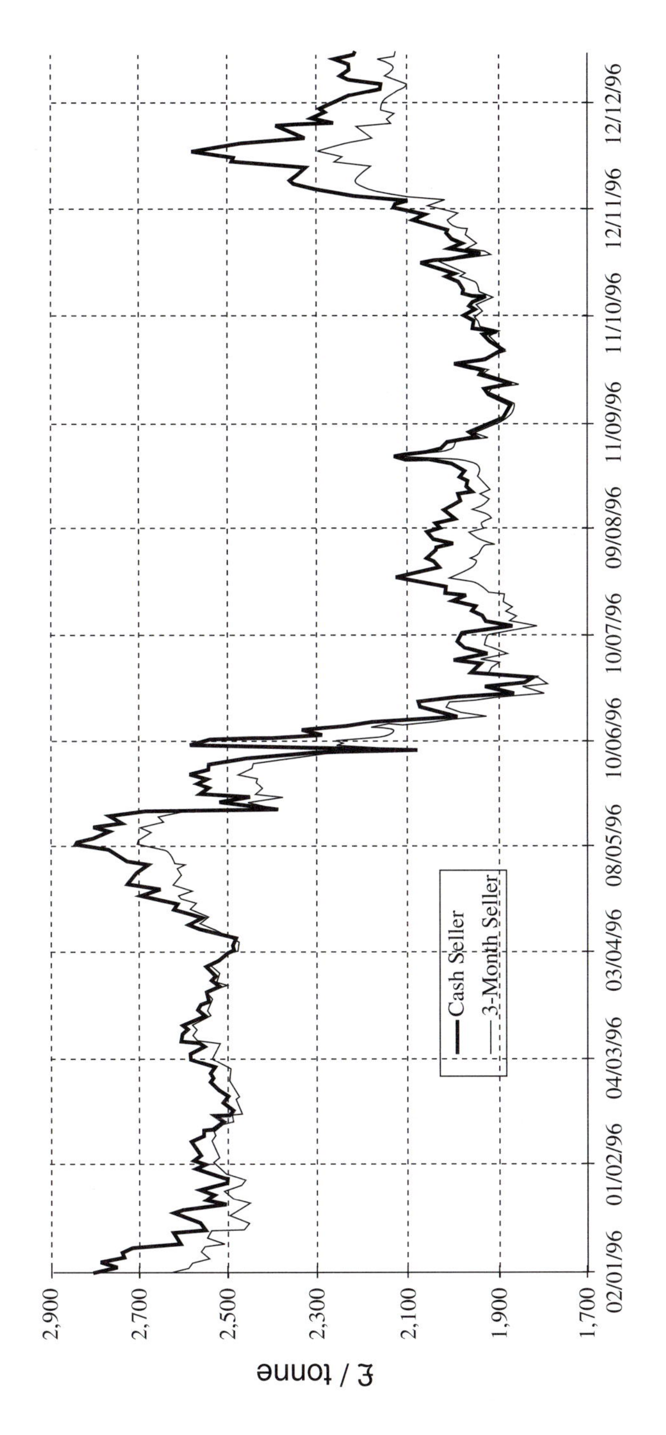

Fig. 4.2 How the price of copper varied throughout 1996 (Courtesy of The London Metal Exchange information department).

pay for it and become the true owner. I was amazed when recently visiting a large copper manufacturer to find the stockyard empty and the plant about to be shut down for some days. This was the result of rough weather at sea preventing the landing of copper. Evidently the 'just-in-time' philosophy has its dangers.

If you would like to know more about copper, why not use the World Wide Web? Try:

http://www.copper.org
http://www.lme.co.uk

I think you will find them interesting.

Aluminium, silver and gold also have to be dug up and extracted. For aluminium (a base metal) the ore is cheap, but the extraction expensive. For silver and gold it is the other way round (see Section 2.8).

Shaping copper

The next stage in the manfacturing process is to *shape* the copper ingots. This is described in the next three Sections 4.2–4.4.

4.2 Making large cross-section conductors

Typical examples of large cross-section conductors include busbars, winding strips, rotor bars and commutator sections. These are normally *extruded* hot from a cast bar which is ~750 mm long and 200 mm across. Figure 4.3 shows a schematic picture of how an extrusion press works with a photograph of the real thing underneath.

4.3 Small cross-section conductors: wires

Extrusion is not very suitable for producing small cross-sections, and so wires are made by *drawing* i.e. instead of *pushing*, we *pull*.

Typically the copper is cast continuously (Fig. 4.4) and then rolled while still hot down to 7–8 mm diameter rod (see Fig. 4.6 for an example of rolling (not part of the same line)). There are several sets (e.g. 12) of rolls which alternate between vertical and horizontal. This rod is cleaned chemically ('pickled'), abraded and waxed and then wound round a drum up to a weight of several tons before being cut off and another drum started. The coil of copper is fed into a wire drawing apparatus: the metal is pulled through a sequence of dies (see Fig. 4.5) and its size reduced by a few % each time. The large dies are made from cemented carbides and the smaller ones from industrial diamonds. After every few reductions the wire has become so hard and brittle because

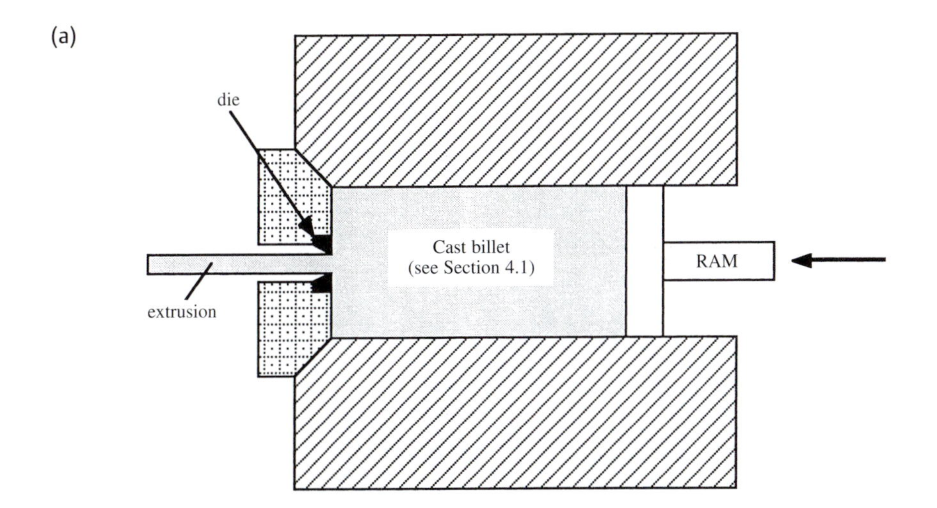

Fig. 4.3 (a) Schematic diagram illustrating the extrusion of a large cross-section conductor. (b) The business end of a real extrusion press (courtesy Cerro Extruded Metals Co).

of work hardening that it has to be *annealed* to soften it and make it ductile. This is done by passing it through a furnace. At the end of the process the wires may be twisted together to form multistrand wires and then insulated by coating with a polymer of some sort (e.g. PVC). I will deal with this in Chapter 8 on plastics.

Fig. 4.4 Continuous casting of copper. The copper is melted over to the left. The liquid copper is conducted to the continuous casting unit on the right. You can see the bright stream of copper hitting and running round the water cooled and slowly rotating wheel. The solid (but still hot) copper strip is led off to a series of rolls. For an example of rolling, see Fig. 4.6. The whole sequence is called the Southwire process (from the Southwire Company in Georgia, USA). (Directing a jet of molten metal at a spinning wheel does have its dangers. I was standing innocently in front of the wheel, which was just being started up, when the foreman suggested I might be more comfortable† about 20 m further away. When the operation got under way, I could see what he meant. (See cover of book)(† Actually, this wasn't *exactly* what he said.)(Courtesy BICC Rod Rollers Ltd.)

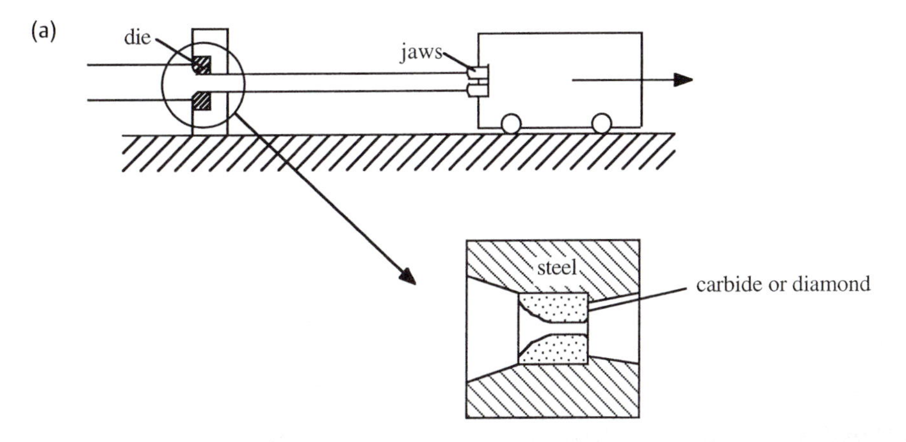

Fig. 4.5 (a) Wire drawing (schematic) (b) An early stage in drawing wire (courtesy Thomas Bolton Copper Products Ltd).

4.4 Flat conductors: rolling

Many conductors are used in the form of sheet. The copper is continuously cast (see Section 4.3) as slab, *hot rolled* down to ~10 mm and then *cold rolled*. A photograph of rolling is shown in Fig. 4.6. Hot rolling requires less force (most things become softer as the temperature rises) and the copper does not work-harden (become harder): it

Fig. 4.6 Rolling copper strip. (Courtesy Thomas Bolton Copper Products Ltd.)

continually *recrystallises*, i.e. new perfect grains nucleate and grow into the deformed copper, replacing material with high densities of dislocations with new, perfect (and soft) crystals—see Fig. 4.7. Cold rolling requires greater force, but gives a better surface finish and a springier product (if this is what is wanted) because recrystallisation has not occurred. (This is because recrystallisation is thermally activated: it needs temperatures of the order of $\frac{1}{3}$ or $\frac{1}{2}$ of the melting point (in K).) The final component is *stamped* and/or *pressed* and/or *machined* from the sheet.

Finally, to complete our roll-call of shaping operations, Fig. 4.8 shows *open-die forging* of a copper ring.

In Section 4.2–4.4 I have described some simple **forming operations** appropriate to a soft ductile metal like copper. In real life, of course, the exact method used may be a mixture of all these, and the final choice will probably be influenced by what is available on the spot. There is a tremendous amount of science, technology (and art!) involved in forming operations which I haven't even touched upon. As you might guess, the balance between tensile and compressive stresses is crucial and, for example, the design of the shape of a wire drawing die, the lubrication, the temperature and the reduction ratio (after/before) all have to be correct to ensure the wire does not snap. Also, although I have used copper as an example, because it is the most common metal used for electrical conductors, most of what I have said about forming operations would be relevant to aluminium. For more details, please see the recommended reading list at the end of the chapter.

Fig. 4.7 New grains for old: a heavily deformed piece of copper is recrystallising. The fresh new grains (the clean areas) are replacing the old deformed ones (the mangled looking areas) which are jam pack full of dislocations. The new grains are much softer. (The grain size is about 10 μm.) (Courtesy Professor F. John Humphreys.)

4.5 Joining conductors: soldering

Sometimes, if a relatively simple shape is required, the manufacturing process finishes with one of the operations just described. Often, though, equipment in electrical engineering and electronics is not made in one piece, it is made up from lots of components, which have to be joined together. Sometimes this is a mechanical connection—for example, a wire screwed into a terminal. Sometimes a more permanent or less bulky connection is required. In this case electrical conductors are

Fig. 4.8 Open-die forging of a copper ring. The operative on the right has just positioned the ring (still glowing red) under the hydraulic hammer. (Courtesy Thomas Bolton Copper Products Ltd.)

soldered together. The other main method of joining metals—welding—is used for steels. It will be described in the next chapter, Chapter 5.

In soldering, a low melting point alloy—the solder—is melted and run in between the two parts to be joined (Fig. 4.9). The solder then solidifies and holds the two components together. These are hardly affected, although a little bit of each surface dissolves in the solder. Solder is like a metal glue. In welding, however, the components to be joined are very much altered, close to the join—for example melted. Welding is a much higher temperature procedure, giving much greater strength. Soldering in the sense I am using is sometimes called *soft soldering*. In between soldering and welding there are *brazing* (brass-ing) and *hard soldering*. So we have:

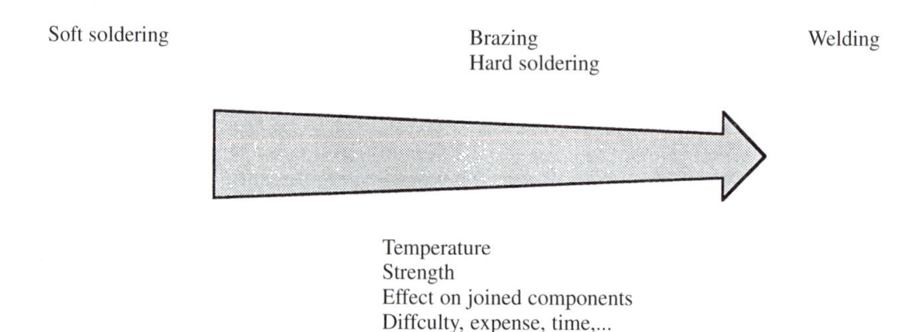

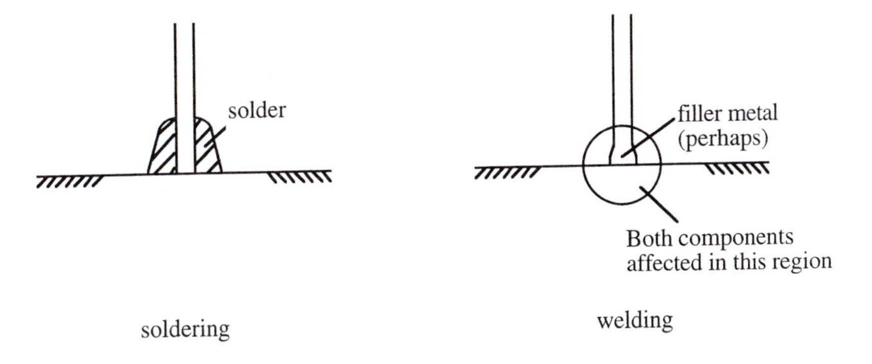

Fig. 4.9 Soldering and welding (but see Fig. 4.13!).

This section is about soft soldering, which I will call simply 'soldering'.

A soldered joint can have three objectives:

Electrical connection

Moderate mechanical strength

(Provide a seal against liquid or gas)

It is the first two which concern us.

Which alloy is used for solder? The most common solder, used for joining copper components together, is made from tin (Sn) and lead (Pb), which are both low melting point metals. The exact composition used is 62 wt% tin and 38 wt% lead, where wt% means 'percentage by weight'. (See Question 4.2 at the end of this chapter.) The reason for this exact choice of composition can be seen from Fig. 4.10, which is a metallurgical *phase diagram*† for tin and lead. The phase diagram shows for any combination of temperature (y axis) and composition (x axis) which *phase* or *phases* are present, and their compositions and fractional amounts. With two elements there is a maximum of two phases under any circumstances: sometimes just one. At high temperature, for example, there is one phase, which is a liquid. At low temperature the alloy is solid, but it may consist of one or two phases, depending on composition. What, then, is a phase? A phase is a region of material which has its own composition and crystal structure. Notice that a 'phase' is different from a 'state of matter'. A state of matter is *solid, liquid, gas* etc‡ but within a solid object there can be several different solid phases.

† Some of you who have studied chemistry may have met 'phase diagrams' where pressure was one of the axes. For liquids and solids, pressure is of minor importance and is usually ignored.

‡ This is a little over-simplistic: we already know there is crystalline solid and amorphous solid. There are other, more subtle, distinctions as well.

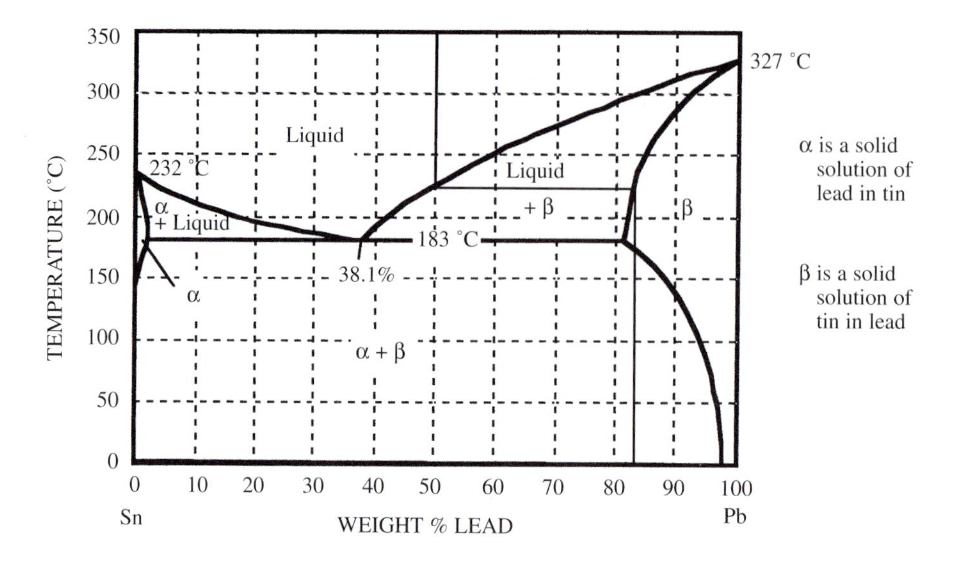

Fig. 4.10 The phase diagram for tin and lead.

For example, imagine, using the procedures described in Chapter 2, cutting up and polishing and etching an alloy containing 98 wt% lead and 2 wt% tin. We would find just one homogeneous phase whose composition is inevitably the overall alloy composition. This is an example of a *solid solution*, which we came across in Chapter 2 (Section 2.5). 2 wt% of the lead atoms are replaced by tin atoms. The crystal structure stays the same as for lead (f.c.c.) and the tin atoms substitute randomly for lead atoms (see Fig. 2.16). This phase is labelled 'β' in Fig. 4.10: solid phases are always labelled from left to right on metallurgical phase diagrams according to the Greek alphabet: α, β, γ etc. The line defining the left hand border of this *single phase field* also defines the maximum solubility of tin in lead. For example, at 150 °C this solubility is about 10 wt%. If we try to put more tin than this into our alloy at this temperature a second phase—α, a solid solution of lead in tin—will form. We have moved from a single phase field to a two phase field. Notice that the solubility of tin in lead increases with temperature (this is quite normal) and that the solubility of lead in tin is quite different from that of tin in lead—again, not unusual.

Let us look at a quite different alloy composition: 60 wt% tin, 40 wt% lead. There are two solid phases present: one is α, a solid solution of lead in tin, and the other is β, a solid solution of tin in lead. Figure 4.11 shows a picture of this structure. The two phases are labelled. If we were to dig a little bit of each phase out and analyse it chemically,† we would find that they had quite different chemical compositions, α being nearly all tin (Sn–0.1 wt% Pb) and β being nearly all lead (Pb–3 wt% Sn). These are

† Actually, this can be done *in-situ* in a scanning electron microscope: no need for a microscopic spade!

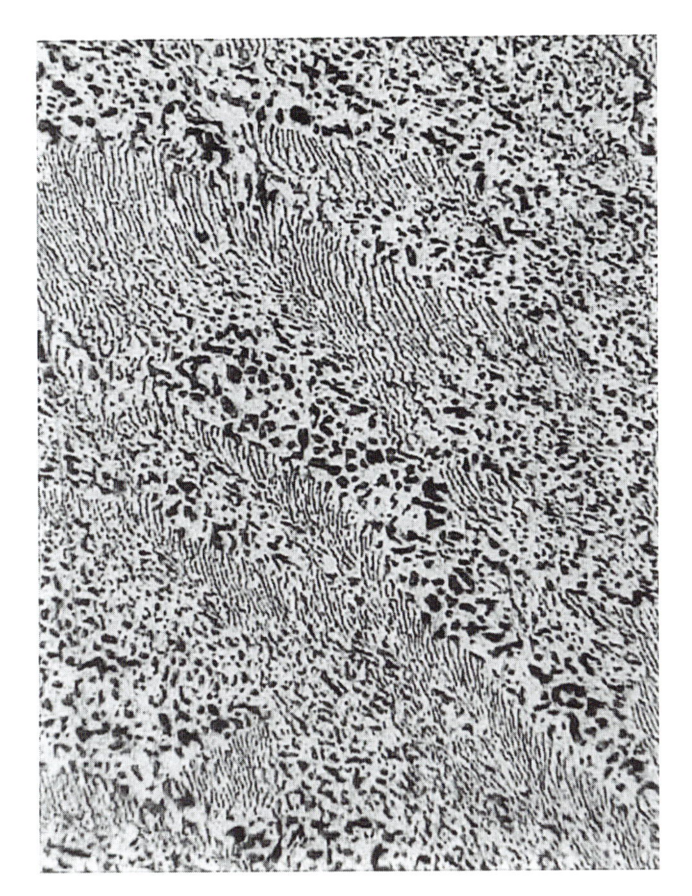

Fig. 4.11 An optical micrograph (a picture taken with an optical microscope) of common solder. The two phases show as dark and light. The light coloured phase is α, a solid solution of $\sim$0.1 wt % of lead in tin. The dark coloured phase is β, a solid solution of $\sim$5 wt % of tin in lead. (The exact amount depends on how quickly the alloy is cooled and when solid state diffusion becomes too slow to reproduce the phase diagram—see Fig. 4.10.) The composition is 62 wt% tin and 38 wt% lead. This is the eutectic composition (see Fig. 4.10) which is used for electrical solder. The microstructure consists of a fine array of alternating plates or juxtaposed particles. (Micrograph courtesy the late Ian Garbett.)

the compositions given by the lines on the phase diagram either side of the two phase region, so as the composition of an alloy changes from Pb–3 wt% Sn across to Sn–0.1 wt% Pb the compositions of the two phases α and β do not change, but the relative amounts of each of them do change, from all α at the left hand end, to all β at the right hand end. You can get some practice at using phase diagrams and reading off compositions and relative fractional amounts in Questions 4.2–4.7 at the end of the chapter. Note that however long we held the alloy, there would always be two phases. They will never turn into one. The phase diagram describes the *equilibrium* situation.

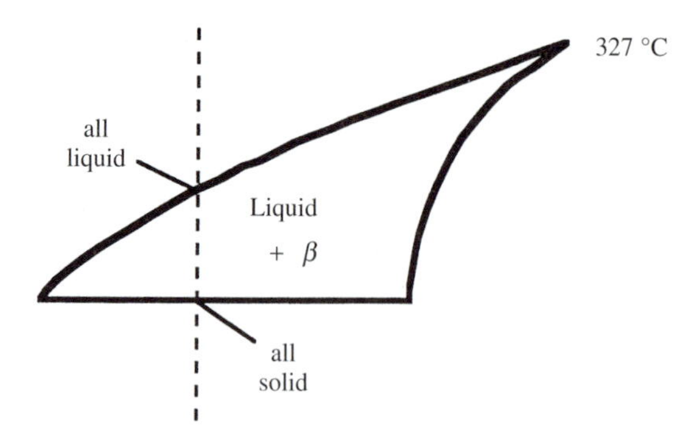

There are two more two phase regions in the phase diagram—those marked 'α + Liquid' and 'Liquid + β' (see above). Here there is a mixture of a solid phase and a liquid phase and the alloy is mushy or pasty. The same rules apply as for the two solid phase region: for example as an alloy is cooled through a two phase region the fraction of liquid in the system changes gradually from 100% to 0%.

So to come back to my original question, why is electrical solder 60 wt% tin and 40 wt% lead? Notice from the phase diagram that this is a very special composition. Only here does liquid turn directly into solid without going through a two phase region. This type of diagram, which is quite common in metallurgy, is called *eutectic* (Greek for 'well melting') and the special composition used for solders is called the eutectic composition. The eutectic composition is ideal for electrical† solders because:

- it represents the lowest melting point (minimum damage to surrounding electronic components)
- melting is very sudden and not protracted (imagine holding a component in place for a long time while the solder passes through the mushy two phase region)
- the eutectic two phase structure is very fine, giving good strength

(for why, see the following section, 4.6 and Chapter 5, Section 5.2)

Notice that the melting point of common electrical solder is 183 °C. In general, solders melt below ~300 °C. Solders suitable for other metals and temperatures are listed in Table 4.1. They are all eutectics, sometimes involving more than two elements.

There is enormous environmental pressure on manufacturers to eliminate lead from solders. This is not a trivial problem and currently nothing as good and cheap as lead–tin has been discovered.

Making a good joint: fluxes

The essence of soldering is that the solder *wets* the two surfaces to be joined. The very accommodating metallic bond does the rest. For the solder to wet the surfaces, these must

† Plumbers' solder is chosen to be more lead rich because, unlike an electrician, a plumber does not want sudden solidification of the solder. He/she wants to be able to manipulate and wipe the joint during the mushy stage. (Also lead is cheaper than tin!)

Table 4.1 Solders for different materials and purposes.

To Join		Solder
Cu, Ag, Au		Sn–40 wt% Pb
Au (better)		Sn–30 wt% Pb–17 wt% In–0.5 wt% Zn
Al		Sn–8 wt% Zn
Low temperature solder	(145 °C)	Sn–32 wt% Pb–18 wt% Cd
High temperature solder	(221 °C)	Sn–3.5 wt% Ag
	(310 °C)	Pb–1 wt% Sn–1.5 wt% Ag

be clean—free from grease and free from oxide. To ensure this, the two surfaces are usually cleaned, and a *flux* is used with or before the solder. The cleaning operation can be simply abrasion, or sometimes the surfaces are *pickled*—dunked in an acid which removes dirt and corrosion layers. For example, copper is pickled in warm dilute sulphuric acid. The flux performs a similar function, but *during* or *immediately preceding* the soldering operation. The most gentle flux is *rosin* (a type of resin, distilled from pine sap) in an organic solvent. Rosin contains mild organic acids which clean the copper (etc.) surface and leave harmless residues. In fact, rosin is often too mild and has to be 'activated'. This involves adding stronger organic acids or chlorides. The problem with this is that the more 'active' the flux, the more harmful the residue, which is often inconvenient or difficult to wash away.[†] The final choice of flux depends on the exact application.

When the flux is introduced at the same time as the solder, a *cored* solder is used, with the rosin based flux in channels in the solder. Figure 4.12 shows how the flux cleans the surface and allows the solder to make a good bond.

The necessity for the solder to bond with a clean metal surface means that soldering is very suitable for the noble metals copper, silver and gold but quite difficult for the base metal aluminium, for which special solders (see Table 4.1) and aggressive fluxes are necessary.

It is quite normal for solder to dissolve the top layer of the components. Usually reaction products are formed between the tin and lead in the solder and the copper, silver or gold. These reaction products are 'intermetallic compounds'—chemical compounds with their own formulae, like Cu_6Sn_5, but formed between two metals and themselves a metal. Intermetallic compounds are very brittle, but are harmless so long as there is not too much of them.[‡] We will return to intermetallics at the end of this chapter in a completely different context.

[†] If you have ever had a copper hot water cylinder start to leak, it nearly always happens round the soldered joint near the bottom, where the flux has not been washed off properly.

[‡] The intermetallic compound formed between gold and tin, which occurs during the soldering of microelectronic circuits and is rather a nuisance because it is so brittle, is called *purple plague*!

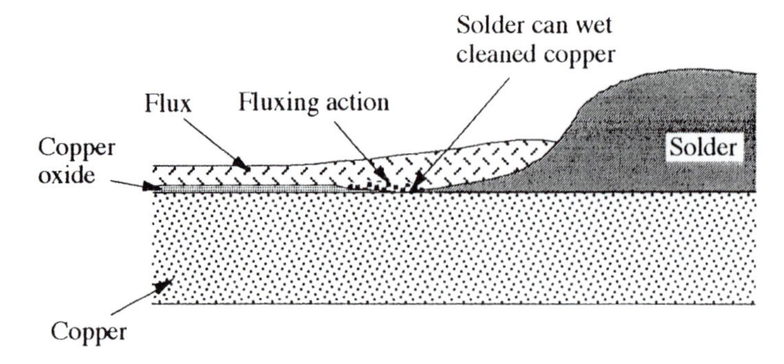

Fig. 4.12 The flux in a multicored solder cleans the copper (or whatever) and enables the solder to bond to it (taken from *Soldering in electronics assembly* by Mike Judd and Keith Brindley, courtesy of Newnes).

The mechanical properties of soldered joints

Solder is a weak material which is used very close to its melting point. If significant strength is required from a soldered joint, then that strength must be achieved mechanically, leaving the solder to make the electrical connection only (Fig. 4.13).

The yield strength of solder is around 30 MPa. Figure 4.14 below shows how the tensile and shear strengths of tin–lead alloys vary with the proportion of tin to lead. Notice that the fine eutectic structure at 60 wt% Sn gives about the strongest alloy. Tensile strength is what I called in Chapter 3 'UTS'. Shear strength I didn't mention, but it is the equivalent parameter for applied *shear* stresses, which are very appropriate for a well designed soldered joint (see Fig. 4.13).

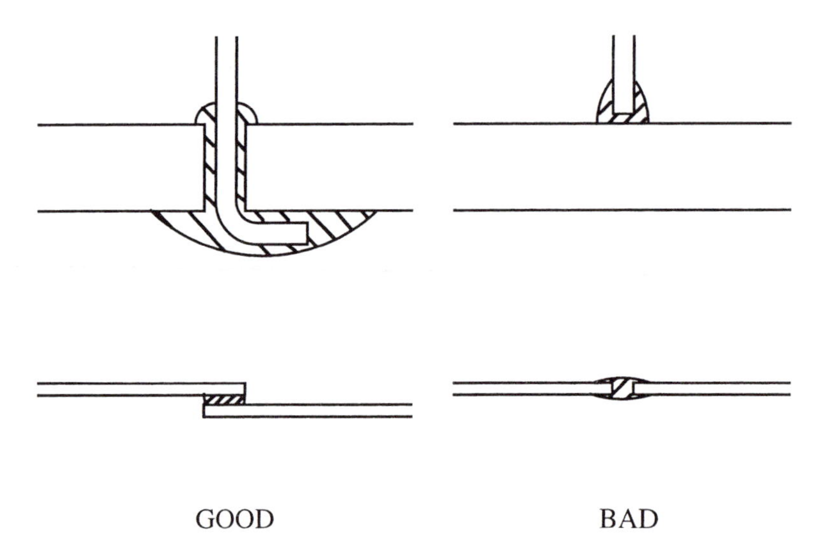

Fig. 4.13 Good and bad examples of soldering practice. (The same considerations apply to adhesives.)

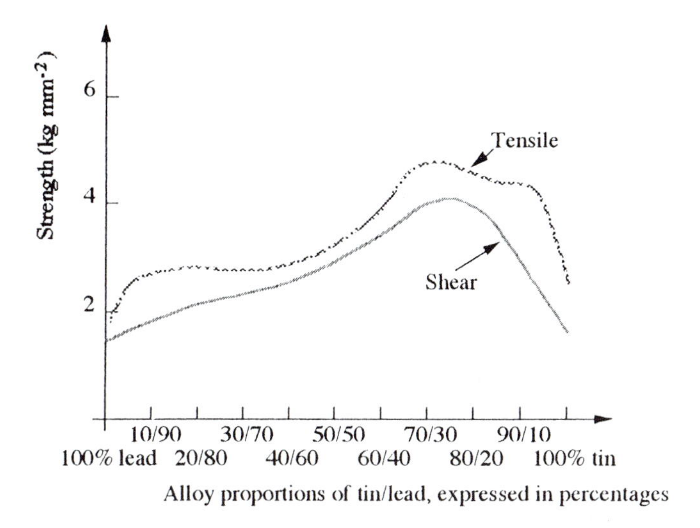

Fig. 4.14 The variation of tensile and shear strengths with composition for tin-lead alloys (taken from *Soldering in electronics assembly* by Mike Judd and Keith Brindley, courtesy of Newnes).

Creep failure (see Section 3.9) occurs after 1000 hours at room temperature at 4.5 MPa and at 80 °C at 1.4 MPa. Solder creeps at room temperature because its melting point is low and therefore room temperature represents a high 'homologous' temperature (temperature as a fraction of the melting temperature, in K). This is not entirely a bad thing, however. One of the chief problems to which solder is exposed is thermal fatigue (see Section 3.8) as current loadings go up and down. The same diffusion processes which give rise to creep also heal the damage which builds up in fatigue, much increasing fatigue life and raising the fatigue endurance limit: solder is quite a forgiving material.

Soldered joints are stronger than bulk solder—for example, Fig. 4.15 shows the variation of shear strength of a copper—common solder—copper lap joint with the thickness of the joint. The increased strength of the joint is because of (a) precipitation of intermetallic compounds in the solder and/or (b) the mechanical constraint imposed by the substrate.† On the one hand the gap needs to be small enough to allow capillary action to draw the solder into the joint, but on the other, not so small that bubbles are trapped in the solder. (In fact, it is usually the bond between the solder and the copper which gives way first (peeling), not the solder itself.) Finally, solder has an electrical conductivity $\frac{1}{10}$th that of copper (and a thermal conductivity $\frac{1}{8}$th) and this in itself means that the joint should be as thin as practicable. Taking all these factors into consideration, and depending on the geometry of the joint, $\frac{1}{20}$th to $\frac{1}{10}$th of a mm is normal.

† The friction between the solder and the workpieces is transmitted into the solder, preventing it from flowing. As the solder becomes thinner, this becomes more effective.

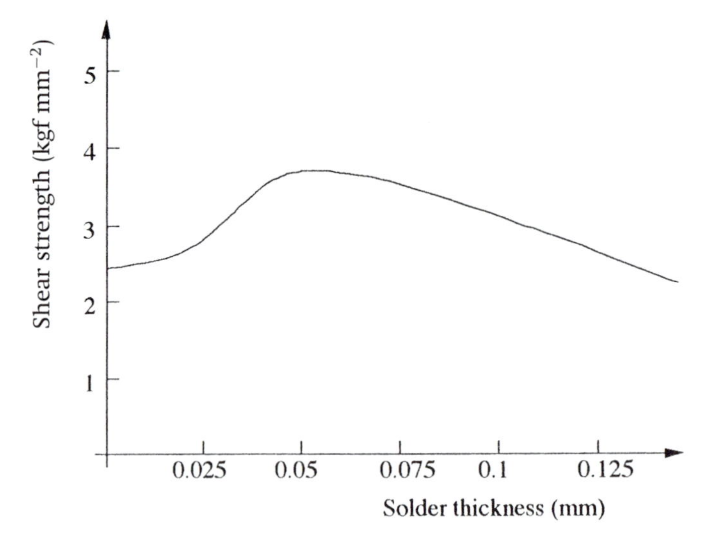

Fig. 4.15 The variation in strength of a copper lap joint with joint thickness (taken from *Soldering in electronics assembly* by Mike Judd and Keith Brindley, courtesy of Newnes). (kgf ≡ kg force ≡ the weight of 1 kg on Earth: an old unit of force.)

Soldering machines

Leaving aside the manual operation of a soldering iron with cored solder, there is a wide variety of automated soldering machines for soldering PCBs.

For CS (CS = position **C**omponent, then **S**older) processes there are dip, drag and wave soldering, illustrated in Fig. 4.16.

In SC (SC = apply **S**older, then position **C**omponent) processes the solder is applied as a paste (= solder + flux + binder) in various ways—for example *screen printing* (see also Chapter 7, Section 7.6 and Fig. 7.23)—and the components may be *surface mounted*. These are illustrated in Figs 4.17 and 4.18 below.

Finally the paste is heated gently to remove the binder and accomplish the soldering operation.

4.6 The limitations of pure elements: strong conductors

So far in this chapter I have been talking about pure elements, principally about copper. Most conductors are indeed made from elements. Metal elements, however, are very weak (if ductile and tough). Electrical engineering is almost unique in employing metal elements. A mechanical engineer would have forty fits at the thought of using

† i.e. with a lead (nothing to do with the element Pb).

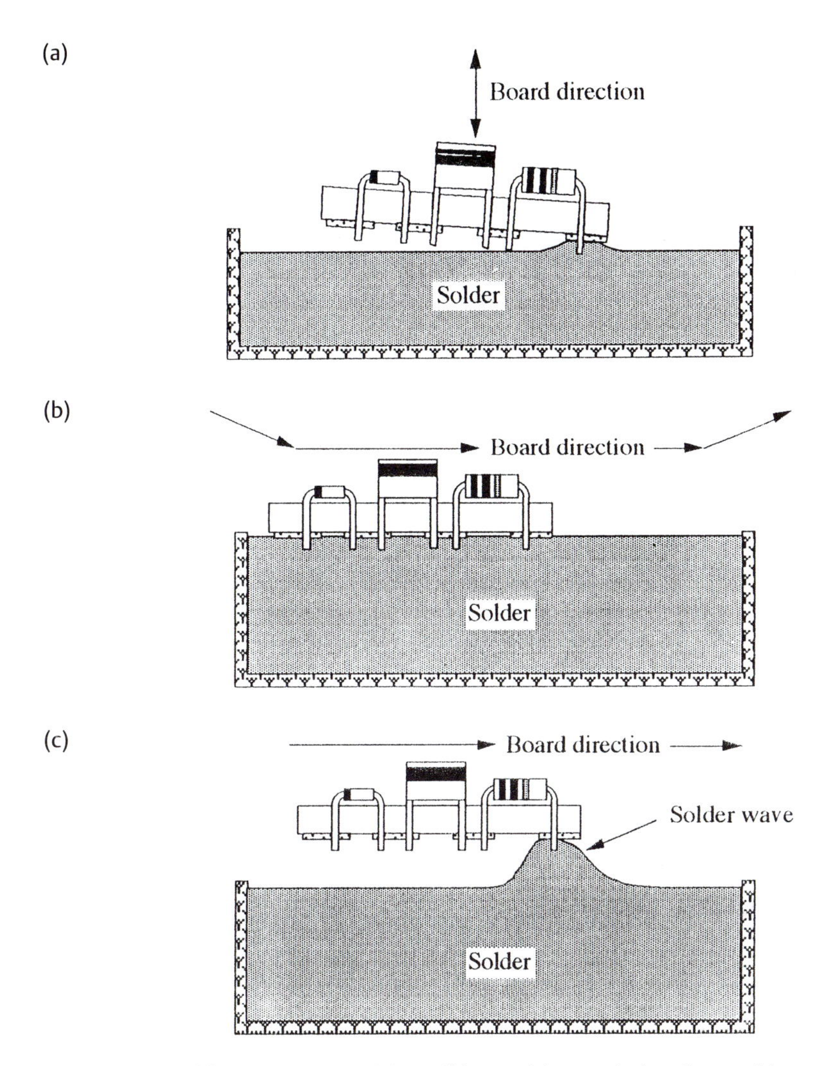

Fig. 4.16 CS soldering processes: (a) Dip (b) Drag (c) Wave (taken from *Soldering in electronics assembly* by Mike Judd and Keith Brindley, courtesy of Newnes).

pure copper or silver for any structural purposes. Sometimes, however, components need to be strong as well as conduct electricity. Examples include:

Overhead power lines for trains and trams
Welding electrodes and holders
Spring contacts
Switch gear contacts
Commutators and rotors on hot running motors (heavy duty)

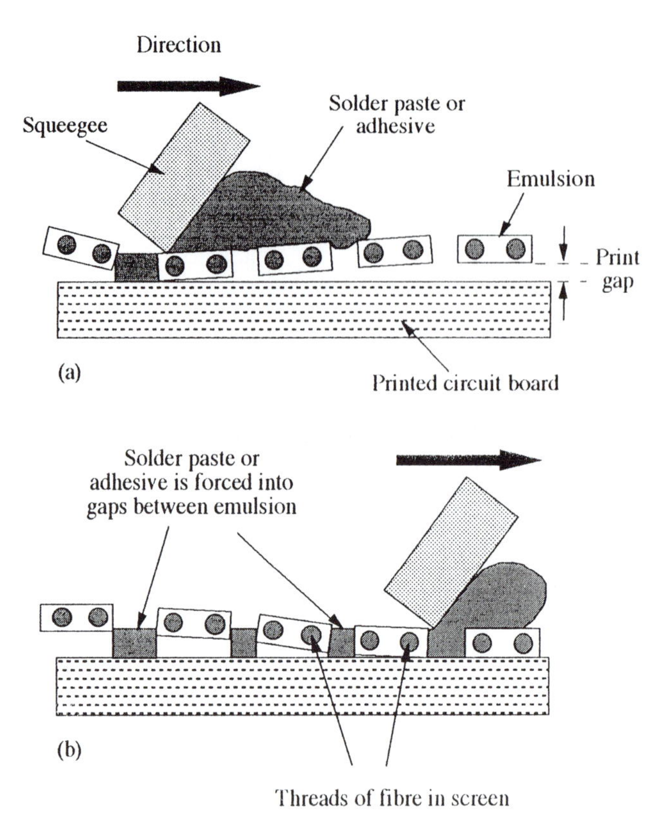

Fig. 4.17 Applying solder paste using the technique called 'silk screening' (not really silk any more!) (taken from *Soldering in electronics assembly* by Mike Judd and Keith Brindley, courtesy of Newnes).

Termination equipment
High strength conductor bars

Notice that often here it is *abrasion resistance* which is being sought, but abrasion is first cousin to hardness, which itself is closely related to yield stress.

In this section I shall explain how conducting metals are strengthened. Unfortunately, the same methods which are used to strengthen metals also decrease conductivity and so again we have to seek the best compromise.

How are metals strengthened?

When we talk about 'strength', sometimes we mean 'yield strength' and sometimes we mean 'tensile strength' (UTS, fracture stress). In this section, when I refer to strength, I mean 'yield strength'.

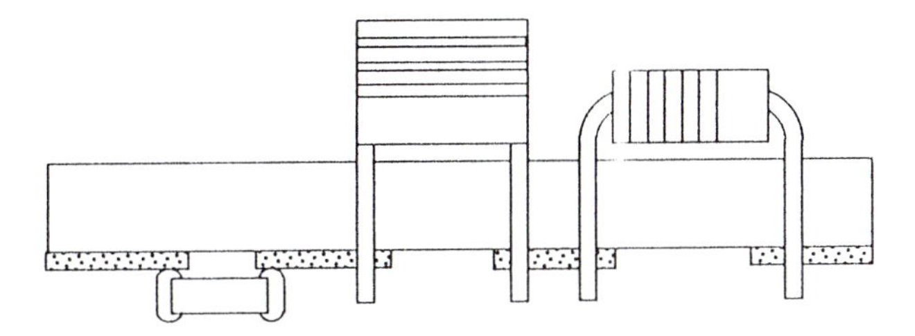

Fig. 4.18 An example of mixed surface mounted and leaded components. (Currently this is called a 'Type II' assembly.) (Taken from *Soldering in electronics assembly* by Mike Judd and Keith Brindley, courtesy of Newnes.)

A metal 'yields' when dislocations start to move. To strengthen a metal, therefore, we need to stop dislocations moving. There are four ways of doing this:

1. Work hardening
2. Grain size hardening
3. Solute hardening
4. Precipitate hardening

Here 'hardening' and 'strengthening' are interchangeable.

1. Work hardening

In work hardening the metal is subjected to plastic deformation. This increases the numbers of dislocations. In an annealed metal there can be as few as 10^8 dislocation lines per m^2; in a heavily worked metal as many as 10^{16} dislocation lines per m^2 (see Question 2.7). The extra dislocations get in the way of each other and so the metal is hardened.

We have already come across this phenomenon as a facet of the tensile stress–strain curve. If we wish deliberately to work harden in order to strengthen a component, however, there are more effective ways of doing it than in a tensile test. We must incorporate a compressive stress to hinder crack formation. Rolling, forging and extrusion, introduced earlier in this chapter, are excellent ways of introducing cold work. Perhaps the best way is wire drawing. Soft annealed copper has a yield stress of 50 MPa. Drawing wire raises this to 400 MPa. Work hardening is an excellent way of raising the strength of an electrical conductor, because it has very little effect on the electrical conductivity. In the example given above, as the strength goes up 8 times the conductivity falls by only 5%.

There is one disadvantage to work hardening: it is sensitive to temperature. Remember how in Section 4.3 the copper wire had to be annealed to soften it before further working could be carried out? Even when wire drawing is used deliberately as a way of strengthening a piece of metal, this effect of heat remains. Look at Fig. 4.19 for the effect of annealing on work hardening.

As we go through strengthening methods 1–4 above we will find that they become more and more resistant to temperature.

2. Grain size hardening

Reducing the size of the grains in a metal strengthens it. This is because the *grain boundaries* act as obstacles to dislocation movement. This method of strengthening is very effective for b.c.c. and h.c.p. metals but not for f.c.c. metals. Thus it is useful for steels (see Chapter 5) which are b.c.c. but is not really very important for electrical conductors, which are f.c.c. This is a pity, because grain boundaries do not have much effect on electrical conductivity.

3. Solute hardening

In Chapter 2 (Section 2.5) and also in Section 4.5 in this chapter I explained that, just as you can dissolve salt into water, so other elements can be dissolved into a metal. Maximum levels of solubility in solids are much smaller than in liquids and the solute (the element which dissolves) can affect markedly the electrical conductivity of the solvent (the metal which does the dissolving).

Dissolving *anything* into a metal strengthens it, regardless of the nature of the solute element. So a soft element like lead will strengthen a harder one like copper (you will see why in a moment). Once the lead atoms are separated out in solution in the copper, the nature of the solid 'lead' becomes irrelevant.

Solid solutions are of two types—*substitutional* and *interstitial* (see Section 2.5 and Fig. 2.16). Most of the solutions of importance for electrical conductors are substitutional. It is the *size* of the solute which is important to strength, more exactly the *difference* in size between solute and solvent. This difference in size is measured via the change in *lattice parameter* as the solute is added. (The lattice parameter is the size of the little cell which is the smallest unit of the crystal—see Figs 2.8 and 2.11.)

$$\frac{\mathrm{d}\sigma_y}{\mathrm{d}c} \alpha \left(\frac{1}{a}\frac{\mathrm{d}a}{\mathrm{d}c}\right)^2 \tag{4.1}$$

where σ_y is the yield stress, c the concentration of solute and a the lattice parameter of the solvent (see Fig. 4.20 which is in terms of the change in lattice parameter, rather than the *fractional* change in lattice parameter). The origin of this strengthening is illustrated in Fig. 4.21. There an *undersized* atom sits above the glide plane of an edge dislocation, where there is a state of compression, and pins the dislocation. An oversized atom would sit *below* the dislocation glide plane and pin it likewise. Pinning the dislocation stops it moving and strengthens the metal. Now you can see why the hardness of the solute element is not very significant. Its hardening effect as a solute derives chiefly from its moving of the solvent atoms, not from any intrinsic hardness of the atom.

(a)

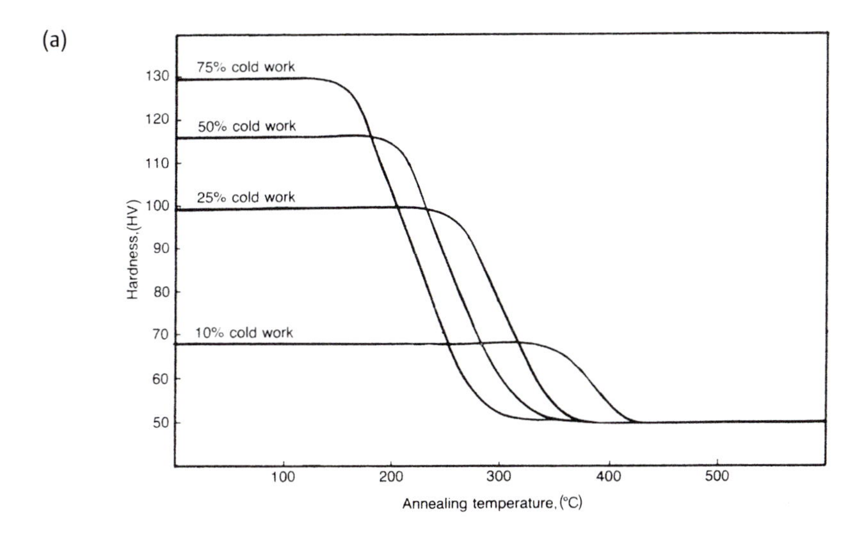

(b)

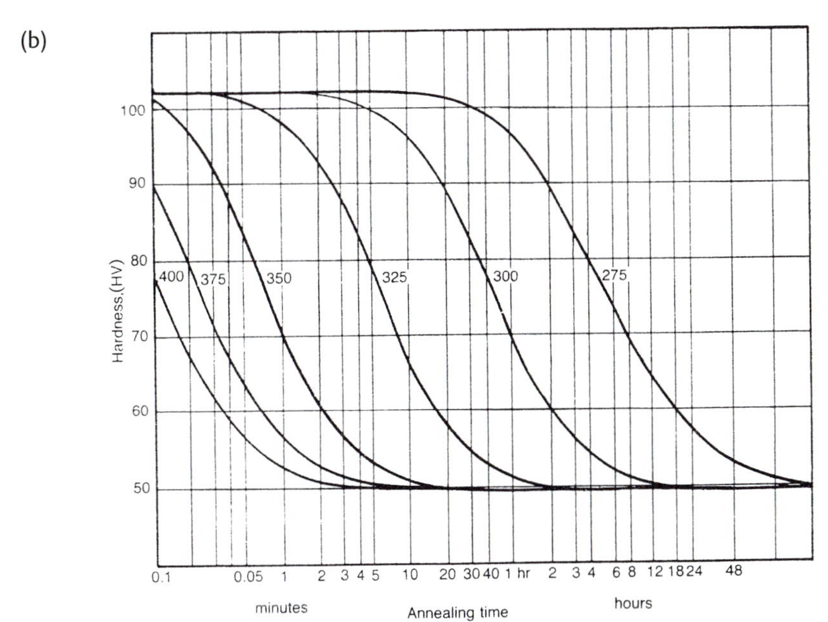

Fig. 4.19 How annealing softens a piece of work-hardened copper. The heavier the work-hardening or the higher the temperature, the faster the softening. At low temperatures/levels of work *recovery* occurs. This is when dislocations annihilate with each other or gather into tidy groups (this is one of the processes in creep (Section 3.9)). At higher temperatures/levels of work *recrystallisation* occurs (Section 4.4 and Fig. 4.7). Effects of (a) amount of cold work (b) temperature. (Taken from *High conductivity coppers*, courtesy of the Copper Development Association (See Recommended reading).

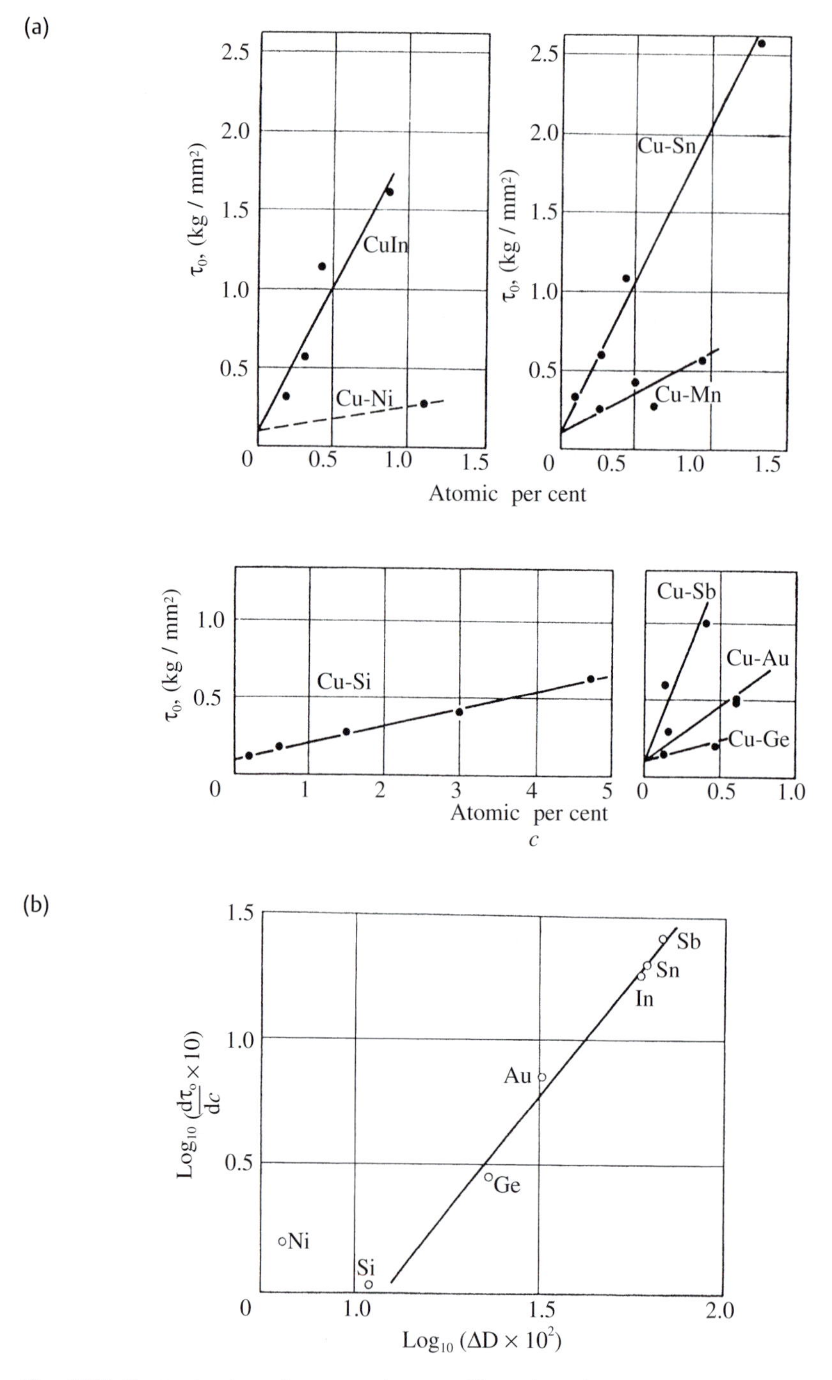

Fig. 4.20 Illustrating how the strengthening effect of a solute atom depends on (a) its concentration and (b) its size in solution. ΔD is a measure of the size difference between the solvent and solute atoms. (From *The plastic deformation of metals* by R.W.K. Honeycombe by kind permission of Edward Arnold.)

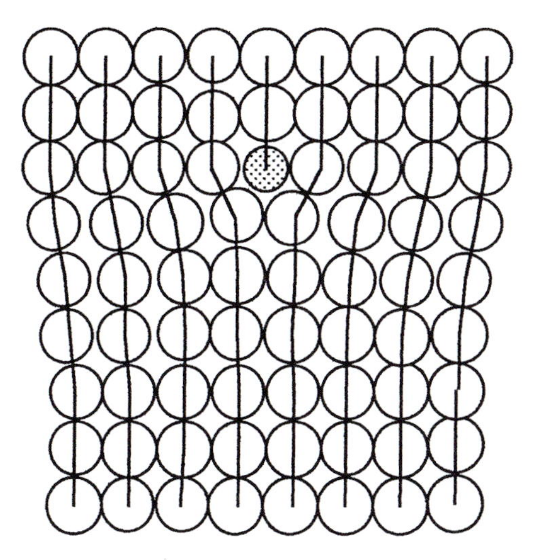

Fig. 4.21 A small atom pins a dislocation by substituting for one of the atoms in the extra half plane. (A large substitutional solute atom or an interstitial solute atom can also pin the dislocation, but this time by sitting *below* the half plane in a region where the metal is under tension.)

On the other hand it is the difference in *valency* between solute and solvent which affects the electrical conductivity. (Valency is the number of potentially free electrons possessed by an atom.)

$$\Delta\rho \; \alpha \; \Delta Z^2 \tag{4.2}$$

where ρ is the resistivity and ΔZ the difference in valency between solvent and solute. The effects can be quite large—see, for example, Table 4.2.

Thus if we want to strengthen copper without ruining its electrical conductivity we need a solute element whose atoms have the same valency as, and a different size from,

Table 4.2 The effects of nickel and arsenic on the resistivity ρ of copper (see also Fig. 4.22). The valency of nickel is similar to that of copper, but that of arsenic is very different (roughly 5 as compared to 1) (see Fig. 1.6).

	ρ (room temperature)
Cu	1.70×10^{-8} Ωm
Cu–1%Ni	2.74×10^{-8} Ωm
Cu–1%As	8.54×10^{-8} Ωm

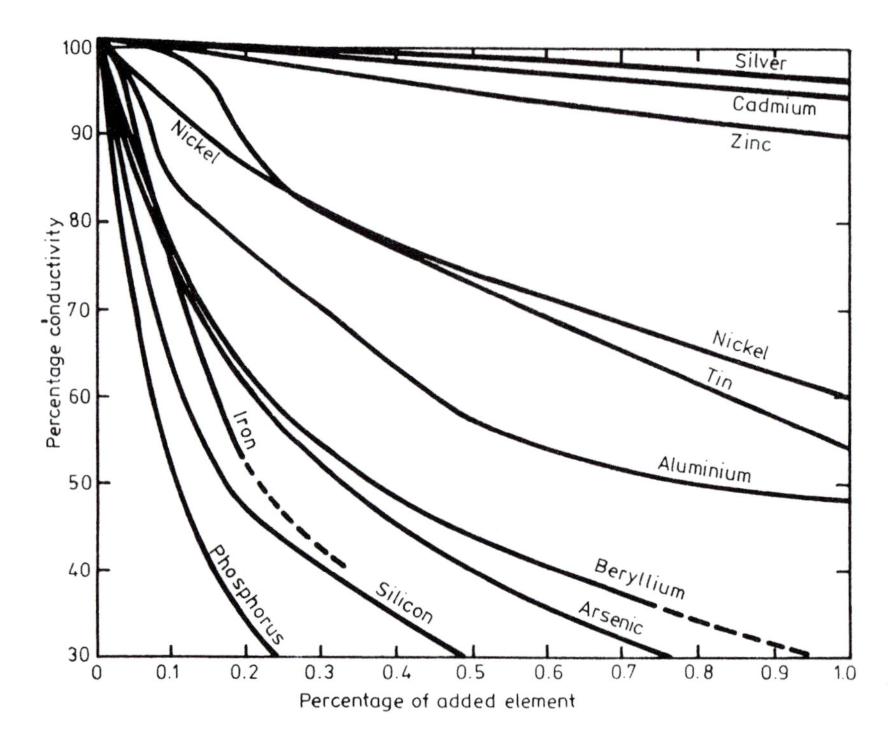

Fig. 4.22 The effects of various solutes on the electrical conductivity of copper (from *Copper and its alloys* by E.G. West, courtesy of Ellis Horwood).

the solvent. Figure 4.22 shows the effects of several alloying elements on the conductivity of copper. Not surprisingly, the three common solid solution strengthened copper alloys are those involving silver, cadmium and zinc. Three typical compositions used in electrical engineering are shown in Table 4.3. Figure 4.23 shows the improved hardness of Cu–0.08%Ag and, in particular, its improved high temperature performance.

I should add that copper–cadmium alloys are sometimes used in a more concentrated form which is closer to Method 4 immediately below and that the

Table 4.3 Uses of solid solution strengthened copper alloys.

Cu–0.08% Ag	used for commutators on hot running motors and for switch gear contacts
Cu–0.7% Cd	used for overhead wires for trains and trams
Cu–30% Zn ('cartridge' or 'alpha' brass)	used for all manner of cheap electrical widgets—terminals in 3-pin plugs, the metal screw on an electric light bulb etc. etc. etc. etc.

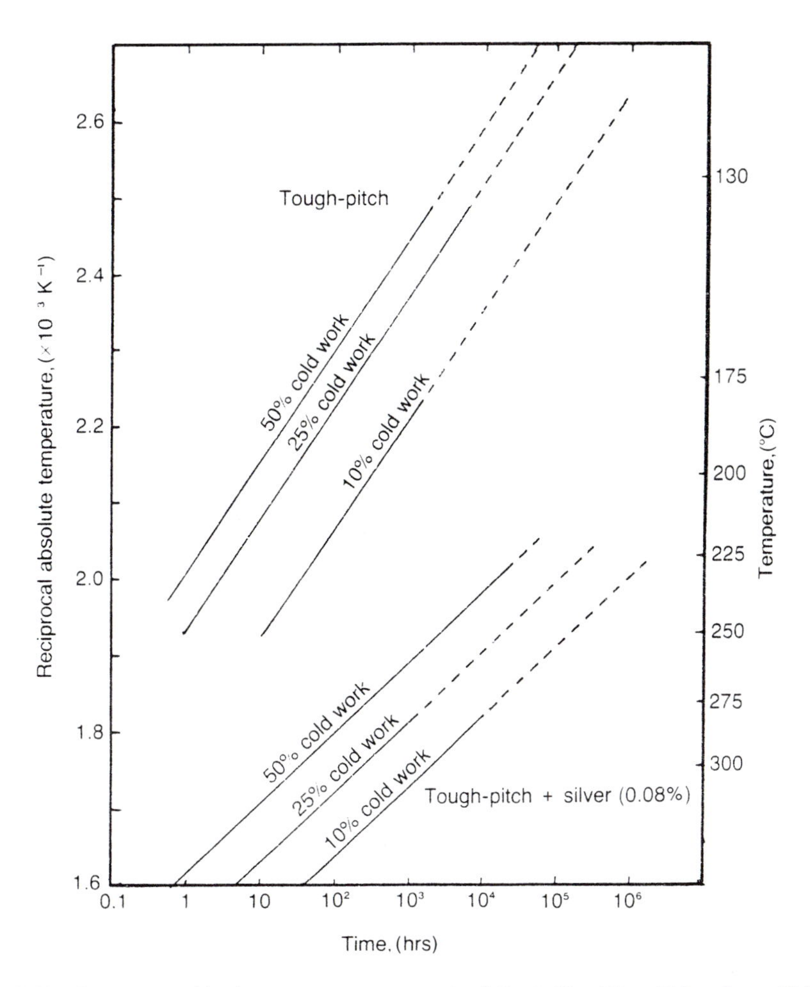

Fig. 4.23 The improved high temperature strength of Cu–0.08wt%Ag. (Taken from *High conductivity coppers*, courtesy of the Copper Development Association (see Recommended reading).)

reasons for using copper–zinc solid solutions (i.e. brass) extend further than just strength. Brass is cheap, corrosion resistant and easily worked. Enormous tonnages are used.

4. Precipitate hardening

This gives the strongest alloys, but has the worst effect on electrical conductivity. Figure 4.24 compares the temperature resistance of work hardened, solid solution strengthened and precipitate strengthened copper alloys.

Recall from the discussion of phase diagrams in Section 4.5 on solder that there comes a point when we are dissolving a solute into a metal solvent where the solubility is exceeded and a second phase appears. Say we dissolve chromium into copper. If we really used *solid* copper and tried to persuade the chromium to dissolve in it by putting

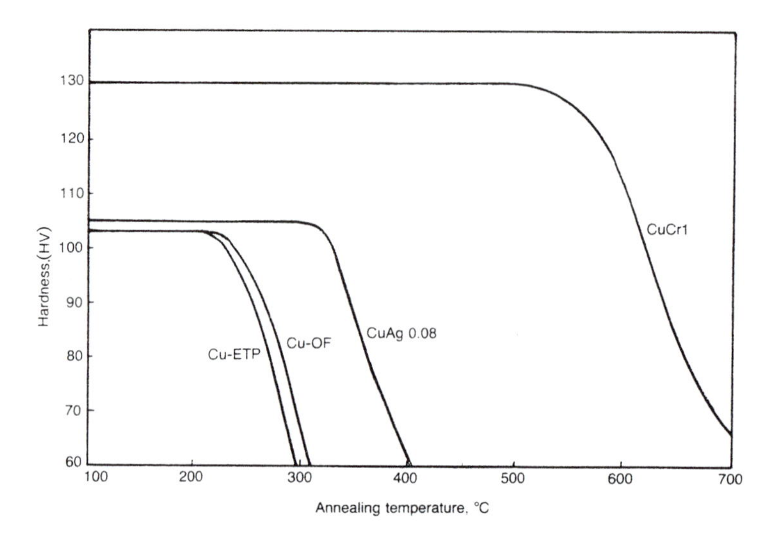

Fig. 4.24 The temperature resistance of strengthening goes up in the order work hardening → solid solution strengthening → precipitate strengthening. (Taken from *High conductivity coppers*, courtesy of the Copper Development Association (see Recommended reading))

two pieces in contact, we would have to wait an awfully long time, because we would have to rely on solid state diffusion, which is very slow, particularly at lower temperatures. What we actually do is dissolve solid chromium in liquid copper and cool to the temperature of interest. If the percentage of chromium is beyond the solubility limit, the excess chromium will *precipitate* out. We therefore now have two phases. They are both *solid* phases, sharing the same state of matter. As the temperature rises, the solubility of chromium in copper rises (just as salt dissolves more easily in hot than in cold water and just as the solubilities of tin in lead and of lead in tin can both be seen to rise in the phase diagram of Fig. 4.10). The phase diagram for chromium and copper is shown in Fig. 4.25. The boundary between the copper–chromium solid solution (called $\alpha^{\dagger}$) phase field and the two phase (α + Cr (β, of course)) field marks the solubility limit of chromium in copper at various temperatures.

The two phase alloy, made this way, is rather useless. It is not very strong and has nearly zero tensile ductility. The reason is that the chromium precipitates in the easiest place—around the copper grain boundaries—providing an easy path for a crack and affecting the dislocations in the copper (i.e. its strength) not at all. We can, however, make a very useful alloy out of copper–chromium. Here is how:

First we raise the temperature of the alloy (an alloy is a mixture of two or more metals—here copper and chromium) sufficiently high that the chromium dissolves in the copper. For example, if there were 0.2 wt% chromium, 900 °C would be sufficient to

† See Section 4.5 following Fig. 4.10.

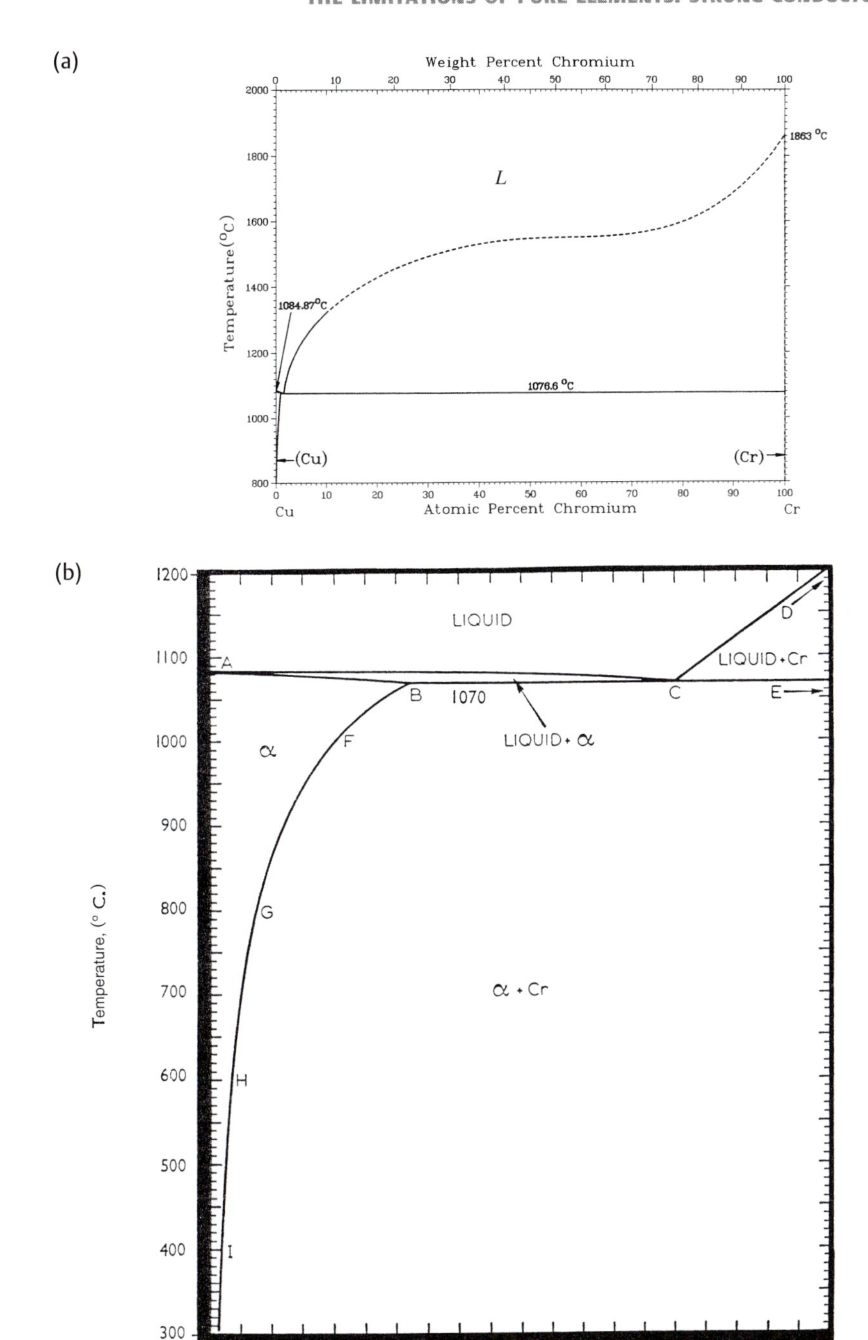

Fig. 4.25 (a) Copper–chromium phase diagram. (b) A magnified view of the copper-rich end of the phase diagram.

see it dissolved. This is called the *solution treatment*. The alloy is then *quenched*—this means it is cooled very quickly by dropping it into water. This prevents the precipitates from forming. The alloy is raised to a temperature intermediate between room temperature and the solution temperature—say about 350 °C here. Under these conditions the chromium precipitates form all through the grains. This is because

(i) the alloy is so far from equilibrium that the driving force for precipitation is enormous and the precipitates will form *anywhere*, not just round the grain boundaries,

and (ii) the necessary solid state diffusion is very slow at this temperature, leading to a high density of very small precipitates.

This process is called *aging*.

The resulting fine dispersion of precipitates immobilises the dislocations and gives great strength (see Fig. 4.26). This and the relatively high concentration of solute cause a considerable decrease in conductivity. Luckily, during the aging, the electrical conductivity of a precipitate hardened alloy improves as the strength also improves (Fig. 4.27). This is because the single solute atoms which are very efficient scatterers of conduction electrons group together to form small precipitates which are very efficient at pinning dislocations. Some figures specific to the very strong alloy Cu–Be–Co are given in Table 4.4.

The reason why the strength in Fig. 4.25 starts to decrease if aging is carried on for too long is that there is an optimum size for the precipitates. If they are too small they are cut by the dislocations, which go straight through them. If they are too large, the dislocations can pass between and around them, leaving a loop of dislocation around each precipitate. This condition is called *over-aged*. The optimum size of precipitate is of the order of 10's of nm: very small by normal engineering standards. Some examples of typical precipitation hardening copper alloys with basic electrical and mechanical

Table 4.4 Electrical conductivity and strength as an age-hardening Cu–Be–Co alloy is solution treated and quenched (ST), and subsequently aged (ST & age). This type of alloy is often cold worked (CW) during production to increase even further the strength (with a conductivity penalty). These strengths would be quite respectable for a steel (e.g. for a mild steel $\sigma_y \sim$ **200** MPa and UTS $\sim$ 400 MPa) (see Chapter 5). UTS is the Ultimate Tensile Stress, ε_L is tensile ductility and VHN is Vickers Hardness Number.

	Conductivity (%IACS)	σ_y (MPa)	UTS (MPa)	ε_L (%)	VHN
ST	16	185	500	45	100
ST & age	30	1100	1160	5	370
ST, CW & age	25	1260	1340	2	400

(a)

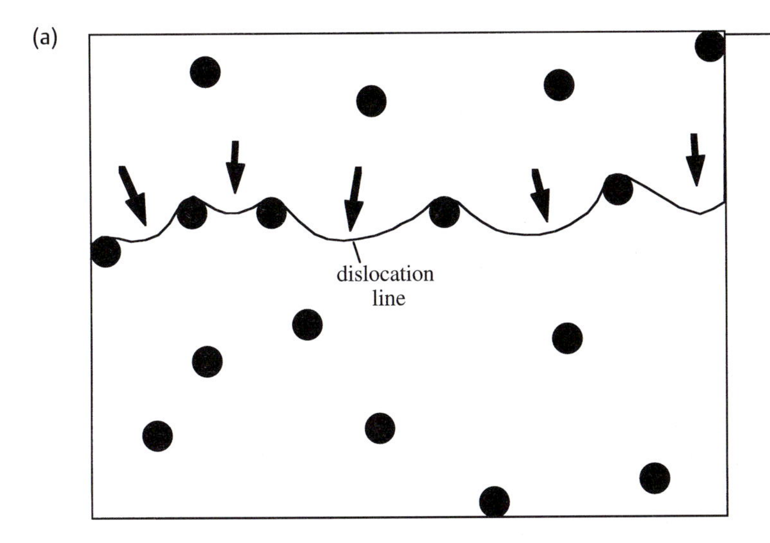

(b)

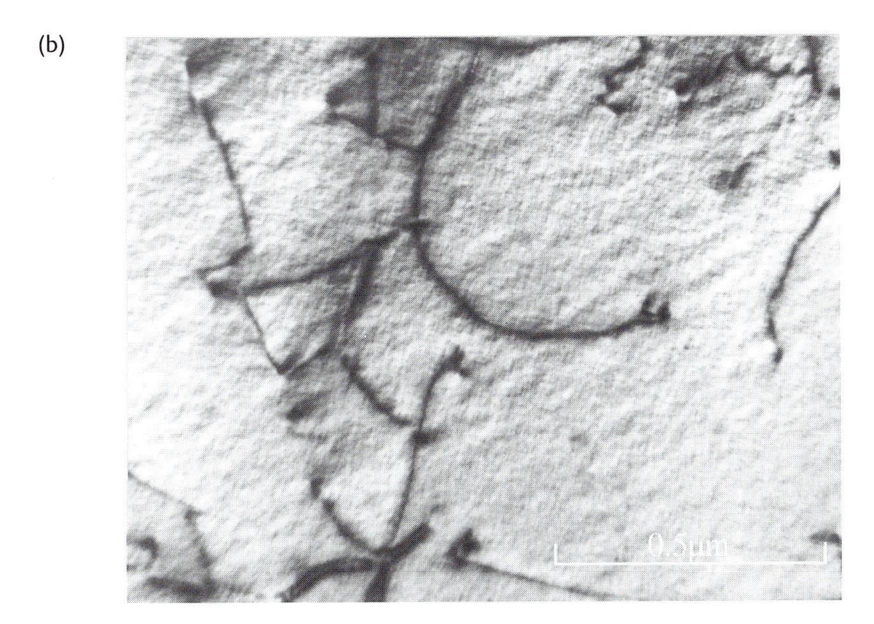

Fig. 4.26 (a) Hard precipitates prevent dislocations from moving and strengthen a metal. (b)–(d) Three *transmission electron micrographs* (i.e. pictures taken in a transmission electron microscope) of Cu-Be-Co.

(b) Solution treated and quenched condition. Some dislocations (the black lines) are visible.

(c) Peak-aged condition: lots of strain contrast. The dislocations are still there, but cannot be seen.

(d) Overaged. This has been annealed for too long. The precipitates have become too large and, more importantly, far apart, and the dislocations can bow between them. (Courtesy Mr A.J. Burbery.)

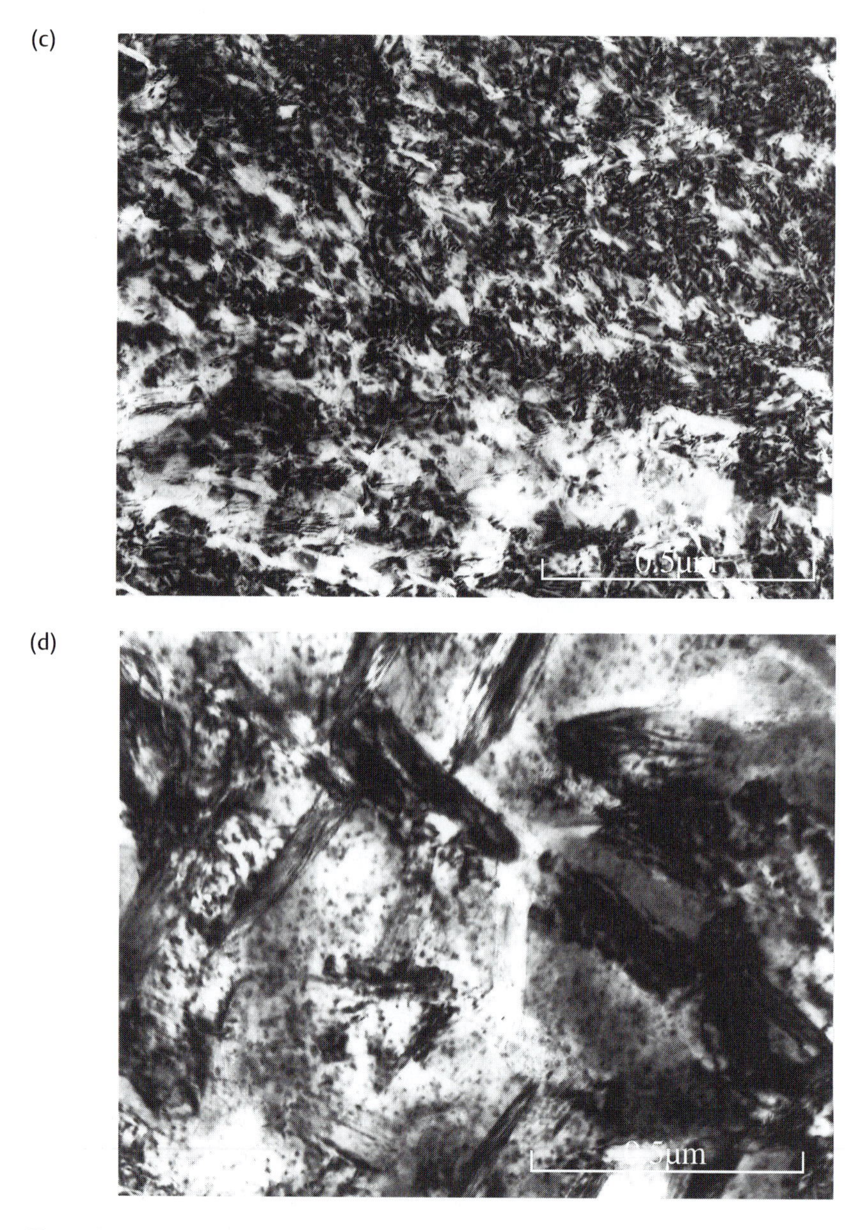

Fig. 4.26 (*continued*)

properties and applications are shown in Table 4.5. For specific applications, fatigue properties, creep properties etc. might be important and would then have to be consulted.

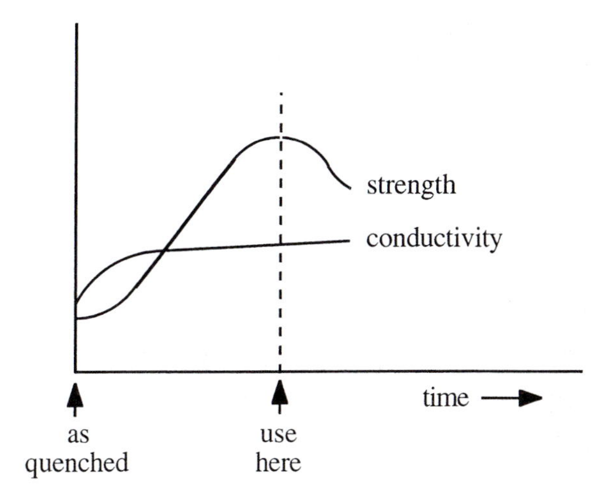

Fig. 4.27 As a precipitation-hardened alloy is 'aged', the electrical conductivity increases as the strength peaks.

4.7 Summary

1. Copper is extracted from its sulphide ore by oxidation and subsequently purified by electrolysis and remelting.
2. Large cross-section conductors are made by extrusion.

Table 4.5 Three common precipitation hardened copper alloys. Note the high strengths and poor electrical conductivities.

Alloy	Precipitate phase	Yield strength σ_y (MPa)	%IACs	Uses
Cu 0.5–1% Cr	Cr	550	80	Spot welding electrodes and electrode holders; high strength conductor bars; termination equipment; heavy duty electric motors (rotors, commutators etc)
Cu 0.15%Zr	$ZrCu_3$	420	87	Alternative to Cu–Cr
Cu 2% Be 0.5%Co	CuBe	1260	25	Springs; contacts; welding electrodes;† heavy duty applications

† Must incorporate fume handling to avoid toxic effects of beryllium oxide (→ berylliosis)

3. Small cross-section conductors (wires) are drawn.
4. Other forming operations include rolling, forging, stamping, pressing and machining.
5. Electrical solders tend to be *eutectic* alloys.
6. Phase diagrams are maps showing the regions of existence of phases in temperature–composition space.
7. Fluxes clean the surfaces to be joined.
8. Solder is weak and creeps.
9. Thin soldered joints have their own intrinsic strength.
10. Metals may be strengthened by (a) work, (b) solid solution and (c) precipitation. Strength, its temperature resistance and electrical resistivity all rise in the order (a), (b), (c).

Recommended reading

Mechanical metallurgy by G.E. Dieter (published by McGraw-Hill)

Engineering materials II by M.F. Ashby and D.R.H. Jones (published by Elsevier)

Manufacturing processes for engineering materials by S. Kalpakjian (published by Addison-Wesley)

Soldering in electronics assembly (2nd edn.) by Mike Judd and Keith Brindley (published by Newnes)

Copper and its alloys by E.G. West (published by Ellis Horwood)

High conductivity coppers for electrical engineering (Copper Development Association, Publication no. 122)

Questions (Answers on pp. 311–316)

4.1 What would be the cost of the material in a kilometre of copper wire of diameter 0.2 mm? (The density of copper is 8.9 Mg m^{-3}. Take the cost of copper from Table 2.5.)

4.2 Common electrical solder consists of 62 wt% tin and 38 wt% lead. What is the composition in at%? (The Relative Atomic Masses of tin and lead are 119 and 207 respectively.)

Use Fig. 4.10 to answer Questions 4.3–4.7.

4.3 How many phases co-exist in Sn—50 wt% Pb at 250 °C
200 °C
150 °C?
in Pb–10 wt% Sn at 250 °C
200 °C
150 °C?

4.4 As liquid Sn–50 wt% Pb is cooled, what is the composition of the first solid to form?

4.5 What are the solubilities of tin in lead and lead in tin at 100 °C?

4.6 In Sn–50 wt% Pb at 100 °C, what are the proportions of the two phases?

4.7 In Sn–50 wt% Pb just below the eutectic temperature, what proportion of the structure is occupied by the eutectic mixture? Of the β-phase, how much is in the eutectic mixture and how much as standalone β-phase?

4.8 Using Fig. 4.19, up to what temperature would it be safe to trust the strength of a work hardened copper wire? Does the load on the wire matter?

4.9 Steel specimens with average grain sizes of 50 μm and 100 μm have yield strengths of 250 and 200 MPa respectively. To what would the grain size have to be changed to give a yield strength of 300 MPa?
(The yield strength σ_y of a metal depends on the grain diameter d via the *Hall–Petch* relationship:

$$\sigma_y = \sigma_0 + k_y d^{-0.5}$$

where σ_0 and k_y are constants.)

4.10 If 1 at% gold raises the yield strength of copper by 16 MPa, what effect would you expect 2% of tin to have? (The atomic diameters of copper, gold and tin atoms in a copper environment are 0.256, 0.291 and 0.313 nm, respectively.)

4.11 The strengthening phase in Cu-Cr is chromium itself (with a little copper dissolved in it). If the densities of copper and chromium are 8.9 and 7.1 Mg m^{-3} respectively, what vol% would the precipitates occupy in Cu–1 wt% Cr? If the precipitates are cubes with edge length 2 nm, what is the average separation between them?

The increase in yield strength due to the bowing mechanism (see Section 4.6, sub-section 4) is $\Delta\sigma_y = \frac{\mu b}{2\pi L}$ where μ is the shear modulus (4×10^{10} Pa for copper), b the magnitude of the Burgers vector (for copper 0.25 nm) and L the separation between the precipitates. Estimate the strength of Cu–1 wt% Cr assuming all the precipitates are 2 nm cubes and that they are not cut by the dislocations, and compare with the values in the table in Question 3.5.

5

Steel

Chapter objectives

- What *is* a steel?
- Making steel
- Plain carbon steels
- Alloy steels
- Stainless steels
- Welding steels together

5.1 Introduction

Steel is not characteristically an electronic or electrical engineering material: its electrical conductivity is not particularly high, although its ferromagnetism does lead to some magnetic applications. So why devote a whole chapter of a book for electrical and electronic engineers to steel? The answer is that steel is by far and away the most widely used structural metal. For example, the annual world production of steel in 1990 was 520 Mtonnes as compared with 13 Mtonnes for copper and 20 Mtonnes for aluminium.

There are two reasons why steel dominates the structural metal market:

1. It is cheap
2. It can be engineered to have an enormous range of strength and ductility.

Why is this?

First of all, what is steel? Steel is mainly iron:

Steel = iron + carbon + other bits and pieces

(Not very much carbon—usually a fraction of a per cent by weight.)

Point number 1: cost

Remember from Chapter 2, Section 2.8 that

$$\text{cost of article} = \text{cost of ore} + \text{cost of extracting metal from ore} + \text{cost of fabrication}$$

The percentages of the earth's crust which different elements constitute are shown in Table 5.1.

Iron ore is very plentiful and very cheap

The cost of extracting a metal from its ore depends on

- how rich the ore is
- how strongly the metal is combined with oxygen (or whatever)—i.e. how base it is

Referring to Table 5.2, aluminium ores are rich in aluminium, but the aluminium is strongly bound to oxygen and requires electricity to separate it out (see Chapter 6).

Table 5.1 Percentage occurrences of the most common elements in the earth's crust* (from Engineering Materials I by M.F. Ashby and D.R.H. Jones (by kind permission of Elsevier)). Notice that copper does not even figure in the list.

Crust		Oceans		Atmosphere	
Oxygen	47	Oxygen	85	Nitrogen	79
Silicon	27	Hydrogen	10	Oxygen	19
Aluminium	8	Chlorine	2	Argon	2
Iron	5	Sodium	1	Carbon dioxide	0.04
Calcium	4	Magnesium	0.1		
Sodium	3	Sulphur	0.1		
Potassium	3	Calcium	0.04		
Magnesium	2	Potassium	0.04		
Titanium	0.4	Bromine	0.007		
Hydrogen	0.1	Carbon	0.002		
Phosphorus	0.1				
Manganese	0.1				
Fluorine	0.06				
Barium	0.04				
Strontium	0.04				
Sulphur	0.03				
Carbon	0.02				

*The total mass of the crust to a depth of 1 km is 3×10^{21} kg: the mass of the oceans is 10^{20} kg: that of the atmosphere is 5×10^{18} kg.

Table 5.2 Cost of extracting three common metals from their ores (1980 prices, adapted from Table 2.5, Engineering Materials I, M.F. Ashby and D.R.H. Jones (by kind permission of Elsevier)). These costs have risen somewhat in the last 20 years, of course, but the comparison is still valid.

	Cost of extraction ($/tonne)
iron	220
aluminium	1320
copper	440

Copper ores are *lean* in copper (i.e. there isn't much) and this raises the price of extraction. Iron ores are rich in iron and the iron is not strongly bound to oxygen:

iron is cheap to extract

Finally,

steel is moderately cheap to fabricate

Thus steel, at a few hundred $/tonne is much, much cheaper than the electrical conductor elements listed in Table 2.5.

Point number 2: wide range of mechanical properties

Steel can be engineered to have strengths varying from 150–5000 MPa (compare copper alloys: from 50–1300 MPa). The fundamental reason for this, which I shall elaborate on below, is that iron changes its crystal structure as the temperature changes (twice in fact):

	iron is...
up to 912 °C	b.c.c.
912 °C to 1394 °C	f.c.c.
1394 °C to 1538 °C (mpt)	b.c.c.

The important transformation is the lower temperature one. These changes in structure are driven by iron's *ferromagnetism* (see Chapter 10). Together with the effect of carbon, they enable an enormous range of structures to be produced which in their turn permit the wide range of mechanical properties I mentioned above.

Putting points 1 and 2 together (cost and flexibility), unless there is a very good contrary reason

steel is always the preferred structural metal

So the *structural fabric* of electrical engineering, as opposed to the conducting part, is provided by steel. The framework of all the large machinery used in power generation, the turbines themselves, motor bodies, power transmission towers (pylons) and cables, conduits, much domestic and laboratory-scale equipment—television sets, washing machines etc etc., right down to the electronics world: the wires coming from your transistor are likely to be gold plated steel; the little metal can from which they emerge will be steel; a major market for British Steel is the backings to solar cells ... steel is everywhere.

I shall spend the rest of this chapter elaborating on point number 2—how the structure of steel can be engineered—with examples of where steel is used in electrical and electronic engineering and the occasional glance at point number 1: cost.

5.2 The structure of steel: plain carbon steels

There are three main types of steel:

- plain carbon steel = iron + carbon
- heat treated, or alloy, steel = iron + carbon + alloying elements
- stainless steel = iron + carbon + chromium (a lot of)

Plain carbon steels are the downmarket cheap ones. Alloy steels are strong and expensive. Stainless steels are used in aggressive (corrosive or oxidising) environments.

Plain carbon steels

Iron ore is iron oxide.

To make steel (of *any* sort, actually):

- *reduce* iron oxide → iron by heating with coke (carbon) in a *blast furnace*. This gives *pig iron* (= iron + 5% carbon) which is uselessly brittle

and then
- *oxidise* pig iron until carbon has desired level (< 1 wt%).[†] This is done in a steel making furnace.

Notice the strange 'forwards' and 'backwards' procedure: first, reduction in the blast furnace and then, *controlled* oxidation in the steel making furnace. If we weren't careful with the oxidation, we would end up back with iron ore!

[†] All percentages in this chapter will be by *weight*, because this is how industry does it and it is how you will always see them quoted. 1 wt% C ≡ 4.6 at% C in steel. (See Question 5.1 at end of chapter.)

To help us understand plain carbon steels let us look at the iron–carbon *phase diagram* (Fig. 5.1), which works in the same way as the diagrams in Chapter 4. We only need the iron-rich end—the rest is of no commercial significance.

Notice the three forms of iron on the left hand axis

α iron (b.c.c.) up to 912 °C

γ iron (f.c.c.) up to 1394 °C

δ iron (b.c.c.) up to 1538 °C

These temperatures are changed significantly by the addition of carbon.

Those of you who know your Greek alphabet will be wondering where 'β iron' got to. Iron changes from *ferromagnet* to *paramagnet* at 768 °C (see Chapter 10). β iron is an old term, not now used, for b.c.c. paramagnetic iron.

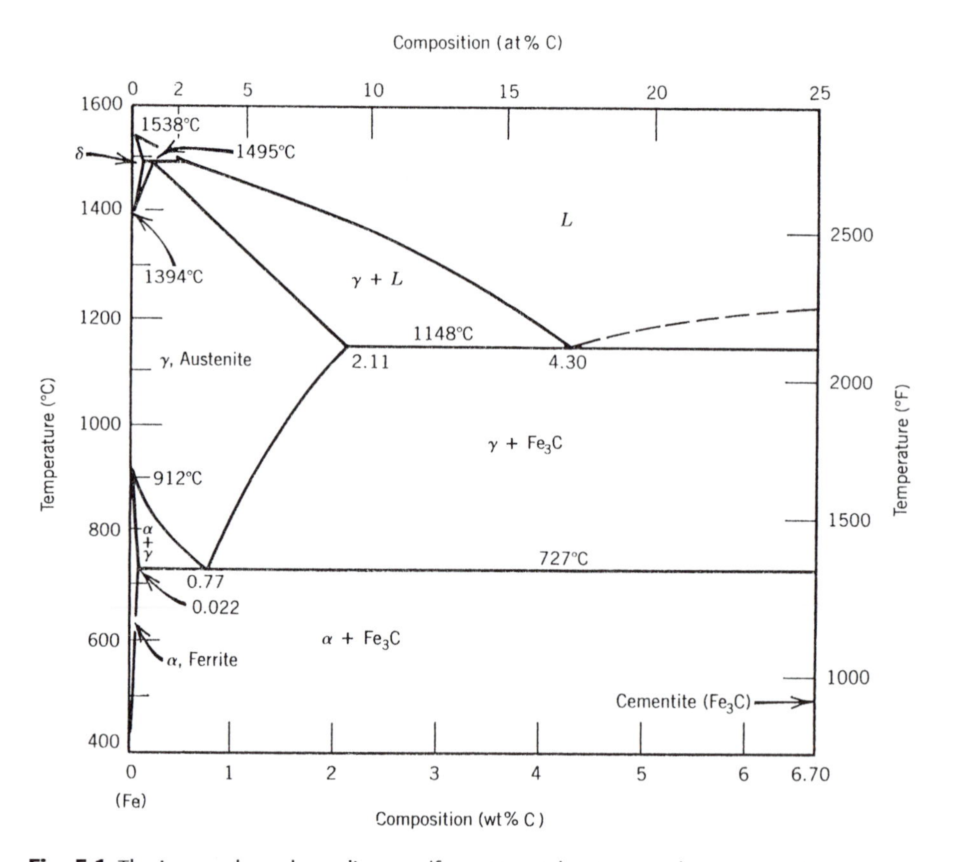

Fig. 5.1 The iron–carbon phase diagram (from *Materials science and engineering* by W.D. Callister, courtesy of John Wiley & Sons Inc.).

α iron is called 'ferrite' (the Latin for iron is ferrum) and γ iron is called 'austenite' (after a Professor Roberts-Austen). The other important name in the diagram is 'cementite', Fe_3C. This is iron carbide and is present in steel as very small particles. It is cementite which gives steel its strength. Plain carbon steels consist of a ferrite *matrix* (base, or framework) reinforced with small cementite plates. The ferrite is soft and tough (see Chapter 3) and the cementite is hard and brittle.

All structural engineering alloys consist of a soft tough matrix reinforced by hard brittle particles

This is true of steels, it was true of the copper alloys of Chapter 2 and it is true of all other structural metal alloys. The matrix provides the ductility and resistance to cracks; the reinforcing particles provide the strength. The matrix by itself would be too weak for structural purposes (see Chapter 4—pure copper, silver and gold); the reinforcement *by itself* would be too brittle.†

Back to the phase diagram. The next thing to notice is that carbon dissolves much more easily in f.c.c. iron—austenite—than in b.c.c. iron—ferrite. This is because carbon dissolves into iron *interstitially* rather than *substitutionally*, as most metal elements, for example, would. F.c.c. is a more ordered structure than b.c.c. (see Chapter 2): although the packing density is greater, the size of the holes or 'interstices' in the structure—though there are far fewer of them—is greater (see Question 5.6 at the end of this chapter). To create a microstructure in the steel which will give satisfactory engineering properties the steel is raised to a high temperature to dissolve all the carbon (called *austenitising*) and then cooled back to the b.c.c. region whereupon the carbon precipitates out in the form of iron carbide, cementite.

Plain carbon steels are cooled slowly.

Heat treated steels are quenched and aged, just like the precipitation hardened copper alloys of Chapter 4.

For the third category of steels, *stainless steels*, the iron-carbon phase diagram is not always very relevant.

Back to plain carbon steels. The part of the phase diagram which controls the microstructure of the steels is the horizontal line, and reaction, at 727 °C, where the steel changes from f.c.c. to b.c.c.. This part is shown magnified in Fig. 5.2. The reaction centred on point 'b' in Fig. 5.2 is called a *eutectoid* reaction. It is just like the eutectic reaction we met with in Chapter 4, but for a eutectic reaction the high temperature phase is a liquid; for a eutect*oid* reaction the high temperature phase is a solid (here austenite, the f.c.c. form of iron-carbon). Just as with a eutectic reaction, a eutectoid

† This principle is very widely applied: for example, a masonry drill bit is a ceramic, tungsten carbide, held together by tough cobalt; sandpaper is brittle glass particles held together by resin.

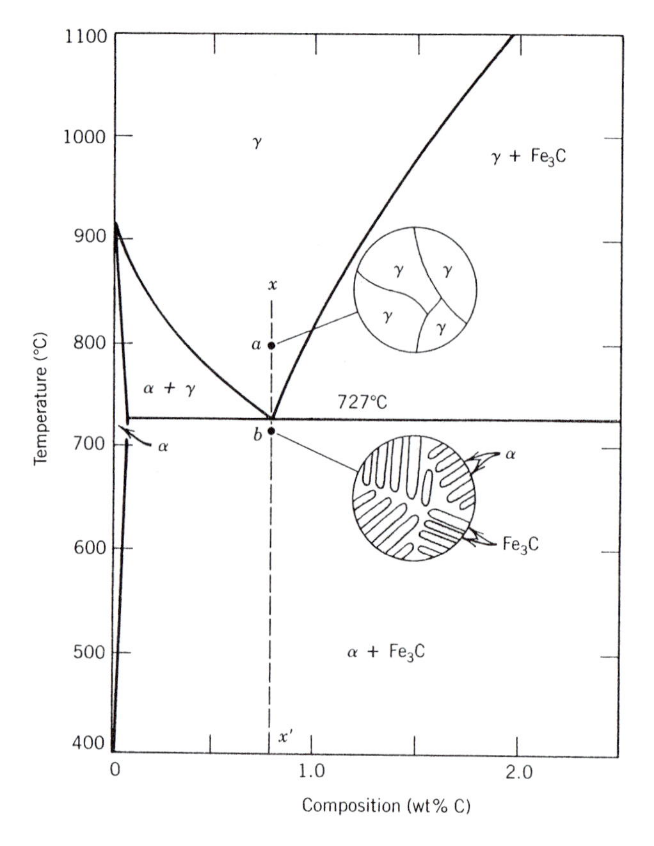

Fig. 5.2 The iron rich end of the iron–carbon phase diagram, shown magnified (from *Materials science and engineering* by W.D. Callister, courtesy of John Wiley & Sons Inc.).

reaction results in an alternating microstructure, here consisting of plates of ferrite and cementite. Figure 5.3 shows a picture of this product. It is called *pearlite*, because the alternating plates diffract light and give the surface of the steel (when properly prepared) an iridescent effect, like mother of pearl (and for essentially the same reasons). Compare Fig. 5.3 with Fig. 4.11, which shows the result of a eutectic reaction in solder.

A steel containing 0.8 wt% C would be all pearlite. Most steels contain a lot less carbon than 0.8%. For a steel with a composition between α and b (see Fig. 5.2) only part of the microstructure will be pearlite; the rest will be pure ferrite. At the α end it is nearly all ferrite, at the b end nearly all pearlite and *pro rata* in between. Figure 5.4 shows a 0.4% C steel, halfway between α and b. Notice that the ferrite occurs in two forms: as relatively large grains in its own right, and as part of the pearlite. The ferrite is almost pure iron and is very soft and ductile (and weak). The pearlite is what provides the strength, but, being brittle, it can also provide cracks as well!

Fig. 5.3 Pearlite makes up one part of a plain carbon steel. It is the result of the *eutectoid* reaction shown in Fig. 5.2. This *optical micrograph* (picture taken with an optical microscope) shows mainly pearlite, with a little bit of ferrite (marked 'F'). The carbon concentration was 0.7 wt%. The marker shows 100 μm.) (Courtesy Dr C.L. Davis.)

Plain carbon steels are usually classified into low, medium and high carbon. Low carbon steels, $< 0.15\%$ C, are soft and ductile and are used for sheet steel, for example for car bodies. Medium carbon steels, $0.15 < \%\,C < 0.4$ are sturdier and are suitable for tubes, girders, plates etc. *Mild steel* is an imprecise term which refers to a low to medium plain carbon steel, typically with $\sim 0.25\%$ C:

Mild steel accounts for 90% of all metal used worldwide

High carbon plain steels are not very common, but do find use where high strength is required with not very much ductility—for example ball bearings, chisels, hammers etc. As we move to higher carbon contents we begin to enter the world of *cast irons*. Initially these contain cementite, like steels (*white* cast irons) but as the carbon content rises towards 2 wt%, the carbon starts to appear as graphite (grey cast irons). Nowadays, cast irons can be engineered to be quite tough.

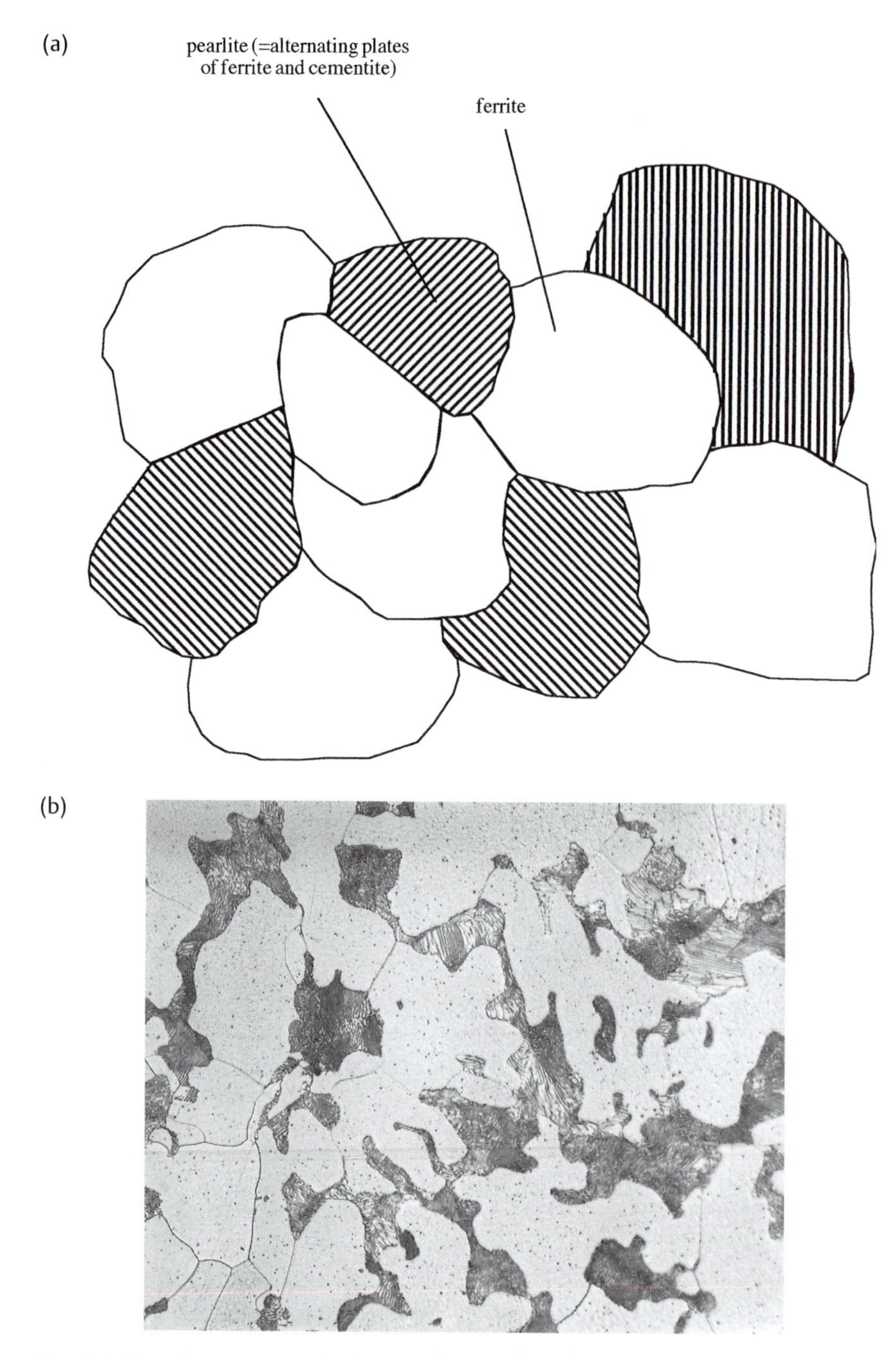

Fig. 5.4 The microstructure of a 0.4 wt% plain carbon steel. The ferrite (nearly pure iron) turns up in two different forms and two totally different shapes: as ferrite, pure and simple, and as a part of pearlite. The ferrite is soft and the pearlite hard (via the cementite in it). There is roughly half ferrite and half pearlite (see Q5.10 and Q5.11 at the end of the chapter). (a) Schematic, (b) a real picture (courtesy Mr. J. Farmer).

Table 5.3 British Steel production for 1995. (Annual report for shareholders.) I have converted from Pound Sterling to US$ using an average exchange rate of 1.6$/£. An 'Engineering' steel (strange terminology!) is a high carbon plain or alloy steel with high strength. (British Steel is now part of Corus.)

	Weight (Mtonnes)	Cost (M$)	$/tonne
Low carbon	7.0	3792	542
Medium carbon	6.1	2976	488
'Engineering'	1.2	826	688
Stainless	0.9	2530	2811

In Table 5.3 British Steel's production for a typical year (1995) is broken down by weight and cost. The cost in $/tonne is higher for low carbon steels because they are supplied as strip, which requires more energy input. The high strength and, particularly, stainless steels are expensive because of the cost of alloying additions (see below). Overall, 5% of steel production goes to electrical engineering as magnetic strip (i.e for specifically electrical purposes, not as structural steels in the electrical engineering industry) (see Chapter 10).

Although I have presented plain carbon steels as iron–carbon alloys, in reality they also contain manganese. Manganese is always present in iron ores, so it is merely left in, rather than added. It fulfils two functions: it removes harmful sulphur as harmless MnS particles, and it strengthens the ferrite somewhat via solute hardening (see Chapter 4). 1% Mn is typical, contributing 30 or 40 MPa to strength.

The fineness of the microstructure of the steel is increased by an increased rate of cooling after austenitisation. Thus

- 'Annealing' is slow cooling in the furnace. It gives large grains of ferrite and pearlite, and a relatively soft but easily machined steel.
- 'Normalising' is air cooling, which is faster and gives a more fine grained and therefore stronger product.

Mechanical properties of plain carbon steels

A typical low carbon steel has a yield stress of 150 MPa, a tensile strength (UTS) of 300 MPa and a ductility of 35%. For a medium carbon steel the equivalent figures are 400 MPa, 550 MPa and 20%.

Figure 5.5 shows where the strength in some plain carbon steels comes from. Solid solution and grain size strengthening are discussed in Chapter 4. Grain size strengthening is much more important for steel (b.c.c.) than for copper (f.c.c.). The contribution from pearlite is because it is a hard constituent of the alloy.

One characteristic of steels (not for some austenitic stainless steels—see Section 5.4) is that they go suddenly from ductile at high temperature to brittle at low temperature.

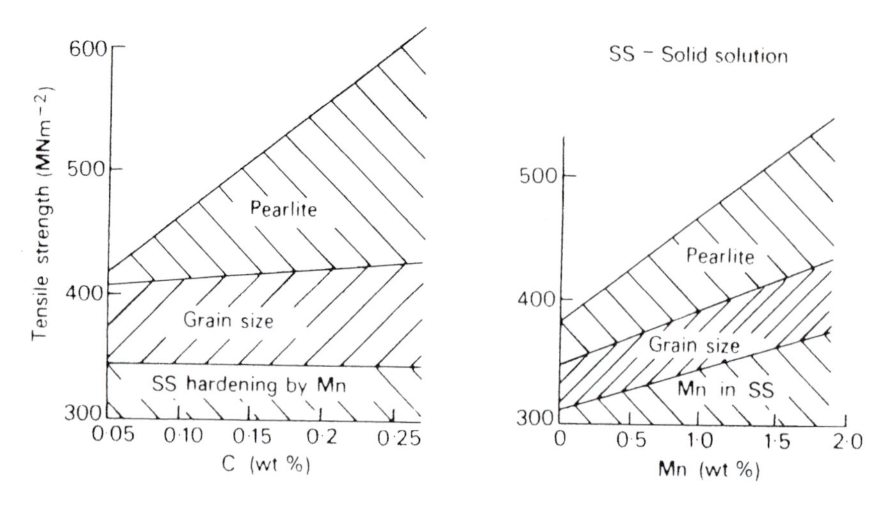

Fig. 5.5 Where the strength of plain carbon steels (with a little bit of manganese) comes from. (From Irvine *et al.*, (1962). *J. Iron and Steel Inst.* **200**, 821, via *Steels* by R.W.K. Honeycombe and H.K.D.H. Bhadeshia (Edward Arnold: London.))

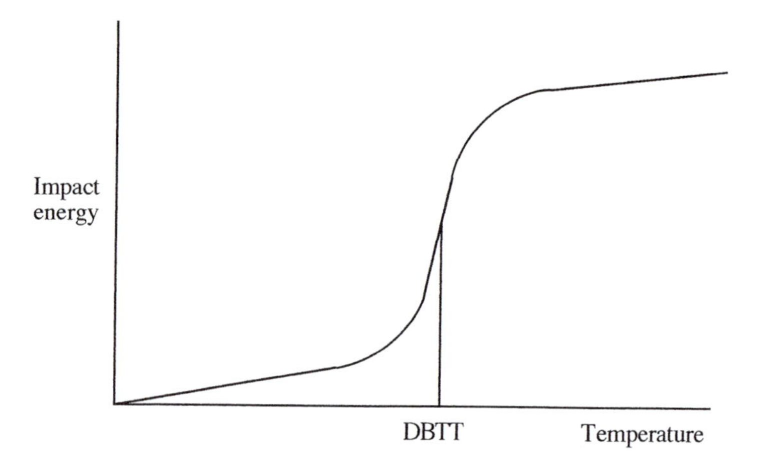

Fig. 5.6 Steels become very brittle very suddenly as the temperature drops. This graph is of impact energy, but the same shape would be obtained for a graph of fracture toughness, or of ductility.

Thus their impact energies vary abruptly (Fig. 5.6). The transition temperature is called the *ductile–brittle transition temperature* (DBTT). For a mild steel the DBTT is about 0 °C. Using a steel at a temperature much below its DBTT can be dangerous.

The representative figures for mild steel presented throughout this section can be much changed by using cleverer heat treatments and small alloying additions. All this, of course, costs money.

5.3 Heat treated, or alloy, steels

Recall the precipitation strengthened copper alloys of Chapter 4 (Section 4.6). There, a matrix of copper was strengthened by a fine dispersion of small precipitates, giving an enormous increase in σ_y and UTS. There was an optimum size of precipitate, in that case, of a few nm—any smaller and the precipitates were cut by the dislocations; any larger and the dislocations could pass between the precipitates.

The pearlite which strengthens plain carbon steels is neither the right size nor the right shape to give optimum strengthening. It is far too coarse, and its platelike shape is a wonderful way of starting off cracks. We can produce a *much* stronger steel by quenching and aging in exactly the same way as with the copper alloys. This is what a heat treated steel is. There are one or two differences in detail between the steel situation and that of the copper, which I shall highlight as we go through. The basic principles are the same: the carbon is taken into solution at high temperature; there is then a quench, followed by an intermediate temperature aging treatment which forms small precipitates.

The first differences which arise are with the quench:

1. Steel has a lower thermal conductivity than copper (or aluminium) and so the quench is both less efficient and physically more violent (because of temporary thermal gradients),

but more importantly

2. the carbon which redistributes itself in steel is an *interstitial* solute (see Chapter 1) and moves very quickly—1000 or 10 000 times faster than the *substitutional* solutes in copper. Therefore the equilibrium phase transformation in steel is more difficult to avoid via the quench.

For both these reasons, alloying elements are added to the steel. This is why for the sorts of steels you are likely to meet as electrical engineers

a heat treated steel = an alloyed steel

The alloying elements work by slowing down the pearlite transformation. This makes it easier for the quench to produce the metastable supersaturated solution (*but see below*). The reason for this is that the alloying elements go into substitutional solid solution. They partition differently between the cementite and ferrite (some like the cementite, some prefer the ferrite). The pearlite phase transformation has to wait for them to diffuse into position and hence it is slowed down. The slowing down of the pearlite transformation means that oil quenching can be used instead of the more violent water quench. Oil quenching is less likely to deform and crack steel components—particularly large ones.

LIVERPOOL
JOHN MOORES UNIVERSITY
AVRIL ROBARTS LRC
TEL. 0151 231 4022

Thus although the alloying elements in a heat treated steel do provide some extra strength (like Mn in a plain carbon steel), the primary reason they are there is a *negative* one: they are added in order to inhibit the pearlite transformation.

So the first difference between producing precipitation hardened copper alloys and a heat treated steel is one of *kinetics*. The second difference involves the product of the quench. In the case of the copper, this is simply a metastable supersaturated solid solution. In the case of the steel, a new phase is produced, called martensite. It has the same chemical composition as the austenite from which it comes, and its formation therefore does not require chemical diffusion, like pearlite. Instead there are small local movements of the iron and carbon atoms (and alloy elements) as shown in Fig. 5.7.

Stage I distorts face centred cubic austenite to a body centred cubic phase. Carbon then occupies one of the cube axes preferentially, distorting slightly the cubic structure

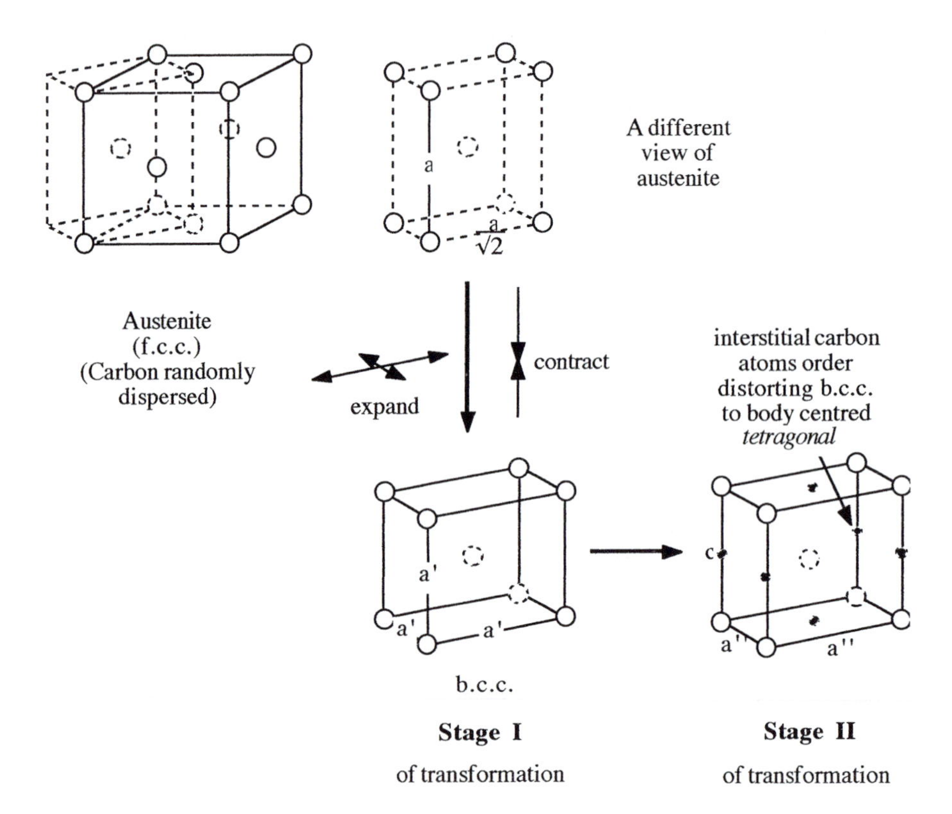

Fig. 5.7 The transformation from austenite to martensite. Really, Stage I and Stage II happen together. The large distortion associated with Stage I is relieved by lots of dislocations. Both these and the carbon give additional strengthening. The label 'c' in Stage II is one of the lattice parameters: by chance I put it in a position occupied by a carbon atom (C). N.B. Not all the four positions shown are occupied by carbon atoms—only a fraction of them—but enough to distort the cubic unit cell to tetragonal.

to tetragonal (square prism). I say *then*, but really Stages I and II happen simultaneously. The strain associated with the enormous shape change of Stage I is relieved by the passage of huge numbers of dislocations, many of which are left behind and contribute substantially to the strength. Martensite is a very hard, strong and brittle phase and is not of much use as it stands. In the final aging treatment, which in steels is called *tempering* (because it improves the toughness or 'temper' of the steel), the martensite decomposes into the equilibrium b.c.c. phase ferrite and small carbide precipitates. Figure 5.8 compares the strengthening of steel with that of copper:

The two products (of copper alloys and steel, that is) are basically similar, consisting of a solution hardened matrix containing a dispersion of fine precipitates. The difference is that the martensite transformation in the steel results in a very small grain size (the transformation takes place in very small lens-shaped packets) and an enormous dislocation density, both of which much increase the strength. (Although precipitation hardened copper alloys are often work hardened, this is nothing like as effective—also grain size strengthening is much more potent for b.c.c. than for f.c.c. alloys.) Which alloying elements are used? Typical ones are Mn (again), Ni, Cr and Mo. Table 5.4 shows some typical mechanical properties. Although the best copper alloys approach the alloy steel figures, higher alloying levels and more sophisticated and complicated thermal and mechanical processing can give alloy steels strengths up to 3000 MPa and beyond.

Figure 5.9 shows a high pressure power generating turbine labelled with the different types of steel used, all of which are alloyed. Most of them, in fact, are stainless (see Section 5.4).

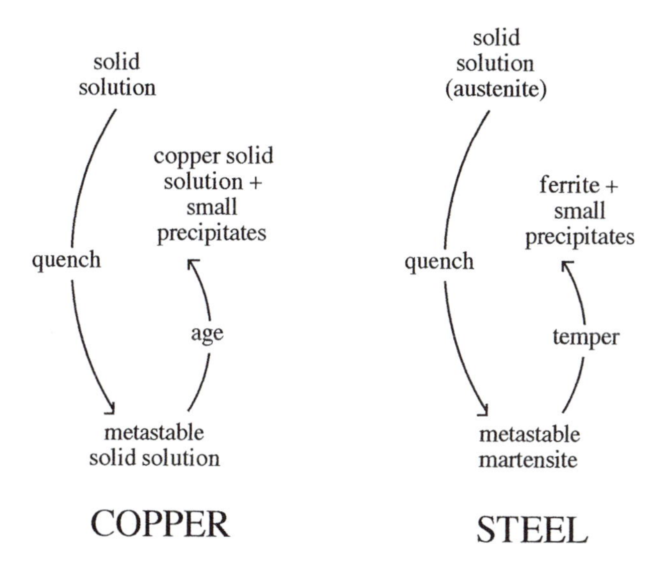

Fig. 5.8 Strengthening steel is very similar to strengthening copper.

Table 5.4 A comparison of the mechanical properties of representative medium carbon plain and alloyed steels.

	Medium carbon steel	Alloy steel (0.4% C, 1.5% Ni, 1% Cr, 0.2% Mo; tempered 1 h at 400 °C)
σ_y (MPa)	400	1400
UTS (MPa)	550	1550
% elongation	20	7

In trying to make the subject of steels accessible to you, and understandable, I am conscious that I have presented a more simple picture than real life warrants. The various categories and processes which I have described separately, in the interests of clarity, are often mixed together in industrial practice. Also, I have done no more than hint at the great sophistication which can be brought to this crucial and competitive area of metallurgy; nor have I, for example, mentioned high strength low alloy (HSLA) steels, which rely on alloy carbides for their strength, rather than cementite. Nevertheless, the concepts introduced in this chapter should give you a basic understanding of this most important of structural materials.

To finish off this chapter, we now turn away from the mainstream of structural steels and move to

5.4 Stainless steels

One of the problems with steels is that they *rust*. Rust is the familiar brown deposit which is unique to steels and which I will deal with at greater length in Chapter 6. It is formed in the presence of oxygen and water and consists of a mixture of iron oxide and hydroxide. The main problem with rust is that it is very porous and does not protect the steel from further corrosion. This brings us to the third and last type of steel: *stainless steel*. Stainless steels are stainless because they contain a large amount of chromium. In the presence of oxygen the chromium forms a thin layer of Cr_2O_3, chromium oxide, on the surface of the steel. This is a very perfect and protective oxide layer[†] which prevents further attack by oxygen and water.

The simplest sorts of stainless steel are iron with between 12 and 30% Cr. They may or may not contain carbon. Round about 12–13% Cr those which do are called 'martensitic stainless steels'—they undergo the same martensite transformation I

[†] Another very perfect oxide is that which forms on aluminium, which is why we don't bother to paint it and why it stays quite bright, even outside, exposed to the elements.

Fig. 5.9 Alloy steels in use: the high pressure section of a power generating steam turbine. The diameter of the turbine is about 1 m. (The low pressure part later on is much bigger.) The various components are identified below. Most of the steels, in fact, qualify as 'stainless' because of their high chromium content, but all the steels, nevertheless, stainless or not, are quenched and tempered (see text). (Courtesy Alstom Power, Mr R.W. Vanstone.) Key to materials indicated on photograph

1. **Moving** blades: 11%CrMoVNbN martensitic stainless steel bar
2. **Outer** casing: 2.25%CrMo low alloy steel casting
3. **HP/IP** rotor: 10%CrMoVNbN martensitic steel forging
4. **Stationary** blading: 9%CrMoVNbN martensitic steel bar and forgings
5. **Bolting:** 11%CrMoVNbN martensitic stainless steel bar
6. **Inner** casing: 9%CrMoVNbN martensitic stainless steel casting
7. **Bolting:** NiCrTiAl (Nimonic 80A) bar (This is not a steel; it is a nickel-based superalloy.)

described earlier. A stainless steel knife blade is likely to be made from a martensitic stainless steel and so may be a high temperature boiler tube. Figure 5.9 shows some martensitic stainless steels in use in a power generating turbine. These would be quenched and tempered like any alloyed steel. If the carbon is missed out from this type of steel it becomes more ductile and can be fabricated more easily—for example, to make a stainless steel exhaust pipe for a car. As the chromium content is increased the corrosion resistance of the steel increases, but the martensitic transformation disappears. This is because chromium stabilises ferrite at the expense of austenite. The steel is b.c.c. from its melting point right down to room temperature. Steels with 17–30 wt% Cr are thus called *ferritic stainless steels*. In these steels, the carbon content must be kept very low (e.g. < 0.08 wt%), otherwise large

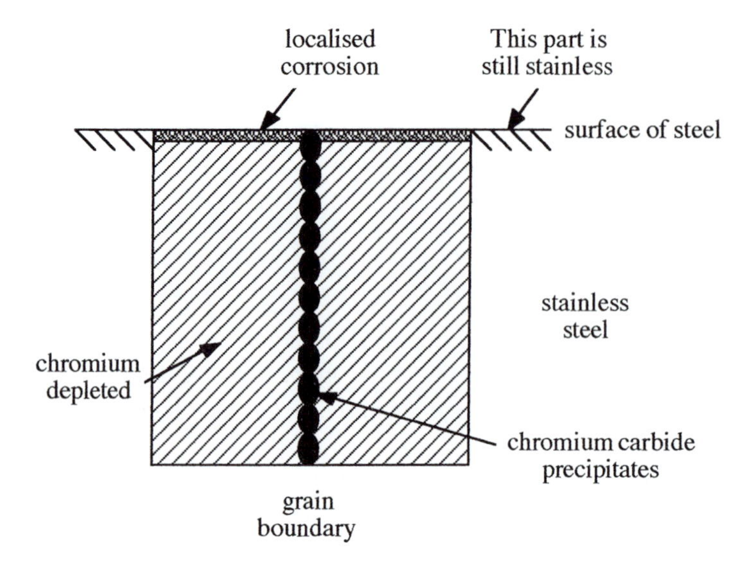

Fig. 5.10 Precipitation of chromium carbides at a grain boundary in a stainless steel causes localised corrosion (sensitisation). The chromium depleted region is of the order of microns (10^{-6} m across). This problem occurs particularly in welding and in the irradiation environment of nuclear reactors.

particles of brittle chromium carbide precipitate at the grain boundaries. Not only is this bad for mechanical properties, but the depletion of chromium in the neighbourhood of the grain boundaries leads to localised corrosion (Fig. 5.10). This is called *sensitisation* by metallurgists and is a particular problem when stainless steels (of any type) are welded (see Section 5.5).

A different sort of stainless steel is obtained by adding nickel to a Fe–Cr alloy. Just as chromium, a b.c.c. element, stabilises b.c.c. iron, so nickel, an f.c.c. element, stabilises f.c.c. iron. As nickel is added to a Fe–Cr alloy, the austenite phase becomes more and more stable until it exists right down to room temperature. Thus an *austenitic* stainless steel is created. Now nickel is *very* expensive (US$/tonne Fe: 200; Cr: 1630; Ni: 7350). It turns out that Fe–18wt%Cr requires the *least* amount of nickel to make it austenitic. Fe–18Cr–8Ni ('18/8') is the most familiar austenitic stainless steel. In fact it is only just an austenitic stainless steel at room temperature. If you place a magnet near your stainless steel sink (which will probably be made from 18/8) you may well discover slight magnetism where the working of the stainless steel has transformed parts of it to martensite.†

† You may also discover that the top of the sink is a martensite stainless steel—to save money. Take a magnet with you when you go to buy stainless steel cutlery etc!

So why have we gone to all the trouble of adding expensive nickel to stainless steel in order to make it f.c.c.? The advantages of austenitic stainless steels as compared with the cheaper martensitic or ferritic ones are better corrosion resistance, workability (ductility) and weldability. Like their ferritic counterparts they suffer from sensitisation, but this can be controlled by small additions of titanium or niobium, which tie up the dangerous carbon as small (and strengthening) precipitates of Ti or Nb carbide and prevent the precipitation of chromium carbide.

Having listed their advantages, I should warn you that although austenitic stainless steels are very ductile, they are actually rather weak (e.g. σ_y 250 MPa; UTS 550 MPa). In fact, 18/8 is only one example of an austenitic stainless steel. 25/20 (Cr/Ni) stainless steels (notice how much more nickel we have had to add) have better corrosion resistance and better strength, particularly at high temperatures. (They are also *much* more expensive.)

Because stainless steels rely on their thin layer of chromium oxide for their corrosion protection, it is important that they are kept in *oxidising* surroundings. If a stainless steel is denied oxygen it will corrode like any other. For example, to return to the stainless steel sink, if you put in a plastic grid to prevent scratching (as I once foolishly did) you will soon find rust where the grid has inhibited oxygen access. Those of you who sail will find the same effect if you use plastic sheathing round stainless steel halyards (there may be safety reasons for having this, of course). The protective layer on stainless steels is particularly sensitive to chloride ions (Cl^-): stainless steel is not much used on sea-going vessels.

Finally I should point out that in a chapter about not-very-good conductors, stainless steels are particularly awful conductors of both electricity and heat (Table 5.5).

5.5 The welding of steels

There are two reasons why the welding of steels is important to electrical engineers. Firstly, welding nowadays is generally an electrically powered process and

Table 5.5 **The electrical and thermal conductivities of copper, mild steel and stainless steel. (Notice that the electrical and thermal conductivities stay roughly in the same proportion. This is because they both rely largely on free electrons.)**

	Copper	Mild steel	Stainless steel
Electrical conductivity $(\Omega m)^{-1} \times 10^6$	59	6.3	1.4
Thermal conductivity $(W\ m^{-1}\ K^{-1})$	0.94	0.124	0.038

resistance welding electrodes, for example, are an important application for the stronger copper alloys (Chapter 4). Secondly, welding is the usual way of joining steel components together, and steel, as we have seen, is the dominant general engineering material. I mentioned soldering and brazing in Chapter 4. In soldering and brazing a low melting point alloy is used to 'glue' two components together. The two components are not really affected by the process. In welding, the two components are joined directly and they *are* affected. There are two sorts of welding: fusion welding and pressure welding. In pressure welding the two components are forced together and the bond, or weld, forms by solid state diffusion.

Fusion welding is the more common process. The surfaces of the two components are melted together. Often some additional material (filler) is supplied. The weld pool needs to be protected from oxidation, either by a molten ceramic or by an inert gas. The heat is supplied in a variety of ways. Four common methods are:

- oxy-acetylene torch
- electric arc
- resistance heating
- electroslag

In *oxy-acetylene welding* (see Fig. 5.11(a)) the heat comes from burning acetylene in oxygen; additional metal is supplied via a filler-metal rod. This is a lightweight mobile method and, for example, would be how your car would be repaired.

In *tungsten inert gas* (TIG) (Fig. 5.11(b)) welding an arc is struck between a tungsten electrode and the weld. The shielding gas comes from round the electrode and is usually argon. Another type of electric arc welding is *manual metal arc* (sometimes simply *electric arc*) welding. The difference here is that the electrode itself is the filler metal. The arc melts the electrode and it slowly dribbles into the weld cavity. The electrode is therefore consumed. The protection is supplied by a flux (a low melting point ceramic: see Chapter 7) which forms the outside of the electrode and melts onto the weld at the same time, forming a protective cover.

Metal inert gas (MIG) (Fig. 5.11(c)) is a combination of the previous two methods.

In *submerged arc welding* (Fig. 5.11(d)) the arc is actually struck under the molten flux. This is for really heavy jobs—thick steel sections and so on.

In *resistance welding* (sometimes called spot welding) (Fig. 5.11(e)) a current is passed from one copper electrode (see Chapter 4, Section 4.6) to another through the two pieces of sheet to be joined. The I^2R heating melts the two sheets together, forming a spot join. This method is used for low carbon steel strip and is typical of the automobile industry.

Finally, in *electroslag welding* (Fig. 5.11 (f)) the heat comes from I^2R heating from passing a current through a molten slag (another name for a protective molten ceramic).

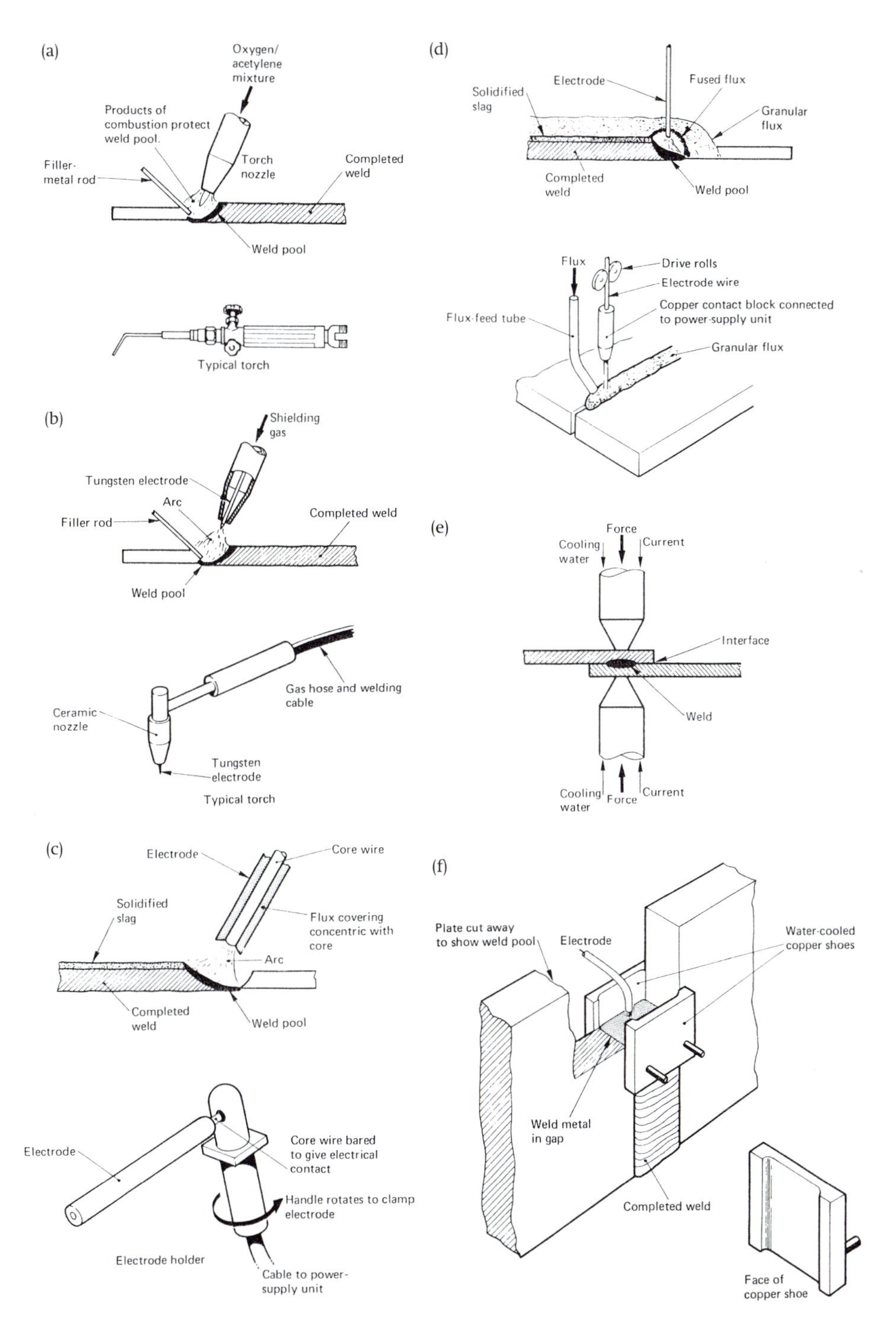

Fig. 5.11 Schematic diagrams of various welding methods. (a) Oxy-acetylene (b) Tungsten Inert Gas (TIG), (c) Metal arc, (d) Submerged arc, (e) Resistance and (f) Electroslag. (From *Principles of welding technology* by L.M. Gourd, courtesy of Edward Arnold.)

Electron beam and, particularly, laser welding are also becoming increasingly popular and common nowadays.

The big problem with steel welds is that they are really little castings made under the worst possible conditions. Hydrogen given off during the welding process embrittles the weld. In a high carbon or alloy steel the molten weld is quenched by the surrounding metal, forming brittle martensite. The *heat affected zone* (HAZ) refers to all that part of the steel (weld and parent metal) whose heat treatment will have been affected by the welding operation. Frequently, post-weld heat treatments are necessary to restore the properties of the steel. In demanding applications, like the fabrication on-site of a nuclear reactor pressure vessel, literally dozens of welding operations are involved. These are defined in tremendous detail and the whole process can take months or years and constitutes the major part of the fabrication process.

5.6 Summary

1. Steel consists of iron, a fraction of a per cent carbon, and perhaps a few per cent of other alloying elements.
2. Iron ore is cheap, iron is cheap to extract from its ore and steel is relatively easy to fabricate. Therefore steel is always the preferred structural metal.
3. There are plain carbon steels, alloy steels and stainless steels.
4. Steel is made by reducing iron ore (iron oxide) with coke (carbon) and then carefully oxidising away most of the carbon remaining in the iron.
5. Plain carbon steels contain iron, carbon and a little manganese.
6. They are heated to 800–900 °C to dissolve the carbon in the f.c.c. iron and then cooled. At $\sim$ 800 °C b.c.c. iron (ferrite) begins to form and at 727 °C the remaining f.c.c.iron–carbon solution decomposes to a mixture of ferrite and iron carbide (cementite) plates called pearlite.
7. Low carbon steels have <0.15 wt% C, are ductile and are used as sheet.
8. Medium carbon steels contain between 0.15 and 0.4 wt% C. They are stronger and are used for tubes, girders and plates.
9. Mild steel encompasses the top end of the low carbon range and the bottom end of the medium carbon range and accounts for 90% of metal used worldwide.
10. Steels suddenly become brittle below their ductile brittle transition temperature (DBTT).
11. Alloy steels are quenched and tempered to form a fine dispersion of iron carbides in ferrite. The alloying elements slow down the pearlite reaction and

enable martensite to be formed by an oil quench. The martensite decomposes on tempering. Alloy steels are very strong and tough.

12. Stainless steels contain chromium. They may be ferritic (b.c.c.) or they may contain nickel which makes them f.c.c. (austenitic).
13. Steels are joined by welding. Most welding involves the melting of the two surfaces and the use of a filler metal. There is a variety of heating methods. Welds are common places for cracks to start.

Recommended reading

Principles of welding technology by L.M. Gourd (published by Edward Arnold)

Steels—metallurgy and applications by D.T. Llewellyn (published by Butterworth-Heinemann)

Steels: microstructure and properties by R.W.K. Honeycombe and H.K.D.H. Bhadeshia (published by Edward Arnold)

Questions (Answers on pp. 316–321)

(Questions 5.1–5.3: The relative atomic masses of carbon, iron and manganese are 12.01, 55.85 and 54.93 respectively.)

5.1 A steel has a carbon concentration of 0.2 wt%. (The 'wt' is often missed out in steel literature.) What is the at% of carbon?

5.2 The strengthening phase in plain carbon steels is Fe_3C. What is the wt% of C in Fe_3C?

5.3 A steel contains 0.1 wt% C and 1 wt% Mn. What are the at% of the two elements?

5.4 What are the names for Fe_3C, the solid solution of carbon in b.c.c. iron and the solid solution of carbon in f.c.c. iron?

You will find that an understanding of Questions 4.2–4.7 (Chapter 4) is very helpful for Questions 5.5, 5.7–5.11. Use Figs 5.1 and 5.2.

5.5 (a) What is the solubility limit of carbon in b.c.c. iron at 727°C?

(b) What is the solubility limit of carbon in f.c.c. iron at 1148°C?

5.6 (a) The lattice parameter of b.c.c. iron is 0.286 nm. Treating the atoms as touching spheres, where are the largest interstices (holes) in the structure? What is their size?

(b) The lattice parameter of f.c.c. iron (extrapolated to room temperature) is 0.365 nm. Where are the largest interstices and what is their size?

(c) A carbon atom in iron has diameter 77 pm. What can you infer about the solubilities of carbon in b.c.c. and f.c.c. iron?

Questions 5.7–5.11 relate to a 0.2 wt% C mild steel.

5.7 Above what temperature must the steel be raised in order to austenitise it?

5.8 On cooling the steel (slowly) at what temperature does the 'pro-eutectoid' ferrite start to form? At what temperature does the pearlite form?

5.9 What is the composition of the first ferrite to form? What is the composition of the ferrite when the pearlite forms? What is the composition of the ferrite at room temperature?

5.10 What would you see in an etched specimen in an optical microscope? What proportion of the microstructure would be ferrite and what proportion pearlite?

5.11 At 727 °C, what proportion of the pearlite consists of ferrite? What proportion of the ferrite constitutes the pro-eutectoid ferrite and what proportion is in the pearlite? What proportion of the carbon is in the ferrite and what proportion in the pearlite?

5.12 A comparison of the two parts of Fig. 5.5 suggests that the Mn content of the steel on the left is 1 wt% and that the carbon content of the steel on the right is 0.13 wt%. For a steel with both 0.13 wt% C and 1 wt% Mn, enumerate the four main contributions to the tensile strength.

You will need to do some factfinding for Questions 5.13 and 5.14.

5.13 What types of steels would the following be made from? In each case give also the main method of fabrication.

(a) Electricity transmission tower (what you may know as a 'pylon').

(b) A disc in the low temperature part of a power generation turbine. (The disc is the central part to which the blades are attached.)

(c) A disc in the high temperature part of a turbine.

(d) The core of a transformer.

(e) The pressure vessel of a nuclear reactor.

5.14 What welding method would you use for the following jobs?:

(a) The steel frame for a building

(b) Repairing some farm equipment

(c) Joining stainless steel tube in a chemical plant

(d) Fixing two car body panels together

(e) Joining two thick plate sections in a pressure vessel

6

Electrochemistry: electroplating and corrosion

Chapter objectives

- The concept of a cell
- Electroplating: why? how?: some examples
- Corrosion: electrode driven and electrolyte driven
- How to prevent corrosion
- Common types of battery
- Fuel cells
- Secondary batteries or accumulators

6.1 Chemistry: some reassuring comments!

This chapter is about chemistry. I know that many of you will not have studied chemistry for many years and even then, not much. DON'T WORRY! I only need a little common-sense-sort-of-chemistry. Believe me, the author is no chemist (as you may soon notice)! Electroplating is a very important part of the electrical and electronic engineering industry and so you should know something about it. Corrosion is electroplating's evil twin. In their natural state most metals exist as chemical compounds called ores: oxides, sulphides etc. As metals, therefore, they are in an unstable situation. Corrosion is nature's way of returning them to their natural stable state. Corrosion costs billions of pounds (dollars, euros ...) each year. Every engineer should know something about corrosion. Electrochemistry is based on the concept of a *cell*. First of all I will explain what a cell is and then I will explain the relationship between electroplating, corrosion, batteries and fuel cells. They are all very closely related. Then we'll take a brief look at each in turn, mainly via lots of examples with many of which you will be very familiar.

6.2 The concept of an electrochemical cell

Imagine we take a glass beaker with hydrochloric acid in it and then drop a piece of iron in. It fizzes around, lots of bubbles come out and as the iron dissolves in the acid, the acid goes green. What has happened? A chemist might write this:

$$\mathrm{Fe} + 2\mathrm{HCl} \rightarrow \mathrm{FeCl_2} + \mathrm{H_2} \tag{6.1}$$

Now this is a *symbolic* equation. Fe stands for one atom of iron, 2HCl stands for two *molecules* of HCl (hydrochloric acid) (i.e. two atoms of hydrogen and two atoms of chlorine), $FeCl_2$ for one molecule of *ferrous chloride* and H_2 for one molecule of hydrogen (which is what the bubbles are). Hydrochloric acid is really HCl *in water*. Because water has such a high *dielectric constant* (see Chapter 7) compounds, if they can, split into *ions* and *dissolve* in the water. Thus a more accurate and informative way to write the above equation is:

$$\underset{\text{metal}}{\mathrm{Fe}} + \underset{\substack{\text{in}\\ \text{soln}}}{2\mathrm{H^+}} + \underset{\substack{\text{in}\\ \text{soln}}}{2\mathrm{Cl^-}} \rightarrow \underset{\substack{\text{in}\\ \text{soln}}}{\mathrm{Fe^{++}}} + \underset{\substack{\text{in}\\ \text{soln}}}{2\mathrm{Cl^-}} + \underset{\text{gas}}{\mathrm{H_2}\uparrow}$$

or

$$\underset{\text{metal}}{\mathrm{Fe}} + \underset{\substack{\text{in}\\ \text{soln}}}{2\mathrm{H^+}} \rightarrow \underset{\substack{\text{in}\\ \text{soln}}}{\mathrm{Fe^{++}}} + \underset{\text{gas}}{\mathrm{H_2}\uparrow} \tag{6.2}$$

I have now written the equation in terms of *ions* (remember the *ionic bond* in Chapter 1?). Fe metal has become Fe^{++} *cat*ions in solution. Thus charge has been transferred. This happens via *electrons*. We can write the part of eqn (6.2) which involves iron as:[†]

$$\mathrm{Fe} \rightarrow \mathrm{Fe^{++}} + 2\mathrm{e^-} \tag{6.3a}$$

The part of the equation involving hydrogen can be written as:

$$2\mathrm{H^+} + 2\mathrm{e^-} \rightarrow \mathrm{H_2}\uparrow \tag{6.3b}$$

Noticing that $2Cl^-$ is common to both sides of (6.2), eqns (6.3a) and (6.3b) added together are equivalent to eqn (6.2). Although it wasn't mentioned in eqn (6.2) or (6.1), electrons have been transferred between atoms. This is fundamental to reactions which take place in aqueous solution (i.e. in water).

When we drop the piece of iron into hydrochloric acid, eqns (6.3a) and (6.3b) are taking place all over the iron and in the acid next to it. What if we separate in space

[†] A chemist would say the iron had been *oxidised*, extending somewhat the use of that word.

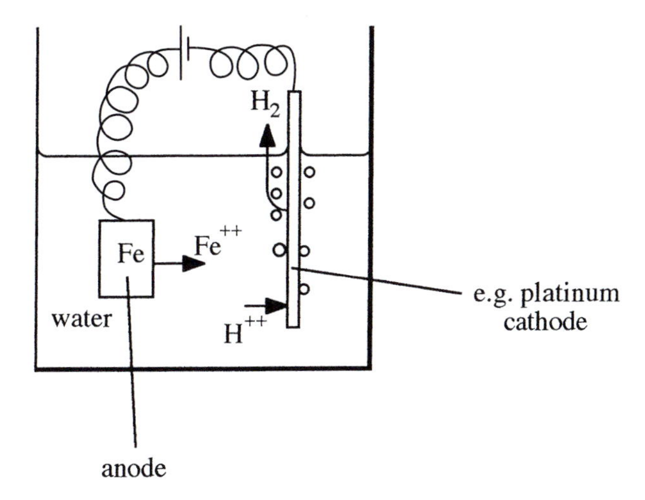

Fig. 6.1 An electrochemical cell.

eqns (6.3a) and (6.3b)?—for example if we provide a more attractive site for equation (6.3b) to take place than all over the iron? The result, shown in Fig. 6.1, is a CELL. We have separated the two halves of reaction (6.1) and enabled the necessary electron transfer to take place via a wire.

All cells are like this. At the *anode*, electrons are given up and, typically, a metal goes into aqueous solution as ions:

$$M \rightarrow M^{n+} + ne^- \tag{6.4}$$

At the *cathode* electrons are consumed. Some examples of how this might happen are:

$2H^+ + 2e^- \rightarrow H_2\uparrow$	This is the one we've already seen
$O_2 + 4H^+ + 4e^- \rightarrow 2H_2O$	In acid solutions
$O_2 + 2H_2O + 4e^- \rightarrow 4OH^-$	In alkaline solutions
$M^{n+} + ne^- \rightarrow M$	The opposite of eqn (6.4) above. This is how copper cathodes are made (see below).

A common choice of cathode is platinum, because it is so unreactive ('noble') that there is no danger of it contributing its own anodic reaction.

If electrons are travelling down a wire, there must be a voltage across it. What *is* the voltage? Going back to eqn (6.1),

$$Fe + 2HCl \rightarrow FeCl_2 + H_2\uparrow \tag{6.1}$$

this equation only takes place, i.e. only goes from the left side to the right side, provided there is an associated energy drop. Every reaction has such a drop in energy. Vigorous reactions are accompanied by a large drop in energy, slow reactions by a small one. In chemistry, the energy is *free* energy—it includes a statistical aspect. It is given the symbol G. So

$$Fe + 2HCl \underset{\Delta G}{\rightarrow} FeCl_2 + H_2 \uparrow \tag{6.1}$$

ΔG is the drop in free energy as the reaction proceeds from left to right. It is measured in eV/atom (for example, eV/atom of iron in eqn (6.1)) or in kJ/mole where a mole is Avogadro's number (6×10^{23}) of atoms. It doesn't matter which—whichever you feel most comfortable with. ΔG is what makes the voltage across the wire in the cell. We can work it out as follows. If one atom of iron reacts according to (6.1), then ΔG eV are released. This equals the work consumed by the two electrons in dropping through a voltage V. So here

$$\Delta G = 2 \times \text{electronic charge} \times \text{voltage in volts}$$

If ΔG is expressed as eV / electron transferred then ΔG is also the resulting voltage in volts. If ΔG is expressed in kJ / mole it is necessary to perform some manipulation of the units—there is some practice at this in the questions at the end of the chapter.

What if I were to attach a battery in the *opposite sense* to that produced naturally by the reaction? What would happen? Common sense—instinct—call it what you will—tells you, quite correctly, that if the applied voltage is greater than the original one, we

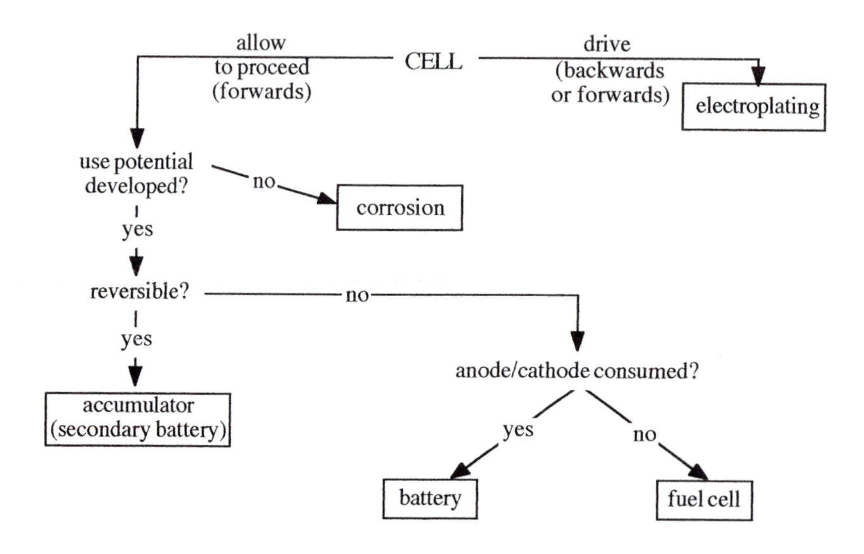

Fig. 6.2 A classification of the main processes and devices in electrochemistry.

can force the reaction to go backwards. (For this particular example we would need to supply the hydrogen, since this has now bubbled away.) So some cells proceed *spontaneously*; others are *driven*. This gives us a useful way of classifying cells: Fig. 6.2.

You are probably familiar with most, if not all, of the names in boxes; I will deal with them one by one. First of all,

6.3 Electroplating

First of all, before I explain *how* electroplating is carried out, I should explain *why* it is carried out. In electroplating a component or object or whatever is coated with a thin layer of metal. This can be to provide

- better *electrical contact*
- *wear resistance* or *lubrication*
- *corrosion protection*
- *soldering*
- *etch masking*
- *catalysis*
- *decoration*

The object to be plated is made the *cathode* of a cell and a voltage is applied with the negative terminal of the voltage supply being connected to the cathode and the other to the anode. Let's look at a simple plating cell and then I shall present you with a variety of examples which will illustrate the wide range of plating applications and methods employed to achieve them.

A simple electroplating cell

Two pieces of copper are placed into an aqueous copper sulphate solution (copper sulphate crystals dissolved in water) (Fig. 6.3). A voltage is applied between the two pieces thus making them into electrodes. That electrode to which is applied the positive connection starts to dissolve, the copper atoms losing two electrons each and going into solution as Cu^{++} ions. This is the anode. The electrons are moved by the battery towards the cathode where two electrons from the wire join with a Cu^{++} ion in solution, neutralising it, whereupon it 'plates out' on the cathode. This might seem to you a singularly pointless exercise, but in fact it is much used commercially to purify copper. The impurities in the copper do not transfer through the solution and fall to the bottom of the cell as sludge, which is periodically removed. Copper is traditionally bought as *copper cathodes* (see section 4.1).

Of course, the cathode and anode in a plating cell do not have to be of the same material—usually they are different—but the idea of an electrochemical cell driven by an

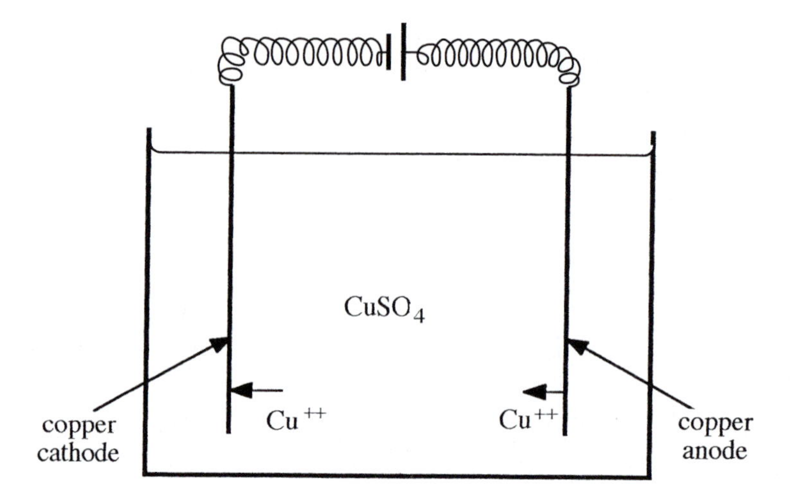

Fig. 6.3 A simple electrochemical cell for plating copper.

externally applied voltage is common to all electroplating. In principle an electroplating cell could proceed in the forward direction under its own steam (as it were), given the right combination of materials. In fact an external voltage is nearly always applied in the interests of control and speed. On the rare occasions when a voltage is *not* applied, the procedure is called *displacement*. It is used for silvering mirrors and for reclaiming copper.

In the following examples I shall try to present you with a range of applications and methods. I should say at the outset that there is still more art than science in electroplating. I remember when I was a metallurgy student in the 1960s being taken around a plant for chromium plating automobile bumpers (fenders). This was in the days when such things were made of metal. The foreman told me that they once cleaned up the plating baths, removing vast quantities of sludge, slime, oil, detritus and goodness knows what. As a result it was months before the process worked properly again. This black magic aspect is because electrochemical processes are very, very complicated: anyway, we must do the best we can!

Example 1 A phone connector (see Fig. 6.4)

The mechanical connection is made at the bottom of the metal connector (where it forks) while it is in place in the white plastic shell. This part needs to be silver plated to give a good electrical connection which can be 'made' many times. The top of the connector is solder plated preparatory to a wire being joined to it.

The connector starts life as brass strip. This is plated on one edge with silver and then it is turned over and the other edge is plated with solder, as shown schematically in Fig. 6.5. The silver plating is carried out in a bath containing silver and potassium cyanides.[†] The solder plating bath contains lead and tin fluoroborates.

[†] In fact, this was the first industrial electroplating process, developed only a kilometre or two from where I am sitting (in Edgbaston, Birmingham, England). Remember this was before mains electricity and A.C.

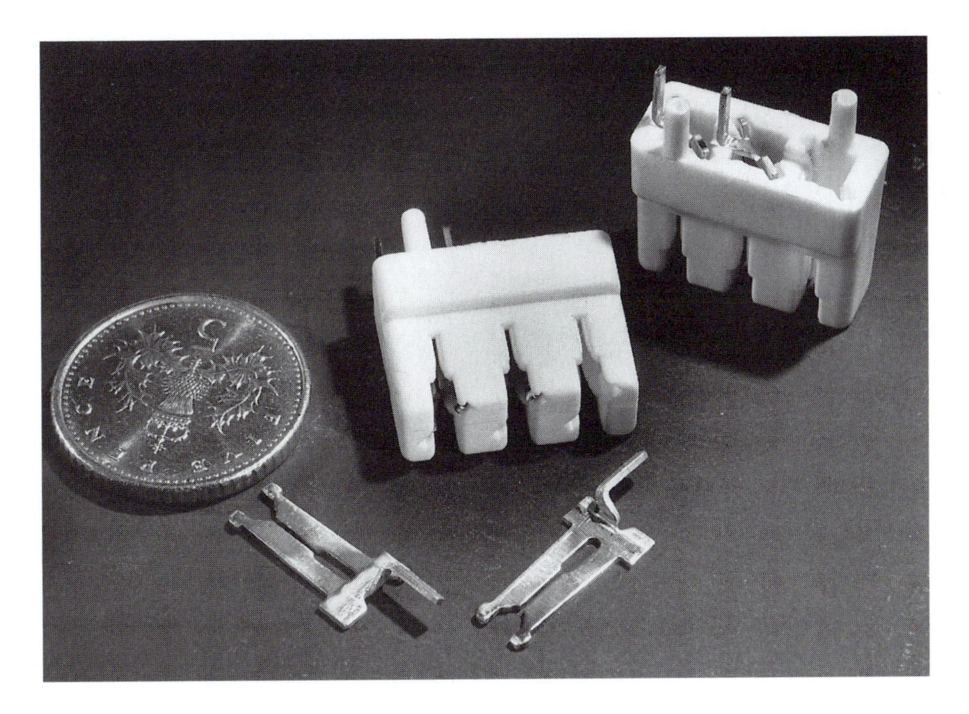

Fig. 6.4 Mechanical phone connectors. Two of the connectors have been removed from their plastic cases and are lying in the foreground.

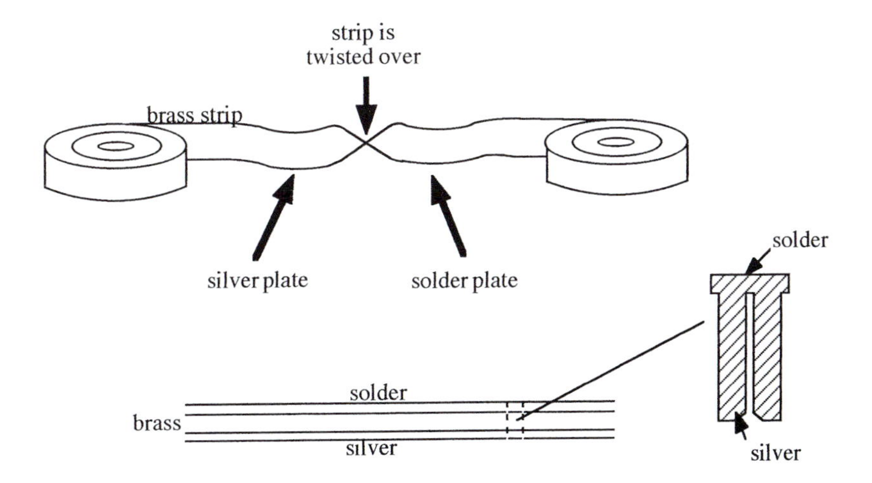

Fig. 6.5 Making a phone connector.

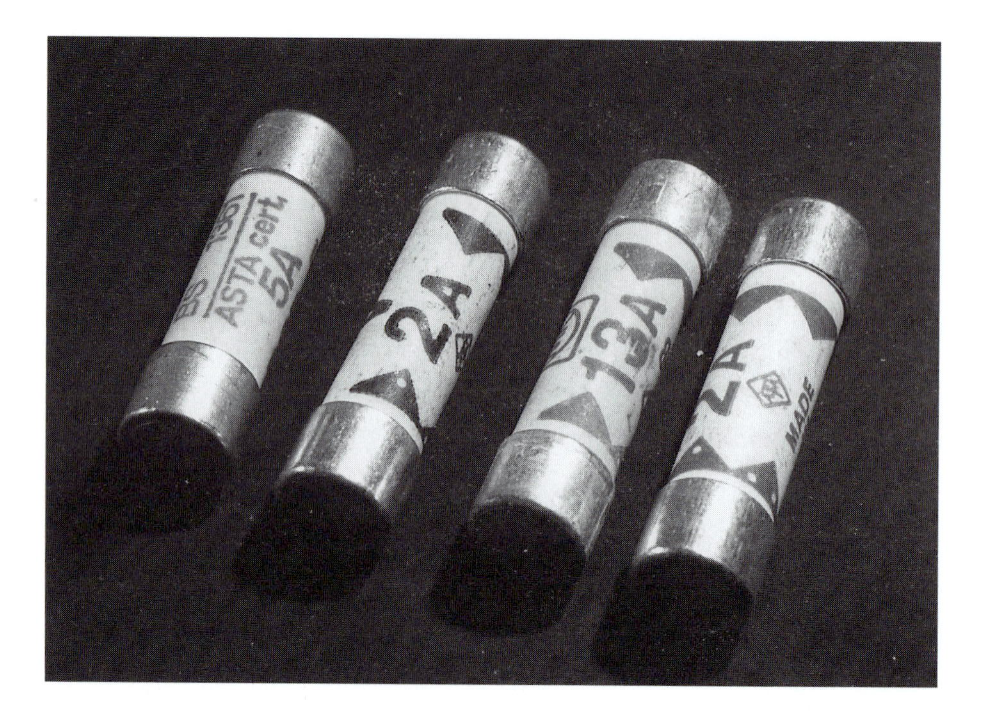

Fig. 6.6 Fuses from domestic plugs. The ends are silver plated.

Finally, the connector is stamped out, assembled with two others in a plastic case (see Chapter 8) and a wire is soldered permanently to the pre-soldered end. The thickness of the silver plate is only ~2 μm: this is why it is practicable to use a relatively expensive metal like silver for a cheap, mass-produced object.

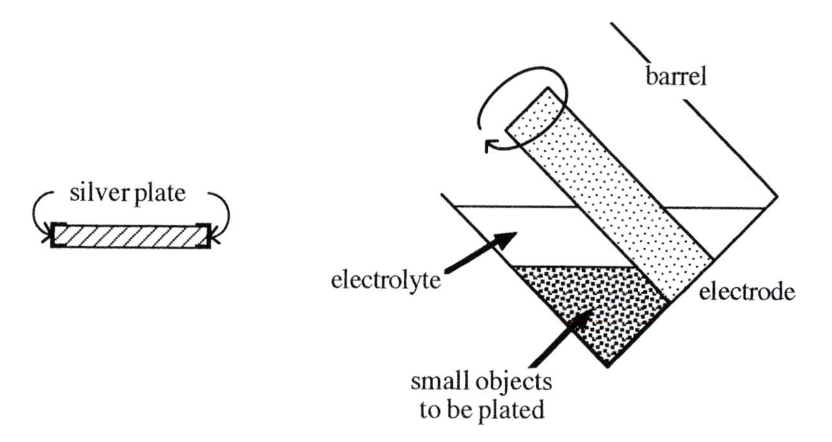

Fig. 6.7 Barrel plating.

For a connector which has to be used in a moist atmosphere, gold plating would be used. The gold is hardened with 2% Co to resist abrasion. A cyanide bath is again used and the thickness of the gold is $\sim 1\ \mu m$.

Example 2 A fuse from a domestic plug (see Fig. 6.6)
This is another cheap, mass-produced article which you might not expect to incorporate silver plate! In fact, all fuses are silver plated, to ensure a good electrical connection in a safety-critical application. The plating process is called *barrel plating* and is illustrated in Fig. 6.7. The electrolyte is cyanide and the electrical circuit is completed via random contacts between the fuse ends.

Example 3 Transistor packaging
Both the cap and the wires are steel. They are successively plated (see Fig. 6.8) with copper (a thin layer called a *copper flash*), nickel and then gold. The nickel provides most of the corrosion and abrasion resistance. The very thin layer of gold is present mainly to ensure reliable soldering. The copper is included because it will stick to steel, but nickel will not.

Example 4 A lead frame (see Fig. 6.9)
As an example of how the economic benefits of mass production control a production sequence, here is a typical route for producing a *lead frame* (frame for supporting the leads connected to a microchip). A cheap mass-produced article travels round the world from one specialist operator to another:

- start with Cu–3%Fe strip (the Fe confers extra stiffness on the copper)
- stamp out frames (e.g. in USA). Frames are still linked together, like a strip of postage stamps
- plate centre of lead frame with $2 \pm 0.1\ \mu m$ silver (e.g. UK, Netherlands). The electrolyte is sprayed at the frame as a thin jet. Several successive baths are used and the plate is built up gradually in a very controlled way. The very tight tolerance is necessary, otherwise wires won't bond thermomechanically to the silver. Any mistake and the financial responsibility will fall on the plater.
- mount semiconductor device (e.g. Korea)
- make the thermocompression bonds with gold wires from device to frame

Example 5 Electroforming copper for a Printed Circuit Board (PCB)
Copper is plated on to a rotating cylinder which is deliberately poisoned so that the copper plate does not stick properly (see Figs 6.10 (real) and 6.11 (schematic)). This makes it easy to skim off with a blade. Producing sheet of this thinness by rolling would be a very difficult and energy intensive operation.

The next (and final) two examples address the problem of plating on to plastic. Plastics, of course, are non-conducting (see Chapter 8) and so some device has to be found whereby the first conducting layer can be attached. Here are two solutions:

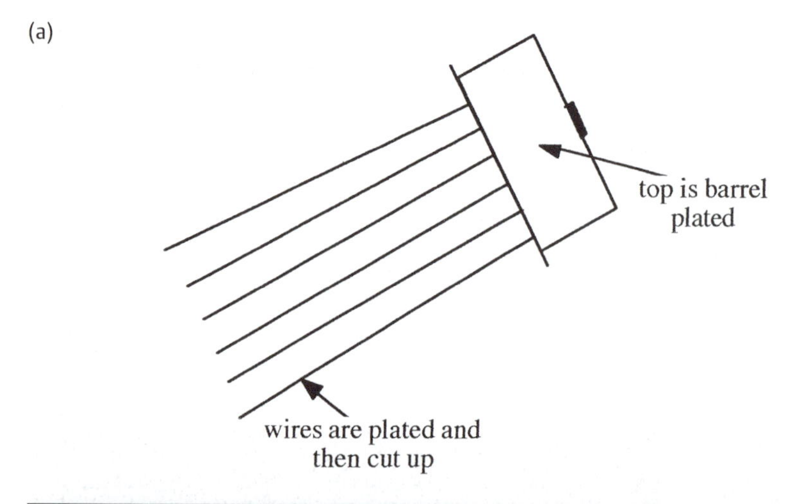

Fig. 6.8 A single packaged transistor: (a) schematic, (b) the real thing. In the example on the left the metal can has been removed, revealing the microchip.

Example 6 Typewriter 'golf-ball'

This is plastic, plated with nickel to provide abrasion resistance. To persuade the first conducting layer to stick, the surface of the plastic is etched in chromic acid. It is then dipped in $PdCl_2$ solution. This 'activates' the surface by adsorbing small clusters of palladium atoms onto it (Fig. 6.12). Nickel is then *electroless plated* onto the surface.

Fig. 6.9 A lead frame in the making.

Electroless, or autocatalytic, plating of nickel consists of depositing metallic nickel from, for example, a metastable solution of metal complex ions and a reducing agent. Electroless plating does not need an impressed potential, but does require a catalytic surface. The solution is stable until the temperature is raised and a catalytic surface (here the Pd clusters) is provided. Once the nickel deposits on the Pd it is sufficiently catalytic itself to maintain the reaction, which spreads across the surface. The plate is not pure nickel; it is Ni–15%P. It is so microcrystalline that it is almost amorphous. It is very hard, also. Electroless plating is expensive, so once full coverage of the plastic has been achieved, the nickel is thickened by conventional electroplating.

Copper also can be electroless plated. This is very good for PCBs, for example, because the plate penetrates into holes, still giving an even covering.

Example 7 Producing a compact disc

This time the plastic is made conducting by *shadowing* it with silver. This is evaporated onto the plastic in a vacuum. (Evaporation in a vacuum is much used in the semiconductor industry: see Chapter 9.) Nickel is electroplated onto the plastic which is then stripped off. The nickel forms the master disc. Because the dots and dashes which make up the contents of the CD are about half a micron wide, by up to 2 μm long and a $\frac{1}{10}$th of a micron deep, the nickel electroplating is done under clean room conditions. A nickel sulphamate bath is used, with small additions of nickel chloride

LIVERPOOL
JOHN MOORES UNIVERSITY
AVRIL ROBARTS LRC
TEL. 0151 231 4

Fig. 6.10 (a) Thin copper sheet for a printed circuit board. (b) Printed circuit board. The copper sheet is so thin that it is cheaper to produce it by electroforming (see Fig. 6.11) rather than by rolling.

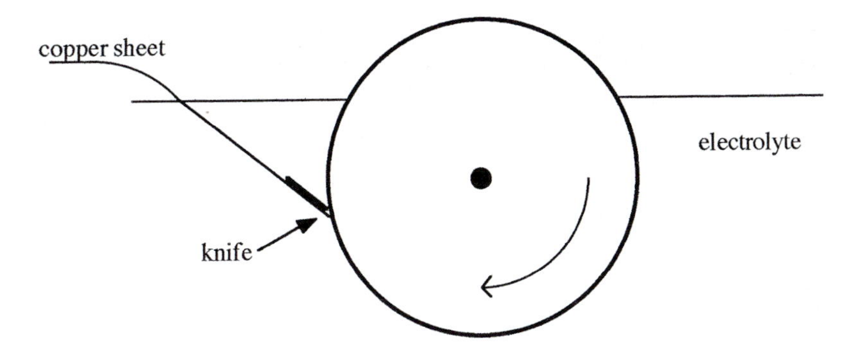

Fig. 6.11 Electroforming copper sheet.

(a)

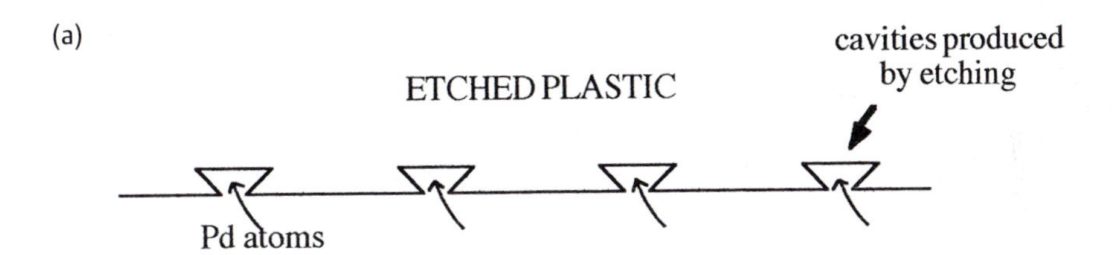

(b)

Fig. 6.12 (a) Activating a plastic for subsequent electroplating. (b) A typewriter 'golf-ball'. (c) Another example of plating onto plastic: an automobile light reflector.

LIVERPOOL JOHN MOORES UNIVERSITY
LEARNING SERVICES

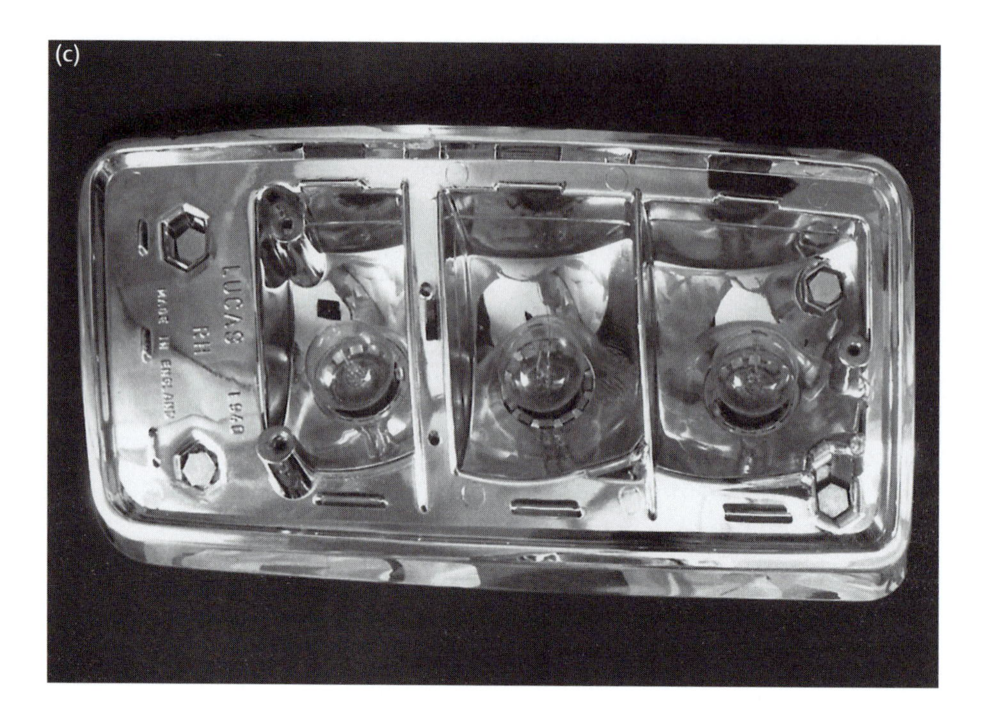

Fig. 6.12 (*continued*)

and boric acid. The 'master' is actually a negative impression. The master is used to produce 'positives', which are used in their turn to produce more 'negative' plates. These are used to produce the compact discs, for example by injection moulding polycarbonate (see Chapter 8).

6.4 Corrosion

In electroplating, an electrolytic cell is driven by an external potential. If you refer to Fig. 6.2 you will see that corrosion is an example of an electrolytic cell which proceeds in the forward direction. The resultant potential is not harnessed and so the energy is wasted. At the same time the anode dissolves and turns into an undesirable inorganic compound (e.g. rust).

Corrosion is enormously costly—typically 3% of Gross National Product. (I am talking here about aqueous corrosion: other forms do exist, but they are nothing like as important.)

Table 6.1 The electrochemical series at 25 °C

Electrode reaction	Standard electrode potential, E_H (V)
$Au^{3+} + 3e^- \rightleftharpoons Au$	+1.42
$Cr_2O_7^{2-} + 14H^+ + 6e^- \rightleftharpoons 2Cr^{3+} + 7H_2O$	+1.36
$Cl_2 + 2e^- \rightleftharpoons 2Cl^-$	+1.36
$O_2 + 4H^+ + 4e^- \rightleftharpoons 2H_2O$	+1.23
$Ag^+ + e^- \rightleftharpoons Ag$	+0.80
$Cu^+ + e^- \rightleftharpoons Cu$	+0.52
$Cu^{2+} + 2e^- \rightleftharpoons Cu$	+0.34
$H^+ + e^- \rightleftharpoons \frac{1}{2}H_2$	0
$Pb^{2+} + 2e^- \rightleftharpoons Pb$	−0.13
$Sn^{2+} + 2e^- \rightleftharpoons Sn$	−0.14
$Ni^{2+} + 2e^- \rightleftharpoons Ni$	−0.23
$Co^{2+} + 2e^- \rightleftharpoons Co$	−0.28
$Cd^{2+} + 2e^- \rightleftharpoons Cd$	−0.40
$Fe^{2+} + 2e^- \rightleftharpoons Fe$	−0.41
$Cr^{3+} + 3e^- \rightleftharpoons Cr$	−0.74
$Zn^{2+} + 2e^- \rightleftharpoons Zn$	−0.76
$Mn^{2+} + 2e^- \rightleftharpoons Mn$	−1.03
$Ti^{2+} + 2e^- \rightleftharpoons Ti$	−1.63
$Al^{3+} + 3e^- \rightleftharpoons Al$	−1.71
$Mg^{2+} + 2e^- \rightleftharpoons Mg$	−2.38
$Na^+ + e^- \rightleftharpoons Na$	−2.71
$Ca^{2+} + 2e^- \rightleftharpoons Ca$	−2.76
$K^+ + e^- \rightleftharpoons K$	−2.92
$Li^+ + e^- \rightleftharpoons Li$	−3.05

(From *Basic corrosion technology for scientists and engineers* (2nd edn.) by E. Mattsson, courtesy of the Institute of Materials.)

Where does the corrosion cell come from? There are two ways. One is when two dissimilar metals are in electrical contact and there is sufficient moisture present to complete the cell. The second way is when the electrolyte varies in concentration from place to place.

Dissimilar metals in contact always leads to corrosion of the baser metal by the more noble. The order of baseness and nobility for metals is given by the *electrochemical series* (Table 6.1). Unfortunately (for us and from this point of view) copper is a very noble metal and will corrode most others. Copper and steel should never be placed in contact with each other when there is likely to be any moisture present.

The other method of creating a corrosion cell is via a difference in the electrolyte. The classic example is crevice corrosion, where a difference in oxygen concentration in water causes, for example, steel to rust (see Fig. 6.13). The potential at one electrode is different from that at the other and as a result they do not cancel out and corrosion proceeds, provided there is an electrical connection between the anode and the cathode. Where the oxygen concentration is high, the cathodic reaction

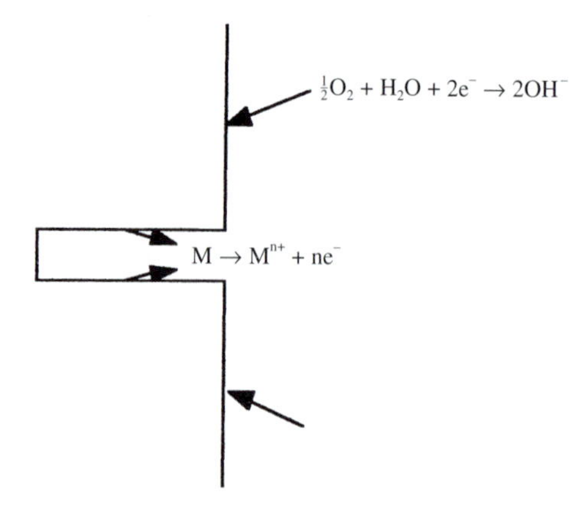

Fig. 6.13 Crevice corrosion. The corrosion cell owes its existence to the fact that the electrolyte concentration within the crevice is different from that without. Here the dissolved oxygen concentration inside the crevice will be different from that outside.

$$O_2 + 2H_2O + 4e^- \rightarrow 4OH^-$$

establishes itself. Where the oxygen is depleted, the anodic reaction is favoured:

$$Fe \rightarrow Fe^{2+} + 2e^-$$

so that the corrosion actually occurs in the crevice.

It is quite good fun looking for examples of corrosion and interpreting it in terms of the mechanisms above. A good place to start is a car or bicycle (not quite such good fun if it happens to be yours).

Prevention of corrosion

Corrosion can be prevented by using a little common sense in design (e.g. not using brass rivets in steel plate) and by the following general methods:

- Paint, etc.
- Inhibitors (e.g. car radiator)
- Metal coatings e.g. galvanising, tin plate
- Cathodic protection (see Fig. 6.14)
- Insulate metals from each other, preventing cell from being set up

6.5 Useful corrosion: batteries

In the examples of corrosion described in the previous section (section 6.4) the electric current generated by the corrosion cells was wasted. In this section I will describe how

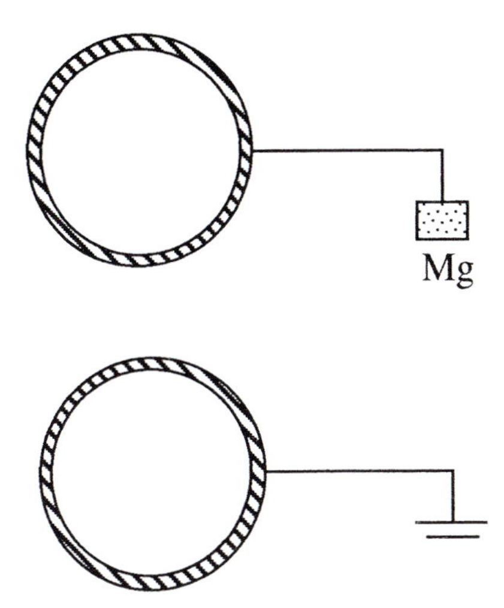

Fig. 6.14 The use of a sacrificial anode to protect a pipe ('cathodic' protection). The magnesium or battery provides an opposite potential and prevents the cell from operating and the pipe from rusting. The magnesium anode or a battery is buried in the soil and connected to the pipeline. Obviously, when an anode has been consumed or battery exhausted it must be replaced (typically years).

this current can be harnessed and put to use in a deliberately engineered corrosion cell. We call this a *battery*.

In principle, *any* corrosion which operates spontaneously can be made into a battery. In practice a host of technical and economic factors intervenes and dictates that out of the hundreds and thousands of possible cells, only a very few are actually made and sold as batteries. In fact, the dry Leclanché cell and its alkaline cousin between them account for a vast majority of the primary battery market. What is this word 'primary' which I have just slipped in? Some batteries are *reversible* and can be reinstated as new by supplying power from an independent source—e.g. the mains, or a generator (like the 'battery' in a car). This type of battery is called a *secondary cell* or an *accumulator* and I will describe it in a separate section, 6.7. Also, batteries where the power comes from chemicals which are piped in to the electrodes and where the electrodes do not take part directly in the reaction are called *fuel cells* and I shall describe those in the following section 6.6. In a *primary* battery, one or both of the electrodes is/are consumed and the reaction is not practicably reversible. It is worthwhile remembering that electricity was understood largely because of batteries, not the other way round.

Figure 6.15 shows schematically how a battery is constituted. At the anode (the negative terminal) electrons are produced and at the cathode (the positive terminal) electrons are absorbed, both processes contributing a share of the total voltage produced by the cell. In Alessandro Volta's (from whom 'Volt') first battery of all (1800) zinc was the anode and silver the cathode. Daniell (1836) incorporated a *depolarising*

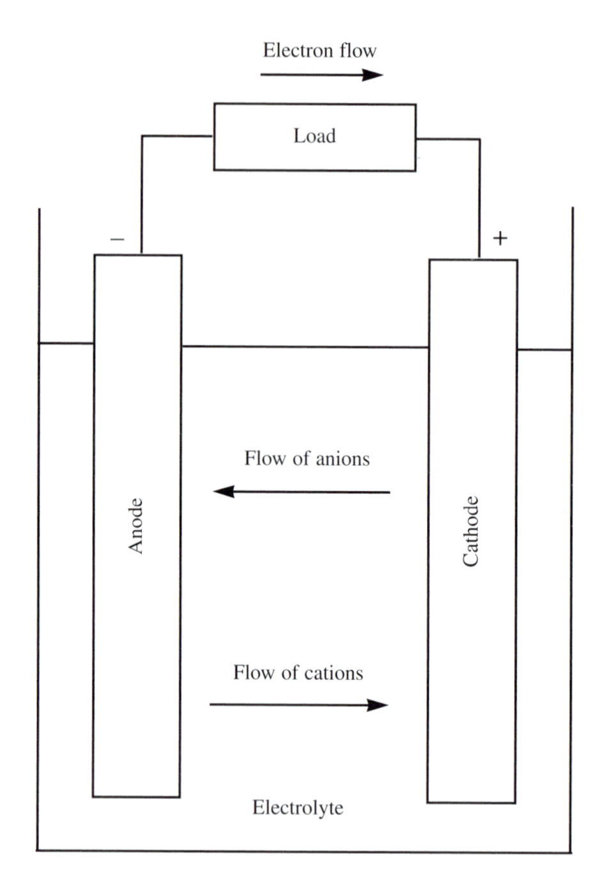

Fig. 6.15 A schematic diagram of a battery (from *Handbook of batteries* by D. Linden (courtesy of McGraw-Hill Inc.)).

agent which prevents *polarisation* of the cathode and much extends the life of the battery. In 1868, Georges Leclanché introduced a cell where zinc was the anode, MnO_2 the cathode and ammonium chloride solution the electrolyte.

The liquid electrolyte is generally replaced nowadays by a paste or a solid electrolyte for convenience, but descendants of Leclanché's cell still dominate the primary battery market, as I described above.

Table 6.2 lists the main battery types in common use today and Fig. 6.16 shows corresponding schematic diagrams. The next time you visit the supermarket try identifying the various types. Which battery is used where depends on a large number of factors—I have tried to include some of the more important. Figure 6.17 illustrates how battery technology has developed over the last 50 years. Finally, notice that battery electricity is much more expensive (thousands of times) than mains electricity. Batteries are only used when mains electricity is impracticable. (See Q6.5.)

Table 6.2 Batteries in common use (adapted from Tables 1–2, 6–2 and others of *Handbook of batteries* by D. Linden (by kind permission of McGraw-Hill)).

Battery	Anode	Cathode	Electrolyte	Reaction	Voltage (volts)	Energy density (Wh kg^{-1})	Typical applications
Primary							
Dry Leclanché	Zn	MnO_2	NH_4Cl + $ZnCl_2$	$Zn + 2MnO_2 \rightarrow ZnO + Mn_2O_3$	1.5	85	Torch; radio; tape recorder; CD player; calculator
Alkaline	Zn	MnO_2	KOH	$Zn + 2MnO_2 \rightarrow ZnO + Mn_2O_3$	1.5	125	Ditto (better, more expensive ...)
Mercury	Zn	HgO	KOH	$Zn + HgO \rightarrow ZnO + Hg$	1.3	470	Hearing aid; camera
Silver	Zn	Ag_2O	KOH	$Zn + Ag_2O + H_2O \rightarrow Zn(OH)_2 + 2Ag$	1.6	500	Watch; camera
Lithium	Li	several e.g. I_2	e.g. solid LiI	e.g. $2Li + I_2 \rightarrow 2LiI$	2.8	800	Heart pacemaker
Fuel cell							
H_2/O_2	H_2	O_2 (or air)	e.g. KOH (low temp. version)	$2H_2 + O_2 \rightarrow 2H_2O$	1.2	3570	Spaceship
Secondary							
Lead–acid	Pb	PbO_2	H_2SO_4	$Pb + PbO_2 + 2H_2SO_4 \rightarrow 2PbSO_4 + 2H_2O$	2.0	70	Automobile
Nickel–cadmium	Cd	NiOOH	KOH	$Cd + 2NiOOH + 2H_2O \rightarrow 2Ni(OH)_2 + 2H_2O + Cd(OH)_2$	1.2	80	Lap-top computer; electric shaver

(a) Dry Leclanché (cylindrical)

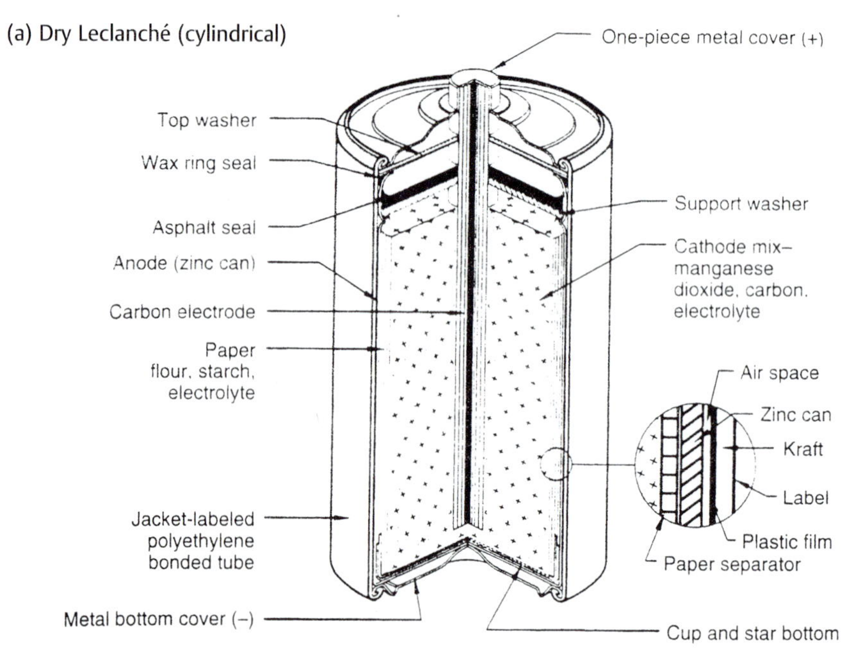

(b) Alkaline (cylindrical)

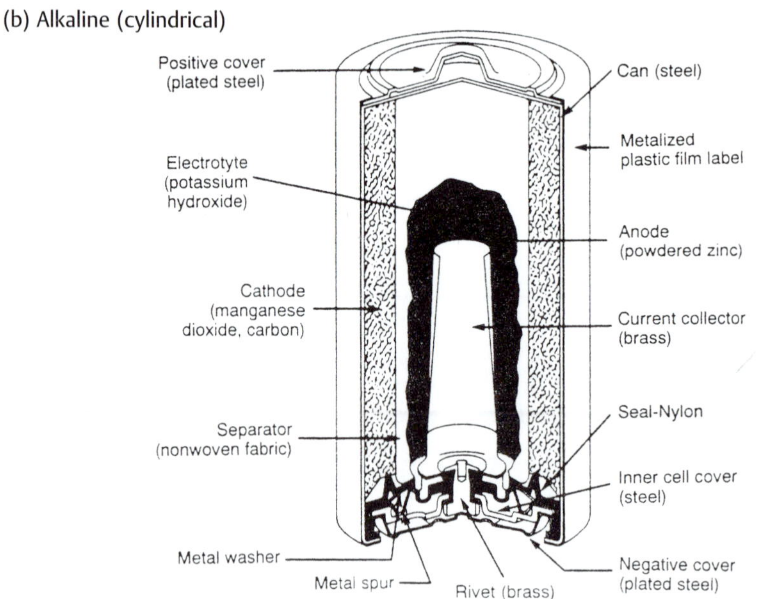

Fig. 6.16 Common types of battery from Table 6.1. (a) Dry Leclanché (cylindrical), (b) alkaline (cylindrical), (c) alkaline (button), (d) mercury (button), (e) silver (button), (f) lithium, (g) fuel cell, (h) lead-acid*, (i) nickel-cadmium (cylindrical)†. *The lead electrodes are actually Pb–0.1%Ca (traditionally Sb). †Nickel-cadmium sealed secondary batteries are facing stiff competition from nickel-hydrogen units.
(a)–(f), (h) and (i) from *Handbook of batteries* by D. Linden,
(g) from *Batteries and energy systems* by C.L. Mantell
(courtesy of McGraw-Hill Inc.).

(c) Alkaline (button)

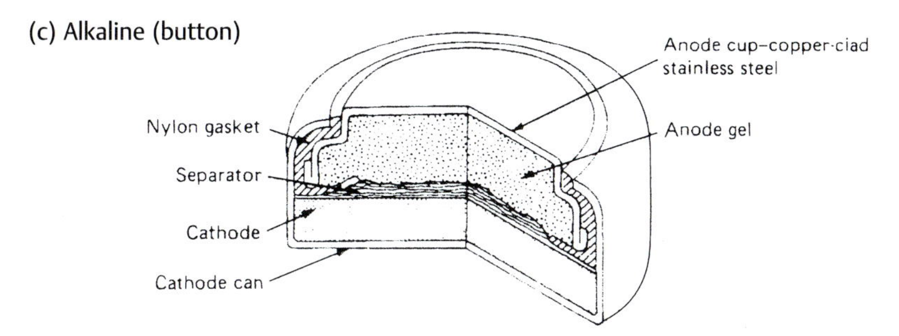

(d) Mercury (button)

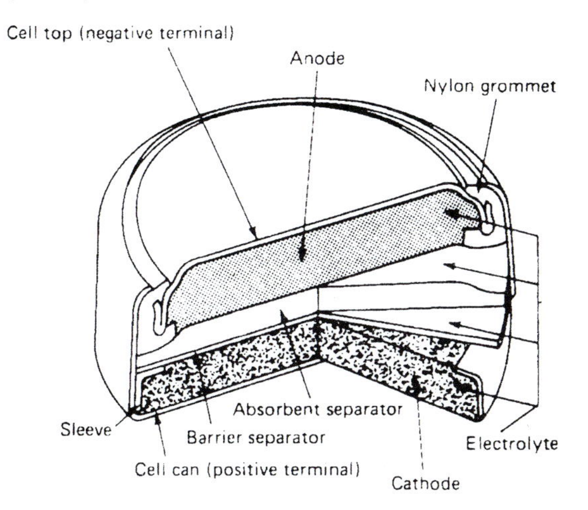

(e) Silver (button)

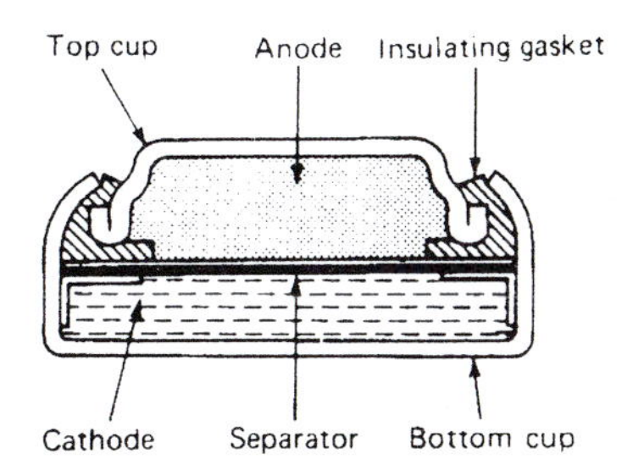

Fig. 6.16 (*continued*)

(f) Lithium

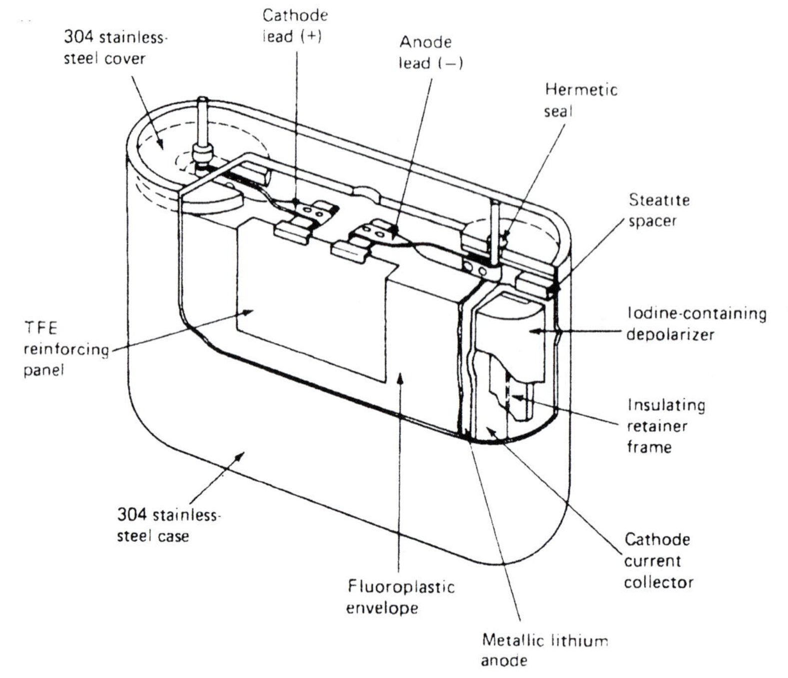

(g) Fuel cell

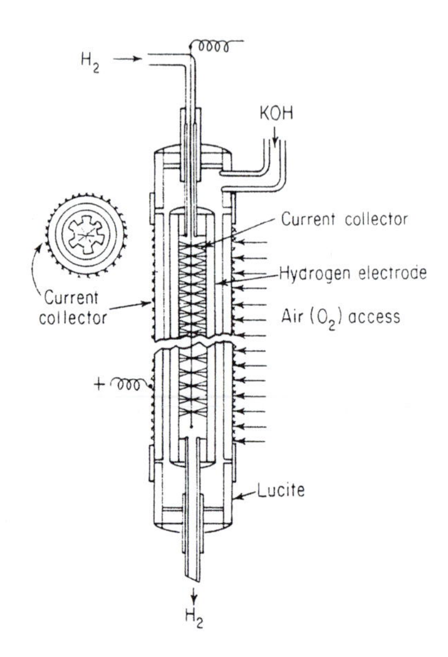

Fig. 6.16 (*continued*)

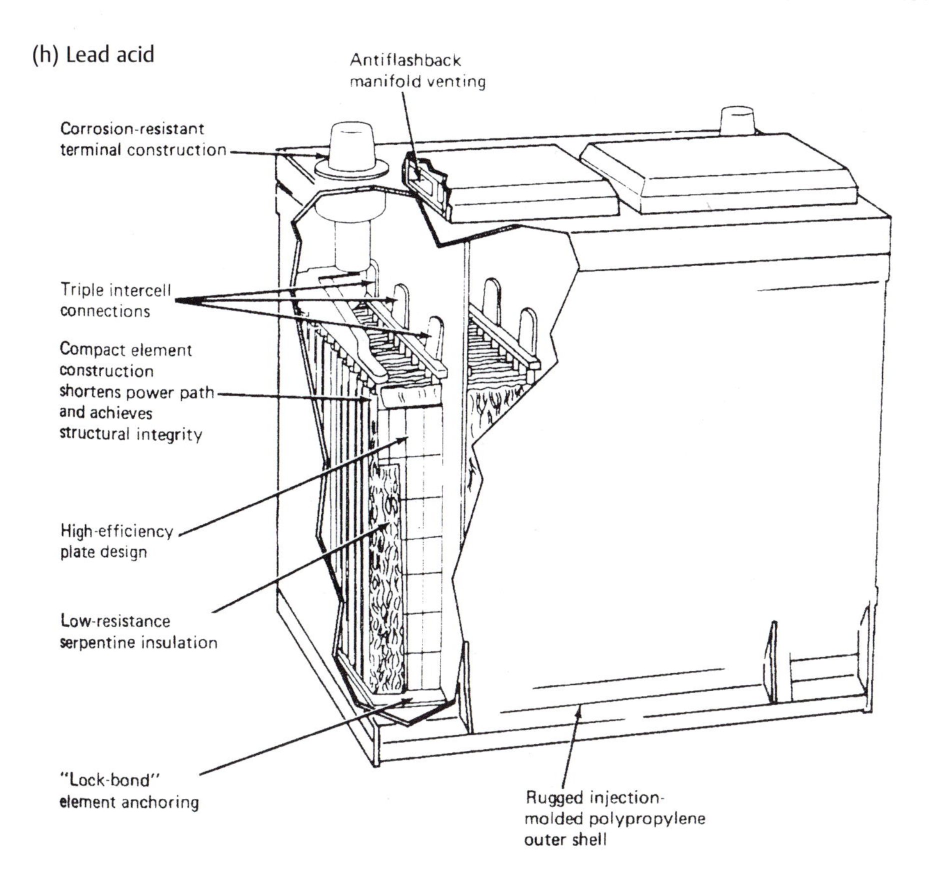

Fig. 6.16 (*continued*)

6.6 Fuel cells

In a fuel cell the chemical reaction providing the power does not involve either of the electrodes. Fuel cells have been developed continuously since they were first conceived of by Sir William Grove in 1839. They remain very expensive and still are used only when longlasting reliable operation is crucial—for example the USA Apollo space missions.

Two common fuel cell reactions are

$$2H_2 + O_2 \rightarrow 2H_2O$$

and

$$C + O_2 \rightarrow CO_2$$

(i) Nickel-cadmium

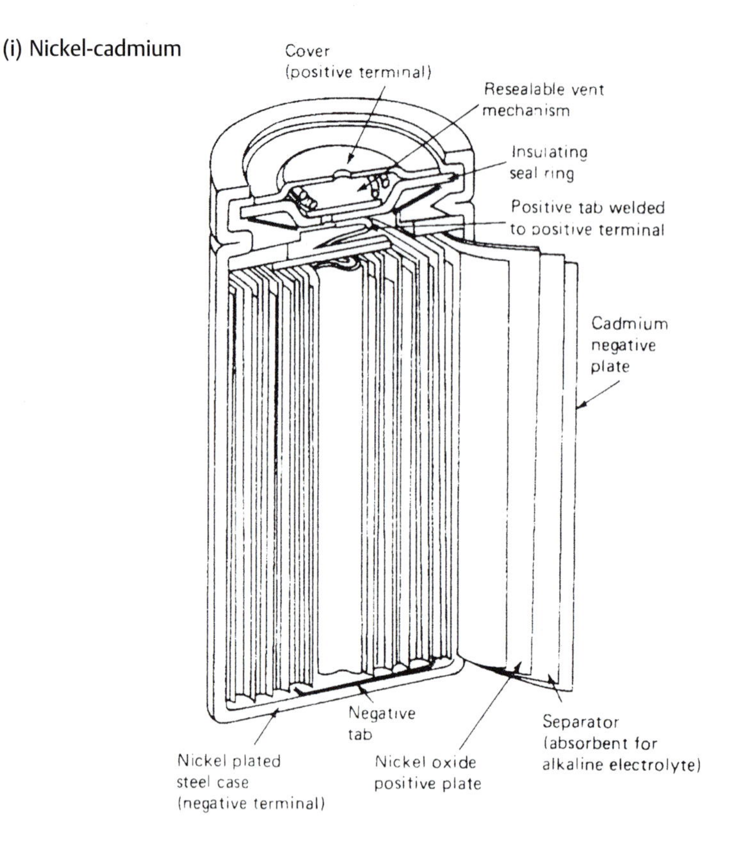

Fig. 6.16 *(continued)*

Figure. 6.16(g) shows one possible way of achieving the first reaction. Notice that the second one is just like a coal or oil-fired power station producing heat which drives a turbine which produces electricity, but the fuel cell is rather more efficient (and convenient)! Unfortunately, the cost of the resulting electricity is still very high. If fuel cells could be made to be cheaper, they would be ideal for an environment-friendly electric car.

6.7 Accumulators (secondary batteries)

In principle, all primary batteries can be *recharged*—that is, a higher reverse voltage can be applied and the cell driven backwards until in its original state. Unfortunately, for various detailed reasons, this does not usually work. Two common systems where the battery *is* reversible are lead/sulphuric acid and nickel/cadmium. The lead-acid accumulator is the battery in a car. This is an enormous market. Figure. 6.16(h) shows how a lead-acid battery works. In the charging cycle, the equation (see Table 6.2) is

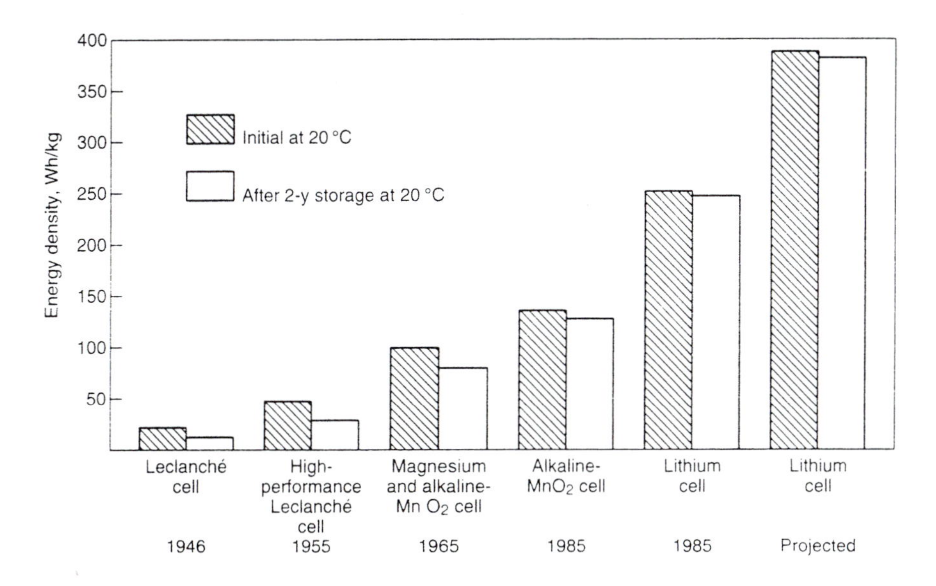

Fig. 6.17 Advances in primary battery performance over the last 50 years (from *Handbook of batteries* by D. Linden (courtesy of McGraw-Hill Inc.)).

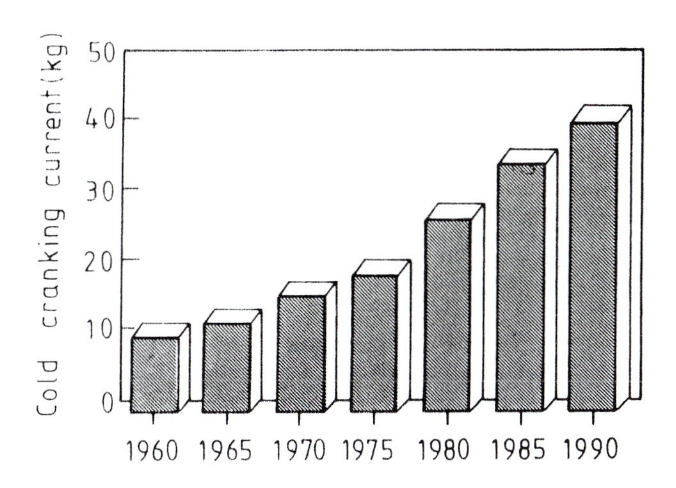

Fig. 6.18 Improvements in automobile, or SLI (Starting, Lighting, Ignition)—i.e. lead-acid automobile batteries—since 1960, measured via the cold cranking, or starting, performance (from *Handbook of batteries* by D. Linden (courtesy of McGraw-Hill Inc.)).

driven to the left and the electrolyte changes from water to sulphuric acid, as revealed by an increase in specific gravity (density). During use the electrolyte changes back to water. Although lead-acid batteries have been around for a long while (1859: Planté) recent improvements to the materials within them (for example the lead electrodes) have meant that car batteries no longer need topping up with water and also last much longer—in the 30 years I have been a car owner the average life of a battery has risen from ~3 years to ~8 years (see also Fig. 6.18).

The other common type of accumulator, which you will find in your lap-top computer or electric razor, is the nickel-cadmium cell. Figure. 6.16(i) shows a schematic of its operation.

6.8 Summary

1. The concept of a cell is basic to electrochemistry.
2. A cell consists of two half-reactions separated in space, one producing electrons and one using them.
3. The voltage produced by the cell is the free energy drop per electron transferred, expressed in eV.
4. In electroplating the cell is driven by an externally applied voltage. The workpiece constitutes the cathode. The surface of the workpiece is modified for decorative or engineering reasons.
5. Silver, gold (both for electrical connections) and nickel (wear) are common electroplates, even for mass produced articles.
6. Plastics may be electroplated after suitable preparation.
7. In corrosion, a cell operates spontaneously, returning a metal to one of its compounds (usually an oxide).
8. Corrosion is driven either by a difference in electrodes (two dissimilar metals in electrical contact) or by a difference in electrolyte (crevice corrosion).
9. Corrosion can be prevented by modifying the surface, modifying the electrolyte, or by applying a reverse potential.
10. A battery consists of a simple cell with a lot of engineering and technology to make it practicable (e.g. to prevent polarisation).
11. Leclanché, silver, mercury and lithium batteries account for most of the market.
12. Fuel cells, where gases combine but the electrodes are not consumed, are very efficient, but very expensive to run. The common reactions are of hydrogen or carbon with oxygen.
13. In an accumulator, the battery can be driven backwards and recharged. The common types are lead-acid and nickel-cadmium.

Recommended reading

Modern electroplating edited by F.A. Lowenheim (published by John Wiley)

Basic corrosion technology for scientists and engineers by E. Mattsson (published by The Institute of Materials (London, UK))

Handbook of batteries (2nd edn.) by D. Linden (published by McGraw-Hill Inc.)

Questions (Answers on pp. 321–322)

6.1 The equation $\Delta G = nFE$ relates the Free Energy change ΔG associated with a reaction to the electrical potential E of a cell based on the reaction. n is the number of electrons transferred and F is called Faraday's constant. If ΔG is expressed in kJ mole^{-1} and E is in Volts, work out F.
For the reaction (see eqn (6.2))

$$\mathrm{Fe} + 2\mathrm{H}^+ \rightarrow \mathrm{Fe}^{++} + \mathrm{H}_2 \uparrow$$

$E = 0.44$ V (see Table 6.1). What is ΔG?
(Avogadro's Number $N = 6.023 \times 10^{23}$ and e, the charge on the electron, $= 1.602 \times 10^{-19}$ C.)

6.2 In a copper electropurification cell, 1 Coulomb of electricity is passed. If the copper cathode has an area of 10 mm^2, what thickness of copper is deposited? [The electronic charge $e = 1.602 \times 10^{-19}$C; the lattice parameter of copper $= 0.3616$ nm]

6.3 Out of the following pairs of metals, which one would you expect to corrode the other?
(a) Copper and iron (b) Iron and zinc (c) Silver and nickel.
(Use the electrochemical series (Table 6.1).)
According to the electrochemical series, iron should corrode aluminium, rather than the other way round, as actually happens. Why does the iron corrode? What solutions to the problem can you think of?

6.4 What type of battery (non-rechargeable) would you expect to be used in (a) a calculator (b) a camera (c) a cardiac pacemaker?

6.5 I recently bought 4 AA/LR6/MN1500 batteries for the equivalent of \$5. In Birmingham I pay about 10¢ per kWh for mains electricity. Using Table 6.2 as a guide, roughly how much more expensive was my battery power than the equivalent supplied through the electricity mains?
(Take the capacity of an AA alkaline battery as 2.5 Ah. (This varies considerably, depending on the mode of use.))

Part III **Electrical insulators**

In Part II, on conductors, I was concerned with making current flow, generally as easily as possible. Now we are moving to a totally different world—that of *insulators*—where the whole point is to *stop* current from flowing. Thus we are moving from the world of the metallic bond to that of the ionic and covalent bond and that of the weak van der Waals bond. With these bonds the electrons are not free to move through large distances: they have to stay roughly where they are.

There are two types of insulator: *ceramic* and *plastic*. Ceramics depend largely on a mixture of ionic and covalent bonds (see Chapter 1) and plastics on the covalent and van der Waals bonds. They are very different types of solid and so I shall describe them in two different chapters, 7 and 8. Some of the basic electrical principles are common to both and I shall try to make it clear where this is true.

First, then...

7

Ceramics

Chapter objectives

- Insulation and capacitance
- What is a ceramic?
- Chemical bonding in ceramics
- Microstructure: crystals and glasses
- How electricity is conducted through ceramics (DC properties)
- Dielectric breakdown
- Mechanisms for capacitance and power loss (AC properties)
- Mechanical fracture and the Weibull distribution
- How insulators and capacitors are made: general methods and some specific examples

7.1 What is a ceramic?

To an engineer, ceramics are NOT cups and saucers. At least, not *just* cups and saucers. Ceramics, in their modern engineering sense, cover most of what, when I was at school, used to be called 'inorganic chemistry'. A pure ceramic phase is a compound of a metal and a non-metal. Ceramics as used in the practical world often consist of a mixture of a number of ceramic phases, some of which may be crystalline and some *amorphous*. (This is our first experience of amorphous phases: metals are nearly always crystalline.) So a bone china cup (to return to the example I was so sniffy about a few sentences ago) consists of quartz crystals bound together by an amorphous feldspar *glass* in which are further fine crystals of mullite. On the other hand, an alumina substrate for a microelectronic circuit consists of one pure crystalline ceramic phase 'alumina': aluminium oxide Al_2O_3. Equally, some ceramics are totally amorphous, or glassy: window glass or a glass bottle are good examples. Figure 7.1 illustrates a selection of ceramic artefacts with electronic or electrical applications.

From this you will deduce that insulators fulfil two main rôles:

1. They prevent the flow of current
2. They can *store* electrostatic charge dipoles (i.e. they can act as capacitors)

Figure 7.1 also shows some more sophisticated applications—for example piezo-electricity, where an applied electrical potential causes a body to change shape. Because these materials do more than merely support a load (*structural* materials) they are called *functional* materials.

It is most convenient to tackle the first rôle, insulation, in the context of direct current (DC) and the second, capacitance, in the context of alternating current (AC), although you shouldn't take these divisions too seriously. Before I talk at all about electrical properties I would like to describe the atomic structure and bonding of ceramics in a little more detail, and also the microstructure, which you may recall from Chapter 2 is the structure of a solid from the atomic scale up to what we can see with the naked eye ($\sim 10^{-10}$ - 10^{-4} m).

(a)

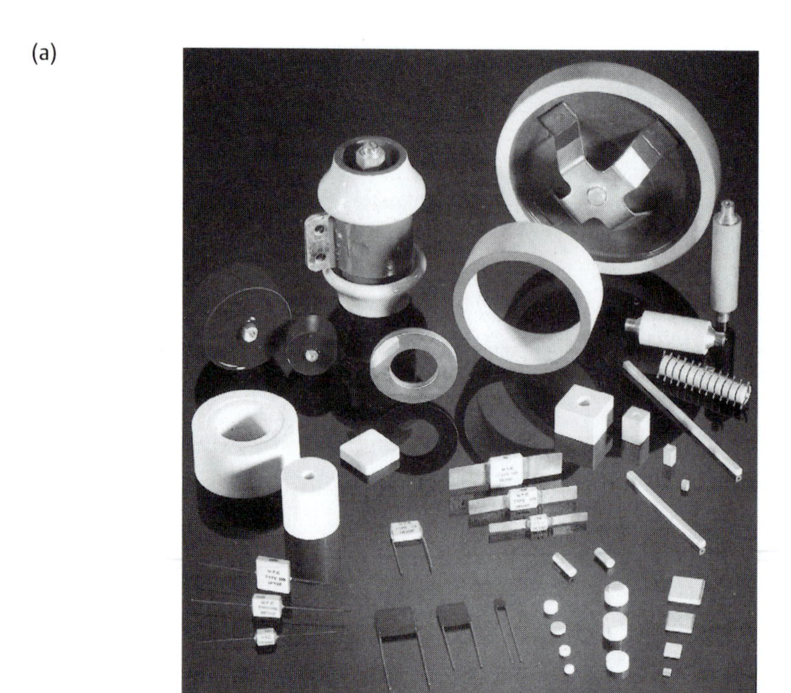

Fig. 7.1 A selection of electrical and electronic ceramics. (a) Starting at the top and working anticlockwise we have RF high power capacitors, high voltage capacitors, microwave ceramics and multilayer capacitors. In the middle are two examples of piezoelectric ceramics.
(b) Common insulators used in the lighting industry (mainly steatite—see Table 7.1). (c) High temperature insulators (MgO based). (Courtesy Morgan Advanced Ceramics Ltd. and Morgan Electroceramics Ltd.)

(b)

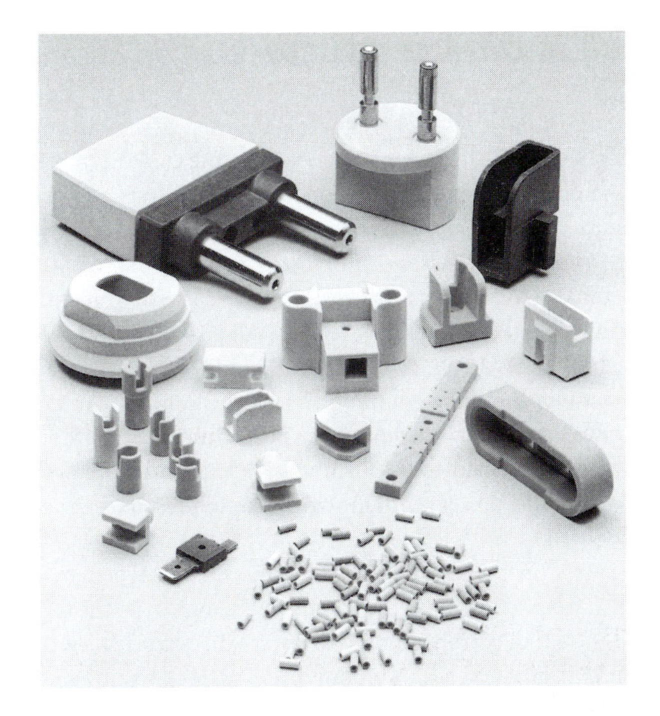

(c)

Fig. 7.1 (*cont.*)

7.2 Bonding and microstructure

The vast majority of ceramics used in engineering consist of crystals mixed up with a glassy phase. The glassy phase helps in the *processing* of the ceramic (as we shall see) and also helps hold the crystals together. The bonding in the crystals is ionic or covalent and that in the glass predominantly covalent.

Most ceramics are oxides, i.e. the non-metal part of the metal + non-metal compound is oxygen. Primarily this is because oxygen is the most plentiful element in the earth's crust and so oxides are cheap. The 'metal' part of the ceramic is, however, quite various, and includes Al, Mg, Fe, Na, K, Ca, Ba and Si.

Table 7.1 lists some of the more common electrical ceramics. The major difference from the metals of Chapter 2 is that some of the phases involved are amorphous rather than crystalline. Some of the names and formulae are long and difficult-looking: try not to let this worry you. All I am concerned with is basic principles: anyone can buy a cookery book.

The *crystalline phases* (in the ceramic product) are of two types: mainly ionic and mainly covalent (Chapter 1, Sections 1.5 and 1.6). In the ionic type the oxygen atoms are charged anions O^{2-}. Anions are generally larger than cations (regardless of atomic number). For example, alumina is aluminium oxide, Al_2O_3. In alumina the aluminium atoms lose 3 electrons each and the oxygen atoms gain 2 electrons each, so alumina is $(Al^{3+})_2(O^{2-})_3$. The negatively charged oxygen anions form hexagonal layers, which are stacked together just like the close packed metals, except that here the oxygen anions do not touch (see Fig. 7.2). The aluminium cations, Al^{3+}, occupy some of the interstices and they are large enough to hold the O^{2-} anions apart. This is quite important for an ionic solid where we want to maximise the electrostatic attraction between the + and - ions, but minimise the repulsion between like + or - ions. Recall that in the close packed metal structures all of B or C were filled. For alumina there are 3 O^{2-} ions for every 2 Al^{3+} ions and so only 2/3 of either B or C are filled. In fact the O^{2-} ions form a hexagonal array ABAB ... etc. and the Al^{3+} ions distort this slightly. Alumina is also known as *corundum*. *Sapphire* is a doped form, which has the same crystal structure. Alumina has very low electrical conductivity, but reasonable thermal conductivity. It is therefore a common electronic substrate material. It is also an important part of alumina porcelain (e.g. spark plugs) (see Section 7.7). A second example of this type of ceramic crystal, barium titanate ($BaTiO_3$), is the basis of microcapacitors, ferroelectrics and piezoelectric materials and the same structure (the *perovskite* structure) is found in the high temperature (HT_c) superconductors (see Chapter 11). The structure (Fig. 7.3) is a mixture of the f.c.c. and b.c.c. structures we encountered in metals. The Ba^{2+} anions form a simple cubic array, the O^{2-} cations occupy the face centres and the very small Ti^{4+} cations the body centre positions. Below 120 °C the barium titanate structure undergoes various slight distortions which render it very useful indeed. I shall come back to barium titanate later on.

Table 7.1 Some common electrical ceramics.

Ceramic		Crystalline phases	Glassy phase = SiO_2+	Made from
Porcelain	Triaxial	Quartz (SiO_2) Mullite $(Al_2O_3)_2(SiO_2)_3$	Al, K	45% clay e.g. kaolin (China clay) $Al_2(Si_2O_5)(OH)_4$ 35% flux e.g. feldspar $KAlSi_3O_8$ 20% filler e.g. quartz: flint or sand SiO_2
	Aluminous	Al_2O_3	Al, Mg	95% alumina Al_2O_3 5% talc (steatite) $Mg_3(Si_2O_5)_2(OH)_2$ (very similar to clay) or as for triaxial porcelain with flint replaced by alumina
Steatite		Enstatite $(MgO)(SiO_2)$ (mainly)	Mg	15% clay 83% talc 2% chalk ($CaCO_3$)
Cordierite		Cordierite $(MgO)_2(Al_2O_3)_2(SiO_2)_5$	Mg	80% clay 20% talc
Alumina		Alumina (Al_2O_3)	—	Bauxite $(Al_2O_3)(H_2O)_2$
Barium titanate		$BaTiO_3$ (= BaO-TiO_2)	—	Barium carbonate ($BaCO_3$) and rutile (TiO_2)

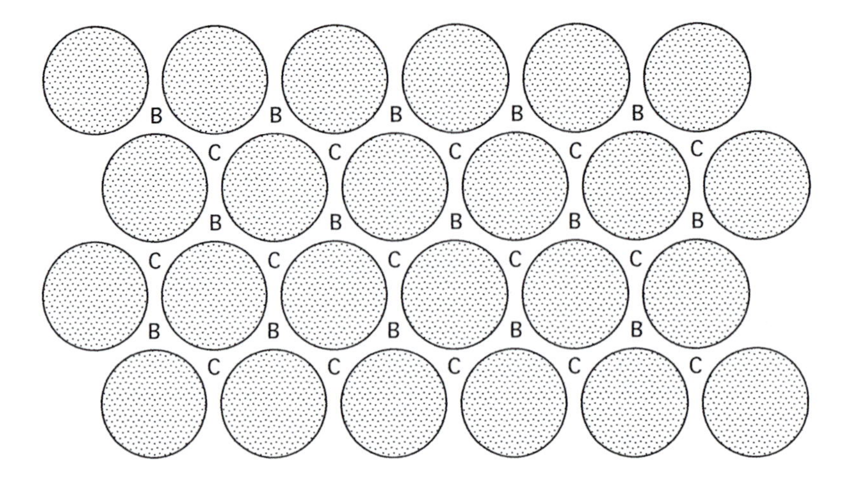

Fig. 7.2 The oxygen anions in alumina form a hexagonal array, but without touching. The aluminium cations fill some of the B and C interstices. (Compare with Fig. 2.6.)

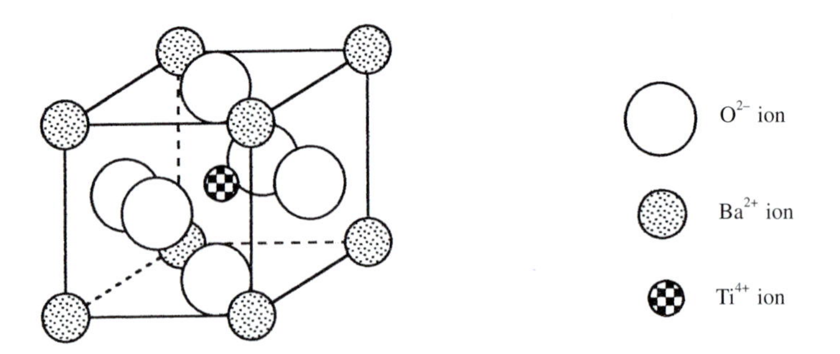

Fig. 7.3 The unit cell of barium titanate, $BaTiO_3$.

The other common type of ceramic crystalline structure is based on linked SiO_4 tetrahedra (Fig. 7.4) where the bonding is mainly covalent. This is true of silica itself (in fact silica exists in many different crystalline forms), mullite, where the SiO_4 tetrahedra are linked by AlO_6 octahedra, and enstatite, where the SiO_4 tetrahedra form chains (see Table 7.1). Cordierite is a *ring* silicate with hexagons in one plane and chains of four membered rings in the perpendicular direction. Generally we will not need to worry about the crystal structures of these ceramics, except perhaps to note that in the clay based minerals, the SiO_4 tetrahedra form sheets which are interleaved with $Al–O–(OH)_2$ sheets. Neighbouring sheets are loosely bound together by weak van der Waals bonds. They can therefore slip over one another very easily. This is extremely important in the manufacture of clay-based ceramics (e.g. some of the porcelains in Table 7.1).

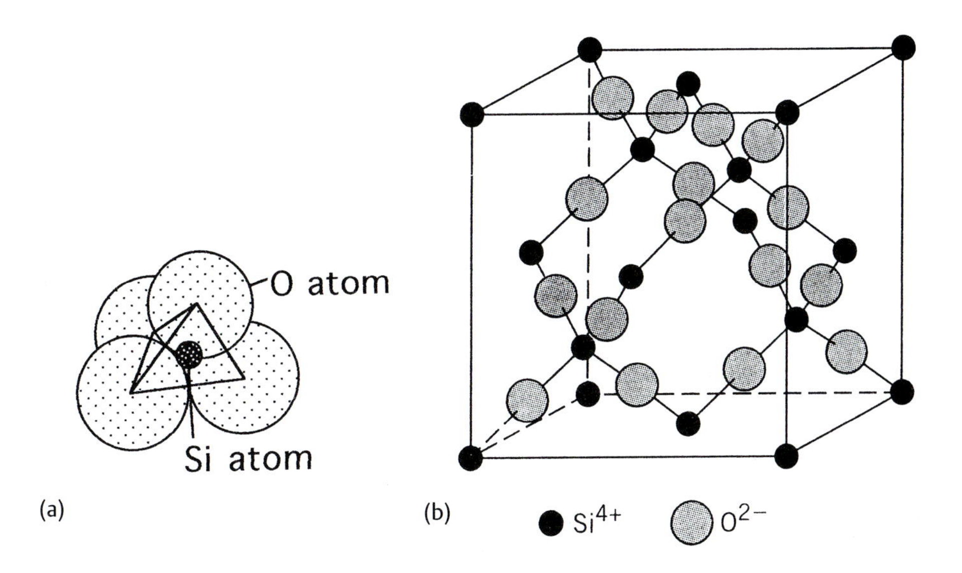

Fig. 7.4 (a) SiO_4 tetrahedra are the basic building blocks of many ceramic crystals and glasses. (b) An example of one of the structures adopted by silica: high cristobalite—you can see the SiO_4 tetrahedra. (From *Materials science and engineering: an introduction* by W.D. Callister, courtesy of John Wiley & Sons, Inc.)

So much for the crystalline part of Table 7.1. Now what about the *amorphous*, or *glassy* part? This is the first time we have come across this type of solid. All the metals I described in Part II were crystalline. It is very difficult to make a metal anything other than crystalline. This is not true, however, of materials where the chemical bonding is ionic or, especially, covalent. When these materials are cooled quickly† from the liquid state of matter they do not form crystals, they form a *glass*. This is a type of solid‡ which is more ordered than a liquid, but less ordered than a crystal. The different types of bonds favour crystals in the order metallic > ionic > covalent. This is simply a question of how demanding they are in terms of spatial arrangement. The requirements of the metallic bond are really very simple: just a given number of atoms in a given volume. At the other end of the scale, covalent bonds are quite picky in that they require neighbouring atoms to be at the right distance and at the correct angle vis-à-vis other neighbours. This is also why the glassy phase in a ceramic tends to be more covalent than the crystalline phase. When a liquid solidifies to a crystal the corresponding temperature is called the melting point T_m. When a liquid solidifies to a glass, the temperature is called T_g (T_{glass}). This transition is more subtle and blurred out than when a liquid forms a crystal. Figure 7.5 shows how the volume of a liquid changes as it is cooled.

† Actually, the same is true of a metal (see Chapter 10), but the necessary rates of cooling are enormous.

‡ I have classified glasses as solids because of their very high viscosity. One could make an argument, however, in favour of their being classified as liquids, on the basis of their atomic arrangement.

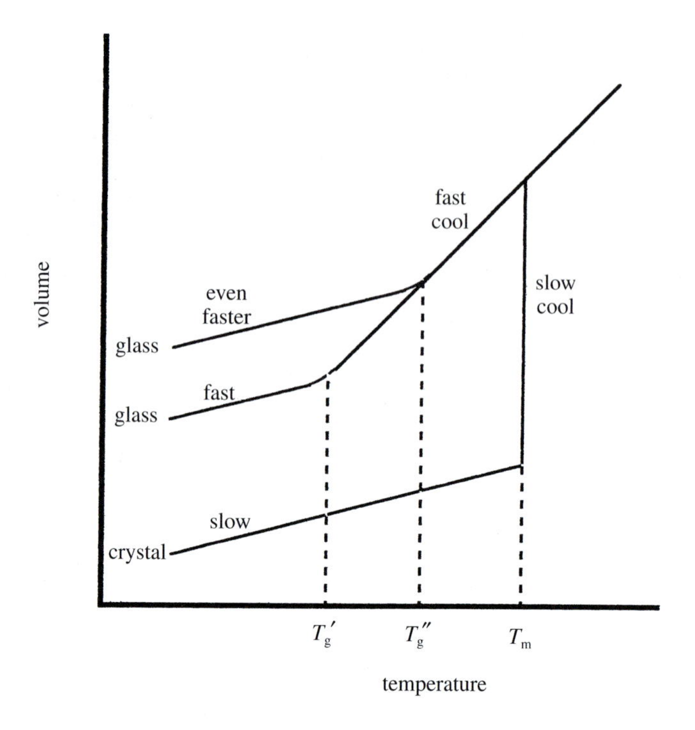

Fig. 7.5 How the volume of a material changes on cooling and how it depends on the cooling rate. Slow cooling encourages crystallisation; higher cooling rates may lead to a glass.

You can see that

1. For the transition to a crystal the volume undergoes a discontinuity, whereas for the transition to a glass it is the *gradient* of the curve (i.e. the *coefficient of thermal expansion*) which undergoes a discontinuity.
2. The transition to a crystal is abrupt, that to a glass more gentle.
3. T_g varies with the rate of cooling.

The most important glasses for us are those based on SiO_2. SiO_4 tetrahedra, which I used above to describe some crystals, come in useful again. Figure 7.6 compares crystalline SiO_2 with glassy SiO_2: you can see that the glass is *quite* ordered—certainly not random—but not as ordered as a crystal. Cations dissolve into the glass in two different ways. If they form glasses easily themselves they substitute for Si^{4+} and are called *network formers*. If they don't, they tend to break up the network and are called *network modifiers* (Fig. 7.7). Some intermediate elements do a little bit of each:

Network formers	Si, B, Ge, As, P, V, Sb
Intermediate	Zn, Pb, Al, Zr, Cd
Network modifiers	Na, K, Ca, Mg

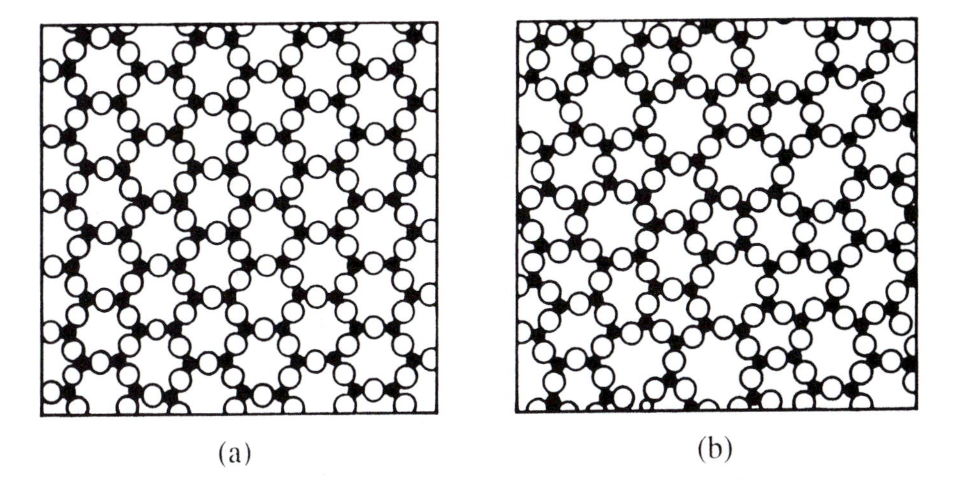

Fig. 7.6 SiO_2 (silica) in (a) crystalline and (b) glassy forms. (From *Introduction to materials science for engineers* by J.F. Shackelford, courtesy of Prentice Hall International Inc.)

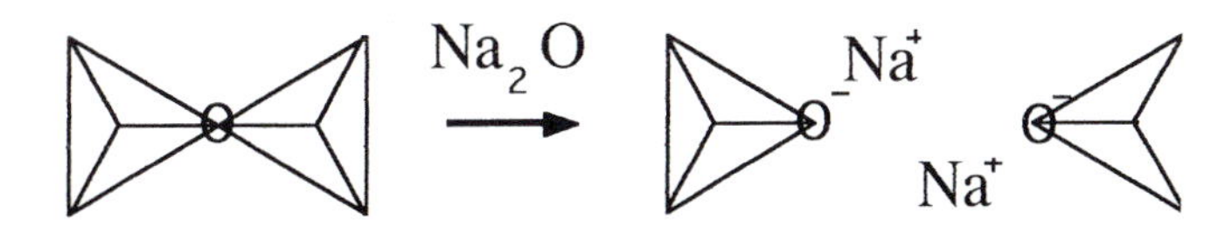

Fig. 7.7 *Network modifiers* like sodium, Na, lower the melting point of a glass and make it less viscous.

Network modifiers reduce T_g and make the glass less viscous and easier to process. Leaving electrical applications aside for the moment, the vast majority of glass made is *soda* or *soda-lime* glass based on the composition $(Na_2O)(CaO)(SiO_2)_6$. This includes window glass, bottles, tableware etc. etc. The raw ingredients are soda ash, limestone and sand (silica). The electrical properties of glasses are distinctly different from those of crystals, as we shall see.

7.3 DC properties

Since the primary purpose of ceramics is electrical insulation, you might think they don't *have* any DC properties. No insulator is perfect, however, and sometimes we even *want* a current to pass through a ceramic component. A respectable ceramic insulator has a resistivity (see Chapter 2, Section 2.7 and Table 7.2) round about 10^{14} Ωm. (In this,

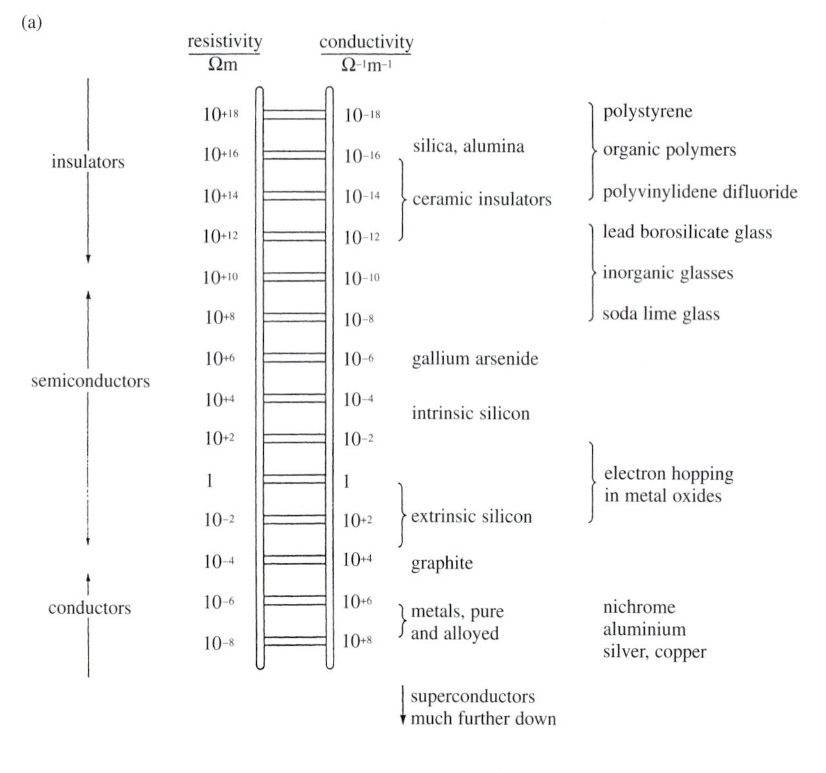

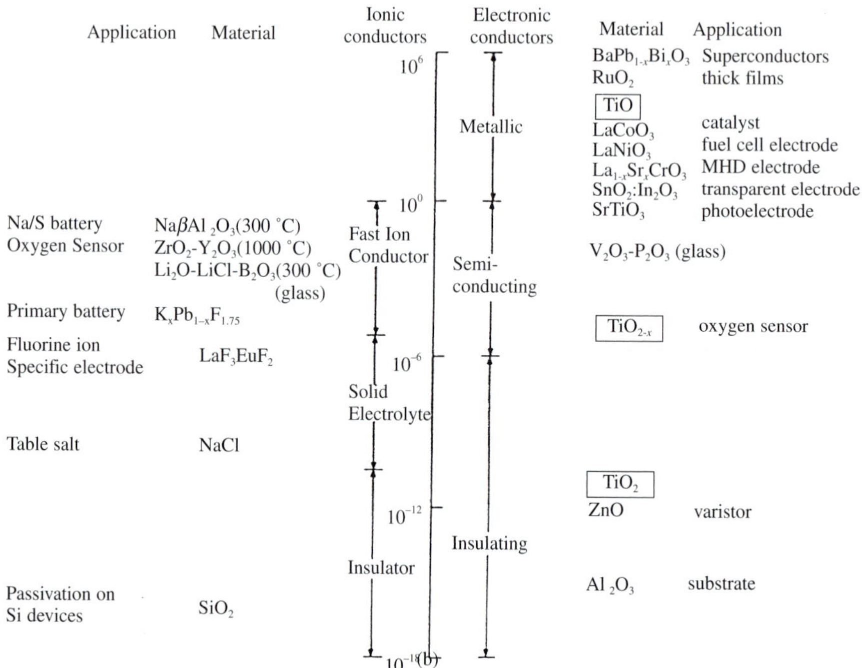

Fig. 7.8 (a) The electrical resistivities and conductivities of conductors, semiconductors and insulators (b) Ceramics in more detail. (These are **D C** resistivities and conductivities.) (Courtesy Professor J. A. Kilner.)

ceramics and polymers (Chapter 8) are quite similar.) Ceramic conductors typically have a conductivity about 100 times worse than the worst metallic conductors (see Fig. 7.8) although the overlap between the two is continually increasing. Figure 7.9 shows how the resistivities of metals, insulators and semiconductors vary with temperature. The electrical resistivity of metals rises gently as temperature increases whereas insulators conduct markedly better as temperature increases. This is because the resistivity of metals is controlled by the *mobility* of the electrons and they are scattered more by the vibrating atoms as the temperature rises. The resistivity of an insulator is controlled by the *number* of charge carriers (including electrons) and this increases with temperature. In metals there is only one effective charge carrier and that is the electron. In a ceramic there are two types: electrons and charged atoms (ions).†

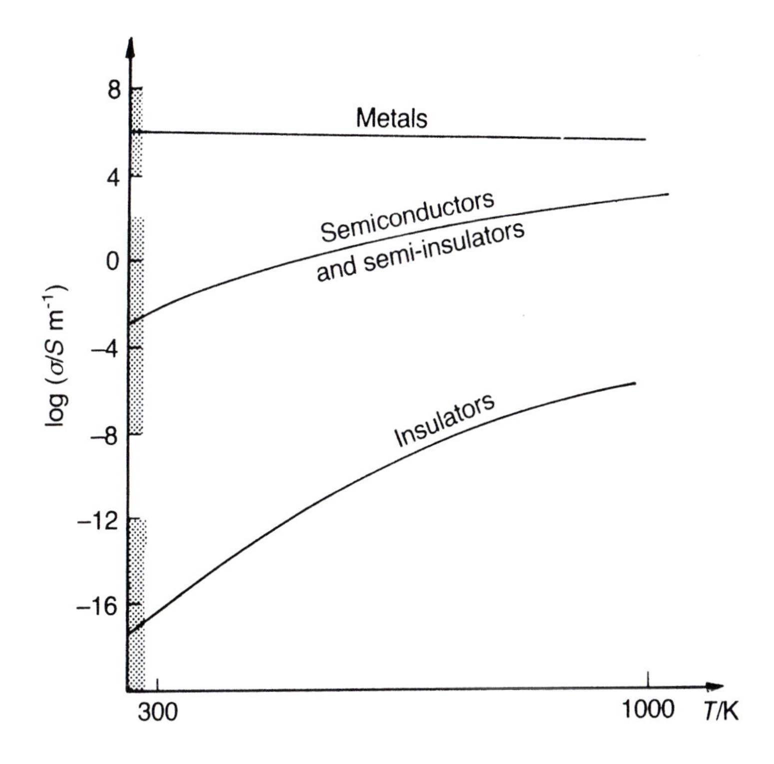

Fig. 7.9 How the electrical conductivities of conductors and insulators vary with temperature. (From *Electroceramics* by A.J. Moulson and J.M. Herbert, courtesy of Chapman and Hall.)

† In principle, ion flow can occur in metals, but its effect is swamped by the much more effective electron flow.

Conduction by electrons: electronic conduction

In a ceramic all the outer (valence) electrons are involved in ionic or covalent bonds and so are restricted to an ambit of one or two atoms. You may recall from Chapter 1, Section 1.10 that the energy necessary to tear a valence electron away from an atom and send it hurtling through the solid (i.e. to promote it from the valence band to the conduction band) is called the *band gap*. The fraction of electrons in the conduction band is $e^{\frac{-E_g}{2kT}}$. A good insulator will have a band gap >5 eV. Remembering that $kT \sim \frac{1}{40}$ eV at room temperature, the fraction of electrons finding themselves in the conduction band at this temperature as a result of thermal excitation is pretty small ($\sim e^{-200}$ or 10^{-80}).

There are other ways of inflencing electronic conductivity in a ceramic which have a far greater effect than temperature (see Fig. 7.10). One is *doping* with an element whose valence is different from the atom it replaces. We have already come across this in Chapter 1 in connection with semiconductors. Semiconductors are the subject of Chapter 9. A semiconductor is just a small band-gap insulator. The doping levels in insulators are generally greater than those used in semiconductors (parts per thousand

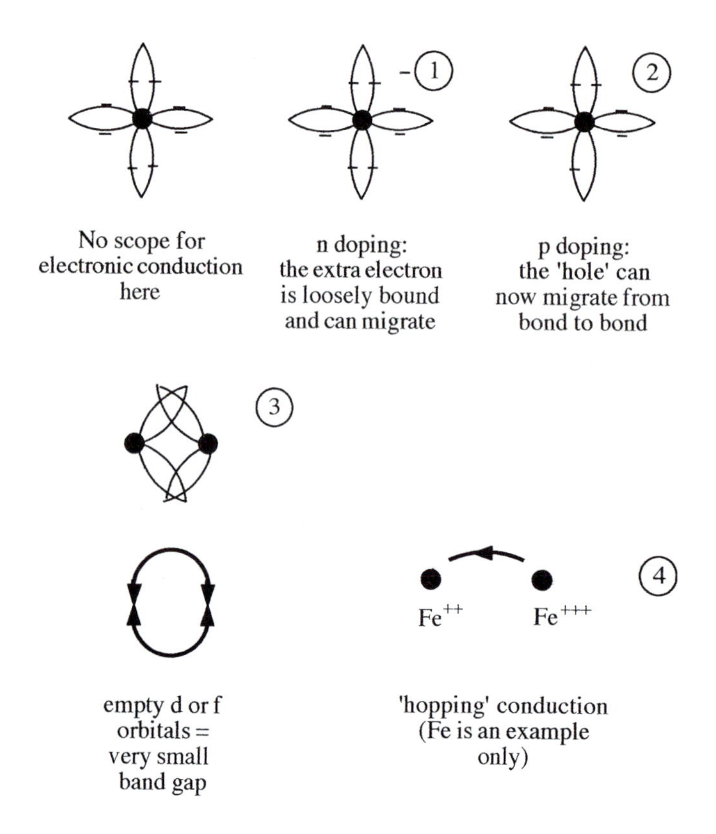

Fig. 7.10 Four ways in which to influence electronic conductivity in a ceramic (apart from thermal excitation).

as compared with parts per million). Of course, turning this around, you can see how important purity is in making a good insulator. Two other ways in which substantial electronic conductivity can be achieved in a ceramic are also shown in Fig. 7.10, both of them involving transition elements. Empty or partially filled d or f orbitals can overlap, providing a conduction network throughout the solid. Finally, if the valence of an ion can be variable—like iron, Fe—so called 'hopping' conduction can occur. (An alternative, up-market term for this is polaron conduction.) Such relatively good conductors are used as resistors and high temperature heating elements (see Section 7.7).

Conduction by ions: ionic conduction

The other method of electrical conduction through a ceramic is by ion migration. Where the bonding is mainly ionic, the atoms are charged, either positively (cations) or negatively (anions). When an electric field is applied the ions can migrate and give rise to an electrical current. The migration (diffusion) of the ions can be via vacancies or it can be interstitial (see Section 2.5). In metals, diffusion *had* to be via vacancies: in close-packed metals interstitials—atoms inhabiting the interstices of the crystal structure—just do not exist (or at least *self*-interstitials don't. *Solute* interstitials exist in metals—see Chapter 5 on steels, Section 5.2—but that doesn't help much here). In ionically and covalently bonded structures, however, there is more free space and interstitials can exist just as well as vacancies. Either way, though—vacancies or interstitials—ionic conduction is *much* slower than electronic. Because the conduction mechanism involves the transport of matter, composition gradients can build up and we talk about electrodes *polarising*. Notice also that the diffusion of ions depends on the presence of point defects: whether vacancies or interstitials. Raising the temperature speeds up ionic conduction both because the concentrations of the point defects rise and because they move more quickly (unlike electrons). The presence of impurities can have a fairly startling effect on point defect concentrations. For example, in a compound A^+B^-, adding an atom of C^{--} will require the formation of a vacancy on the B *sublattice* to preserve electrical neutrality. $\therefore$ 1%C $\Rightarrow$ 1% vacancies. This is an enormous concentration compared with those achievable by temperature alone.

Because glasses are more disorganised than crystals, ionic conduction through glasses can be quite fast. The conduction, of course, is via the charged network modifier ions. The smaller the ion, the more easily it moves and so Li glasses conduct better than Na glasses which conduct better than K glasses and so on. The larger ions, such as those of Ca, Mg, Ba and Pb, are distinctly slower and indeed reduce conductivity by Li, Na, K etc. because they block the conduction channels through the glass. I mentioned in Section 7.2 above (see Table 7.1) that soda ash is a common flux when making ceramics. In the higher quality ceramic insulators soda is replaced by lime to reduce the conductivity of the glass phase. Like electronic conductivity, ionic conductivity in ceramics can be increased to quite high levels. In the so-called *fast ion* conductors, continuous channels through the solid give rise to fast conduction paths for small ions.

An example of this type of conductor is β-alumina as used in Na–S fuel cells (see Chapter 6, Section 6.6).

The manufacture of ceramic conductors will be described in Section 7.6.

Electrical breakdown

At a certain voltage gradient (field) an insulator will break down. There is a catastrophic flow of electrons and the insulator is fragmented. Breakdown is *microstructure* controlled rather than bonding controlled. In this it is just like plastic yielding. The presence of dislocations reduces the yield stress of a metal by several orders of magnitude (see Chapter 3, Section 3.4). Similarly, the presence of heterogeneities in an insulator reduces its breakdown field strength from the theoretical maximum ($\sim 10^9$ Vm^{-1}) to practical values of $\sim 10^7$ Vm^{-1}. Some typical ceramic dielectric strengths are shown in Table 7.2. Heterogeneities can be pores, corners on grains—anything which will act as a field raiser (just as a crack will magnify stress—see Section 3.6).

7.4 AC properties

The second major application of insulators is as capacitors. Capacitance, *per se*, is not specifically an AC property, but, as you will see, there is a complication which is totally AC linked.

First of all, a reminder about capacitance. When a potential V is applied between two plates,† each acquires a charge Q, where $Q = CV$. For the parallel plate capacitor shown

Table 7.2 The resistivities, breakdown strengths, dielectric constants ϵ_R and loss tangents ($\tan\delta$) for some common ceramic insulators and capacitors (compare with Table 8.3).

	Resistivity (Ωm)	Breakdown strength (Vm^{-1})	ϵ_R	$\tan\delta$ at 1 MHz
Steatite	$10^{11} - 10^{13}$	$2.0 - 3.5 \times 10^7$	6.1	10^{-4}
Cordierite	$10^{10} - 10^{12}$	$0.4 - 1.0 \times 10^7$	5.7	10^{-3}
Porcelain	$10^{10} - 10^{12}$	$2.0 - 4.0 \times 10^7$	6.7	10^{-3}
$96Al_2O_3$	$10^{12} - 10^{13}$	$2.5 - 4.0 \times 10^7$	9.7	10^{-4}
Glass	$10^8 - 10^{12}$	$1.0 - 5.0 \times 10^7$	5 – 15	$2 - 20 \times 10^{-4}$
TiO_2	10^6 –	0.5 –	85	3×10^{-4}
$BaTiO_3$	10^{13}	3.0×10^7	230–4770	0.002–0.05

† It doesn't have to be two plates, but this is a simple geometry and one that is much used in real life.

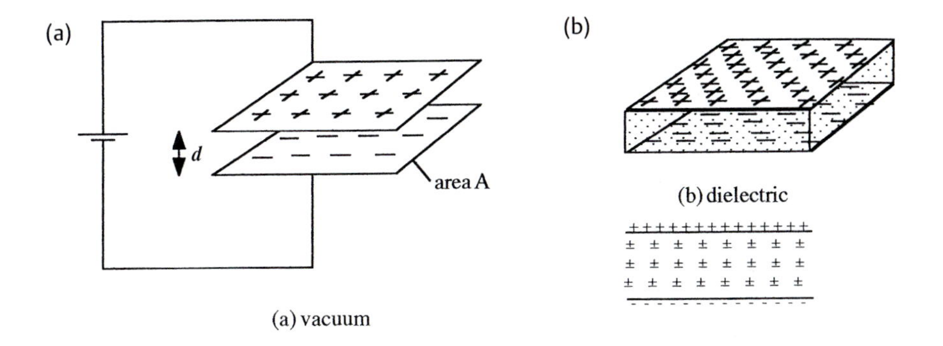

Fig. 7.11 A simple plate capacitor (a) with vacuum between the plates and (b) with a dielectric between the plates.

in Fig. 7.11(a), the capacitance $C = \frac{\epsilon_0 A}{d}$ where ϵ_0 = the permittivity of free space = 8.854×10^{-12} $\mathrm{Fm^{-1}}$.

When the space between the two plates is filled by an insulator (see Fig. 7.11(b)), the charge on the plates goes up. The ratio by which it rises is called the *dielectric constant* ϵ_R of the insulating material:

$$\underset{\text{vacuum}}{C} \longrightarrow \underset{\text{dielectric}}{\varepsilon_R C}$$

A dielectric is just another name for an insulator. The physical reason for the rise in stored charge is that the dielectric *polarises*—see Fig. 7.11(b). The stored charge on the plates rises to cancel out the induced surface charge on the dielectric. Different dielectrics polarise to different extents. Dielectrics which polarise a lot have large dielectric constants and *vice versa*. Table 7.2 includes dielectric constants for some common insulators. Generally speaking, materials with low dielectric constants are used as insulators and those with high dielectric constants are used as capacitors.

How does the polarisation of the dielectric arise? Why do some materials show large polarisation (high ϵ_R) while others show little polarisation (low ϵ_R)? The polarisation arises from induced electrical dipoles within the material. There are four basic mechanisms which are illustrated in Fig. 7.12 in order of their response times. Electronic polarisation is always present, of course. It is mechanism #2, ionic polarisation, which is chiefly responsible for the differences in dielectric constant shown in Table 7.2. Having said this, how on earth can such enormous differences arise? One can imagine alumina having twice the dielectric constant of silica (it is far more ionic) but how can $BaTiO_3$ polarise 1000 times more than either of them? The answer is that the very high values of ϵ_R depend on a new mechanism. When the polarisation

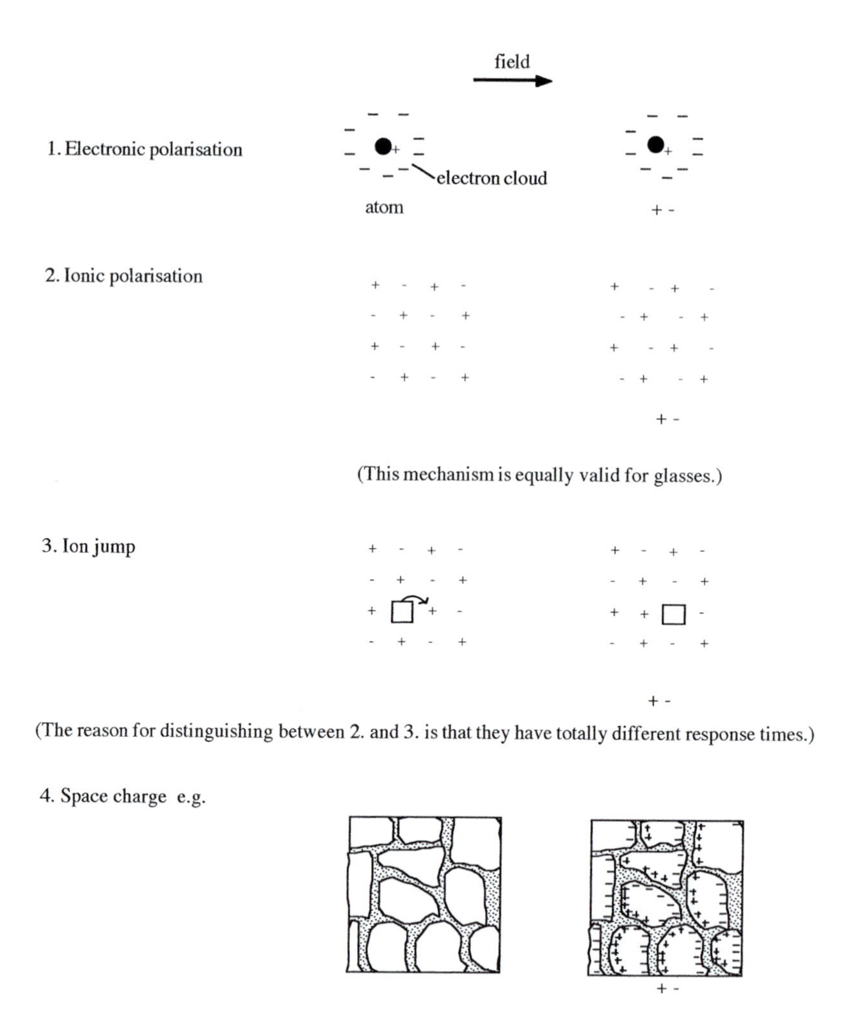

Fig. 7.12 The four basic mechanisms of electrical polarisation.

reaches a certain degree, it can be energetically favourable for all the dipoles to line up cooperatively in the same direction and to become permanent:

$$\begin{array}{ccccccccc} +- & & +- & & +- & & +- & & +- \\ & +- & & +- & & +- & & +- & \\ +- & & +- & & +- & & +- & & +- \end{array}$$

etc.

A material where this can happen is called a *ferroelectric*. Ferroelectricity is analogous in many repects to *ferromagnetism* (which I shall be discussing in Chapter 10) and 'ordinary' temporary electrical polarisation is analogous to paramagnetism. As in ferromagnetism, in a ferroelectric the dipoles do not generally line up across the whole crystal, but form *domains*. This minimises the energy and represents a compromise between domain wall energy and external electrostatic energy. When a field is applied the increase in polarisation is achieved, not by an increase in the sizes of the individual dipoles as for a non-ferroelectric, but by migration of the domain boundaries in such a way that favourably oriented domains grow and *vice versa*. It is the ease of this domain wall migration which is reflected in the large dielectric constant ϵ_R.

One of the reasons $BaTiO_3$ is ferroelectric is because the small Ti^{4+} ion at the centre of the unit cell as drawn in Fig. 7.3 has quite a lot of space to rattle around in and so reacts vigorously to the influence of an applied field, moving a relatively large distance and producing a commensurately large polarisation. Other conditions must be met, of course, before a permanent polarisation is permitted, including conditions on the symmetry of the crystal. The composition of $BaTiO_3$ is tweaked in various ways to produce various proprietary dielectrics. $BaTiO_3$ is the material used for microcapacitors for microelectronic circuits.

So far, so good. You must be wondering, however, when, in a section entitled 'AC properties' I am going to mention AC! Now is the time.

In principle, applying AC instead of DC and measuring an AC capacitance shouldn't make any difference. However, each of the mechanisms illustrated in Fig. 7.12 has its own intrinsic response time and therefore natural frequency. Electronic polarisation will take place at optical frequencies and beyond, ionic polarisation must be linked loosely to atom vibration and IR frequencies ($\sim 10^{13}$ Hz), ion jump frequencies will be ion vibration frequencies times the $e^{-\frac{\Delta G}{KT}}$ factor we encountered in connection with diffusion (Section 2.5) and the space charge mechanism could require some time to build up and therefore correspond to fractions of a Hz. Clearly mechanisms 2, 3 and 4 will be very much affected by temperature.

If the frequency of the applied AC is much lower than the natural frequency of the dipole, then the situation is essentially the same as for DC. When the applied AC frequency is much *higher* than the natural frequency of the dipole, the dipole does not have chance to react and so contributes nothing to the dielectric constant. When the AC frequency is near to the natural frequency of the material dipole, it resonates and absorbs energy from the applied alternating potential. This energy appears as heat. These effects are illustrated schematically in Fig. 7.13.

For a perfect, no-loss dielectric, the current is 90° ahead of the potential. When losses start to occur, because of resonating dipoles, the current lags by an angle δ which is called the loss angle. The charging current is proportional to ϵ_R and the heating (lossy) current to $\epsilon_R \tan\delta$, which is called the *loss factor*. Precisely which of these parameters is critical varies for different applications. Sometimes it is $\epsilon_R \tan\delta$, sometimes ϵ_R, sometimes $\tan\delta$. You may also come across $Q = 1/\tan\delta$ for high frequency circuits.

Certain it is, however, that for a high permittivity dielectric used in a microcapacitor, it is very important to keep δ as low as possible. Figure 7.13 is rather idealised: Fig. 7.14 and Table 7.2 show some real data.

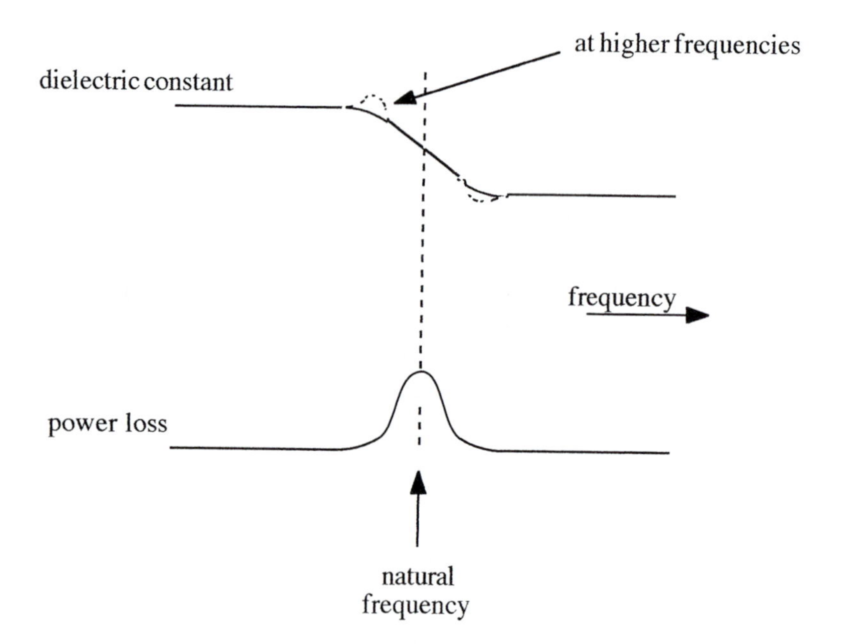

Fig. 7.13 How the dielectric constant and power loss vary around a natural dipole frequency in a material.

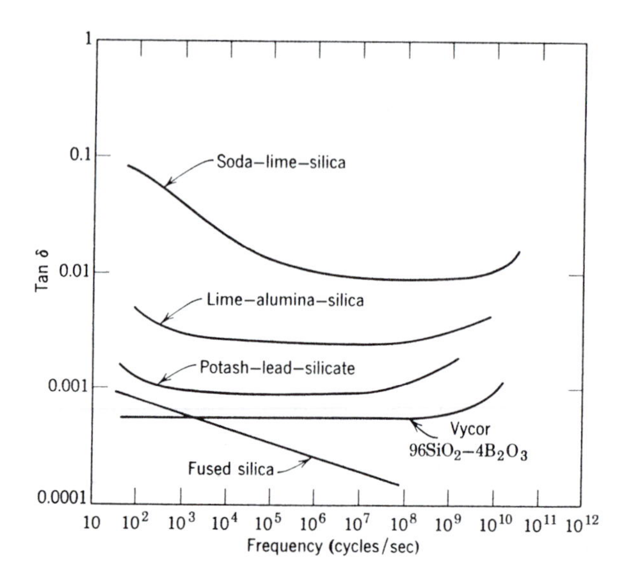

Fig. 7.14 Loss tangent (tan δ) vs frequency for some glasses. The rise on the left is due to ion jumps (mechanism #3 in Fig. 7.12). The rise on the right is the start of ionic polarisation (mechanism #2, Fig. 7.12), which must resonate at the frequency of atom vibration. (From *Introduction to Ceramics* by W.D. Kingery, H.K. Bowen and D.R. Uhlmann, courtesy of John Wiley and Sons, Inc.)

7.5 Mechanical properties

In Chapter 3 I described the mechanical properties of conductors. These were dominated by dislocations, whose copious presence and easy motion confer the plasticity we associate with metals. We are now, however, in a very different world. Ionically and covalently bonded solids do not tolerate or welcome dislocations. There is no plasticity worth talking of, at least not in this sense. This has profound implications for the way insulators are both used and made.

This chapter is concerned with ceramics and the next with plastics. We will find that the manufacturing solution (so far as it exists) to the lack of dislocations is completely different for the two different material types.

Anyway, back to ceramics. We are faced with a material which does not deform plastically, or at least with extreme reluctance. This means, for example, that a tensile test curve for a ceramic ends in the elastic region (Fig. 7.15).

> Ceramics are dominated by brittle fracture

So important is this that I shall recall the important expression from Chapter 3 which tells us when a crack has become critical:

$$\sigma_{\mathrm{f}} = \frac{\sqrt{G_{\mathrm{c}}E}}{\sqrt{\pi a}} = \frac{K_{\mathrm{Ic}}}{\sqrt{\pi a}} \qquad \text{from eqn (3.4)}$$

σ_{f} is the fracture stress, G_{c} is the crack energy per unit area, a is the length of the crack and K_{Ic} is the fracture toughness. The fracture toughnesses of ceramics are ~1 MPa $\mathrm{m}^{0.5\dagger}$—i.e. G_{c} is dominated by surface energy, not, as with metals, by plastic

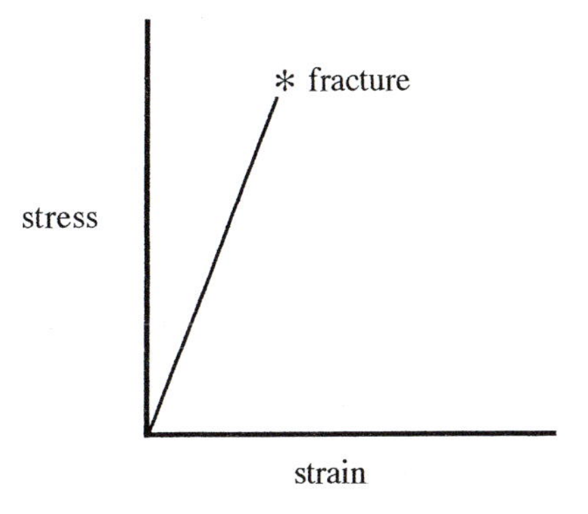

Fig. 7.15 A stress-strain curve for a ceramic.

† Unless we go to a lot of trouble. Tough (and very expensive) structural ceramic composites are now available with a K_{Ic} similar to those of the more brittle of the usable metals.

deformation. Equation (3.4) shows us that it is the large cracks which are dangerous. The methods which have to be adopted to *process* ceramics (see the next section, 7.6) mean that there are generally holes or *pores* inside the body of the ceramic. These control the mechanical properties. It is the largest pore which dictates the strength of the ceramic. Thus large pieces of ceramic tend to be weaker than small pieces (because there is more chance of the larger piece containing a larger crack). (Exactly the same is true of breakdown voltage (see Section 7.3).)

Clearly the mechanical tests I described for metals are totally inappropriate for ceramics. The only mechanical property we need to, or indeed can, know about for a ceramic is at what stress fracture occurs. Because tensile stresses open up cracks and compressive stresses do not, tensile strengths are typically 5–10% of compressive ones. It is therefore useful to have a tensile type of test which will reflect the size statistics of the crack distribution. A typical mechanical test for a ceramic is the *shell test*.

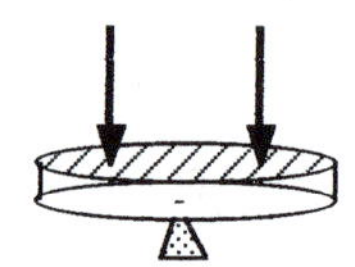

A disc of material is pressed down on a sharp point until it fractures. The stress required is recorded. This is done many times. The fracture stresses follow the *Weibull distribution*:

$$P_s(V) = \exp\left\{ -\frac{V}{V_0}\left(\frac{\sigma}{\sigma_0}\right)^m \right\} \tag{7.1}$$

where $P_s(V)$ is the probability of a given volume of ceramic surviving stress σ. V_0, σ_0 and m are the *Weibull parameters*. By plotting the results in appropriate ways (see Fig. 7.16) the Weibull parameters for the material can be determined. These reflect both the intrinsic strength of the ceramic and the size distribution of pores within it and are the equivalent for a ceramic of a metal's yield stress, ductility etc.

When ceramics are used in electrical or electronic engineering it is clearly important to ensure, as far as possible, that they are exposed to *compressive*, not tensile stresses. The shape of the specimen should not encourage concentrations of tensile stress, for example at sharp re-entrant corners:

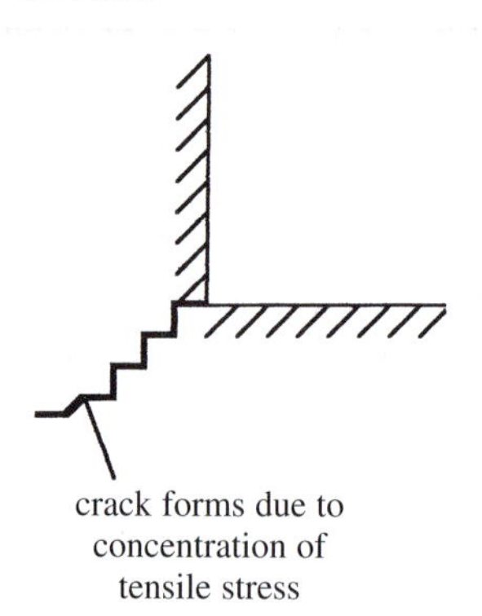

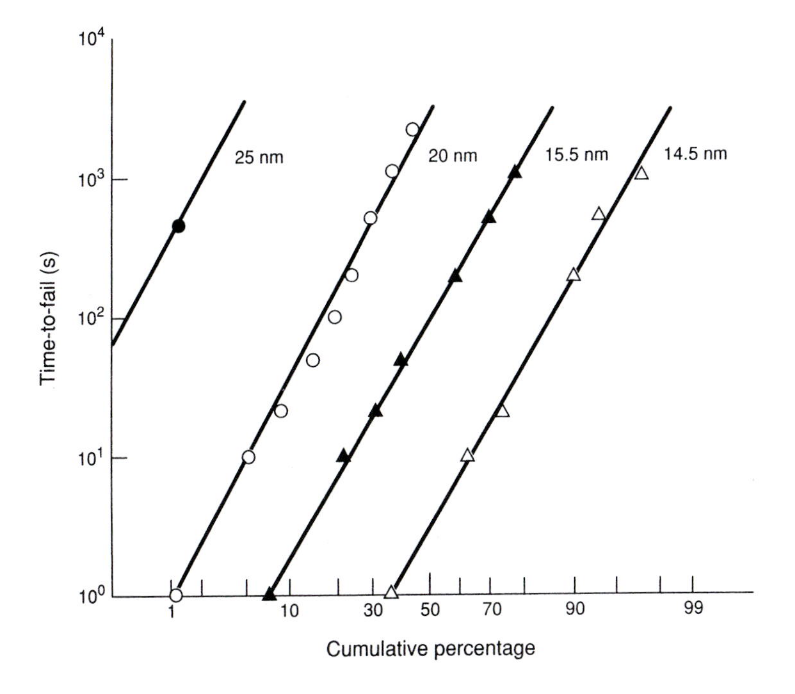

Fig. 7.16 The breakdown of silica films of various thicknesses plotted using the Weibull distribution. (From *Engineering materials science* by M. Ohring, courtesy of Academic Press, Inc.)

What implications does this have for *making* things out of ceramics? This is the subject of Section 7.6.

7.6 Making ceramic artefacts

We are faced with a seemingly intractable problem. We cannot shape solid ceramics, so none of the forming methods for metals which I described in Chapter 4 is possible, at least in the way they were used there. There are two ways round this:

1. With a ceramic which is all or mostly glass we can *cast* the component and shape it by heating the glass.
2. With most† other ceramics a *powder* is used to make the final shape. This is then heated to convert the powder into a single solid body. I shall describe in turn each of these methods in more detail.

† The highest volume ceramic of them all (concrete) does not fit into this pattern.

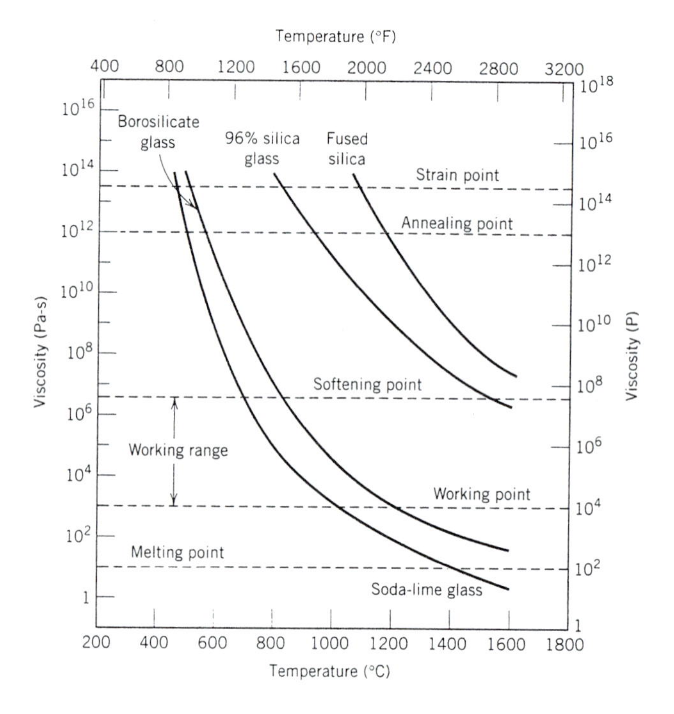

Fig. 7.17 How the viscosity of a glass changes as it melts. The range of viscosity appropriate to annealing (heating gently to remove strain) and working (e.g. pressing; drawing) is not available for a crystalline material, like a metal, where the viscosity drops suddenly on melting from the top of the graph shown to below the bottom. (From *Materials Science and engineering: an introduction* by W.S.Callister, courtesy of John Wiley & Sons, Inc.)

1. Glass artefacts

Glass components which melt at reasonably low temperatures can be cast to the required shape or alternatively cast to an intermediate shape and then formed at an appropriate temperature. Remember that glasses do not melt abruptly to form liquids, as crystals do. Rather they soften gradually, undergoing a rather subtle change from glass to liquid at a temperature which I labelled T_g (see Section 7.2 and in particular Fig. 7.5). If the viscosity (= stiffness or gooeyness) of the glass is plotted as a function of temperature it drops gradually (see Fig. 7.17). When a *crystal* melts its viscosity jumps straight from very high ($> 10^{12}$ Pa s) to very low ($< 10^{-4}$ Pa s). Glasses therefore have a wide and useful range of viscosity which is inaccessible for metals. When the viscosity falls below $\sim 10^6$ Pa s the glass can be worked.

For example, an electric light bulb will be produced by blowing. The light bulb glass is there solely to maintain the vacuum round the hot tungsten light filament so that it does not oxidise. It doesn't need to be an insulator and ordinary soda glass is used. (Soda glass is the 'steel' of the glass industry:[†] 80% of all glass is soda glass.) A glass high tension insulator will be pressed or cast and a glass fibre will be pulled or drawn (see Chapter 11).

† in the sense of dominating usage

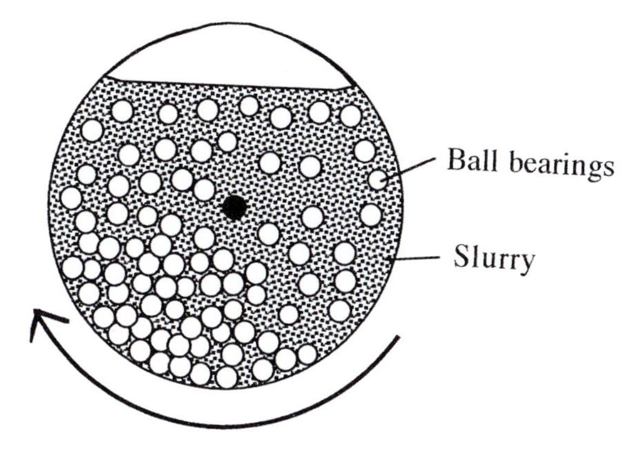

Fig. 7.18 Wet ball mixing (or *blunging*[†]).

2. Non-preponderantly glass artefacts

Here the ceramic is made initially as a powder. It is formed to shape and then *fired* to make the powder into a single solid body.

The raw materials for the lower-cost, higher-volume ceramics are generally just dug up as minerals, concentrated by fairly rudimentary mechanical means and ground into powder (see Fig. 7.18). Where the specification is higher and the purity therefore requires more attention, the raw materials may need to be processed chemically. Of course, this multiplies the cost by a considerable factor. The next step—assembling the powder into the final shape—is very various. Here are some examples. For a fuller list and descriptions, see *Electroceramics* by Moulson and Herbert (see Recommended Reading at the end of the chapter).

In *dry pressing* the powder is pressed without any binder or lubricant at ~75–300 MPa. (see Fig. 7.19). *Isostatic pressing* involves the pressure being applied from all sides, usually

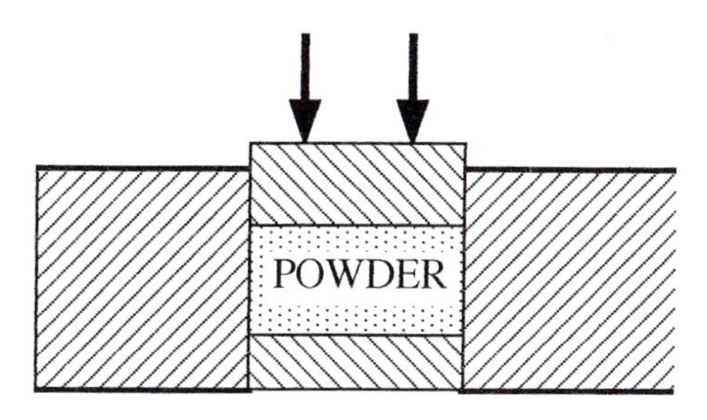

Fig. 7.19 Dry pressing a powder.

[†] What a splendid word! The old ceramics and iron industries are full of mouth-watering terminology like this. I can still remember my delight as a student when I learnt about 'puddling' wrought iron.

via a rubber membrane (Fig. 7.20). Pressures of 20–280 MPa are usual. This method is used to make spark plug bodies. Both these methods employ dry powders. In *jiggering*, however, the powder is mixed with water. Jiggering is the old fashioned way of making cups and plates. It only works if there is a lot of clay in the powder. After the powder is mixed with water, it is shaped on a rotating baseplate and air dried. In an electrical context, this method is used for large porcelain insulators. In *slip casting* (Fig. 7.21) the powder is suspended in water (the slip) and then poured into a plaster of Paris mould. The water is absorbed, leaving a dry layer adjacent to the plaster. The remaining slip is then poured off. This is a very common method.

Three methods which produce ceramic films of various thicknesses are *calendering, band casting* (Fig. 7.22) and *silk screen printing* (Fig. 7.23). In calendering the powder is suspended in a liquid rubber solution and coated on to a tape. A knife trims off the ceramic layer to the desired thickness. The tape is dried and then wound up (e.g. multilayer capacitors or substrates). Band casting (a.k.a. tape casting)

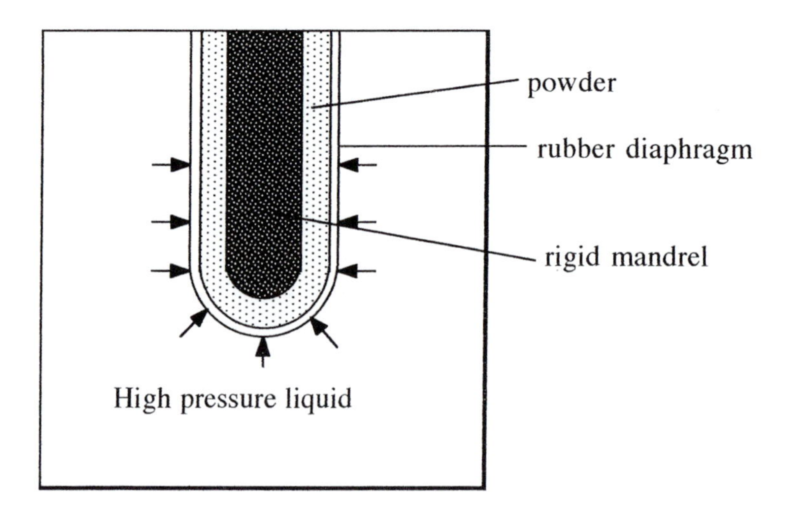

Fig. 7.20 Isostatic pressing of a powder.

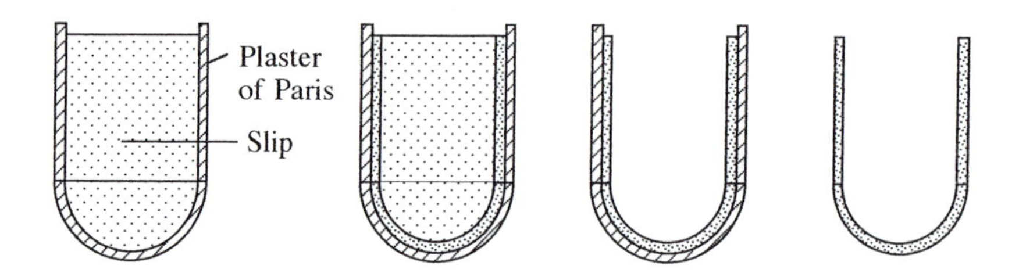

Fig. 7.21 Slip casting.

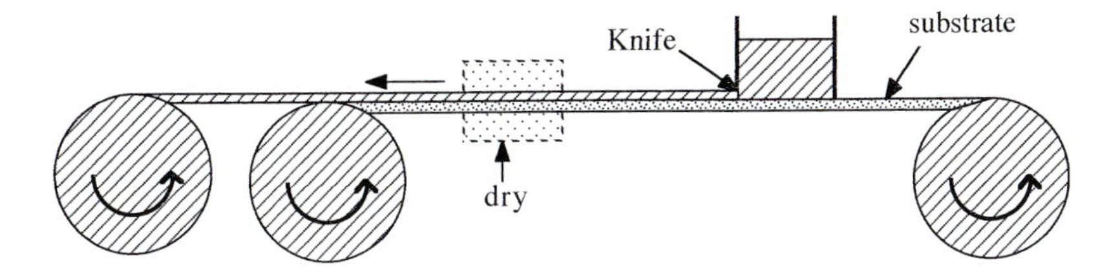

Fig. 7.22 Band casting.

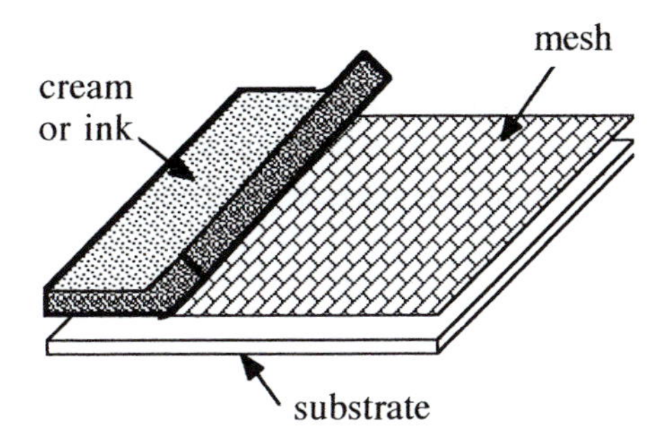

Fig. 7.23 Silk screen printing.

is similar to calendering, but uses a less viscous liquid and is also used to make substrates. Silk screening is used for producing so-called 'thick films' (actually these are quite thin by normal standards). A fine mesh is held a couple of mm above the substrate. A thick cream consisting of the powder in a binder is rubbed across the screen. This is then raised, leaving a film of cream on the substrate. This flows round the mesh, but a pattern of resist can be added to the screen to give, for example, a printed circuit.

Ceramic pastes can be *extruded*, just like a metal (see Section 4.2), but requiring rather less force. The ceramic powder is suspended in water if clay is present, otherwise starch, for example, is added. This method is used for rods, tubes and sheet.

Finally, *injection moulding*, a method usually associated with plastics (Chapter 8) is used also for ceramics. The ceramic powder is suspended in an organic binder and forced under high pressure into a cavity of the required shape (see Section 8.6).

At this stage the component may be *glazed*. Glazing just means 'glassing': covering with a layer of glass. The main purpose is to protect the surface of the ceramic from mechanical damage (think of the glaze on a cup) although the glaze on a high tension insulator is also made semiconducting. It allows a small surface leakage current which prevents build-up of locally high fields and consequent sparking (and thus mechanical damage).

The final step is *firing*. This is performed in a high temperature oven and drives off any liquid binder, accomplishes required chemical reactions and joins the particles together into one solid mass (called *sintering*). The basic driving force for this is surface tension (Fig. 7.24).

The longer the firing time the higher the fractional density of the product (and the more expensive). 95% dense (i.e. 5% pores) is considered quite respectable. The formation of a glass during firing, as in porcelains, is very helpful. The glass flows into the pores and helps stick the crystal grains together. (Remember that the mechanical and electrical properties of ceramics are improved by removal of pores.) The big problem in firing is shrinkage. Common sense dictates that if we are eliminating free space within the ceramic body then the volume of the ceramic must be decreasing. Problems arise when the shrinkage is not uniform and the final shape of the component differs from the compacted (or whatever) one. Sometimes a *calcining* stage is inserted before compaction. This accomplishes partially any chemical reactions and reduces the amount of shrinkage during firing. It can be carried out, for example, in a rotary kiln as shown schematically in Fig. 7.25.

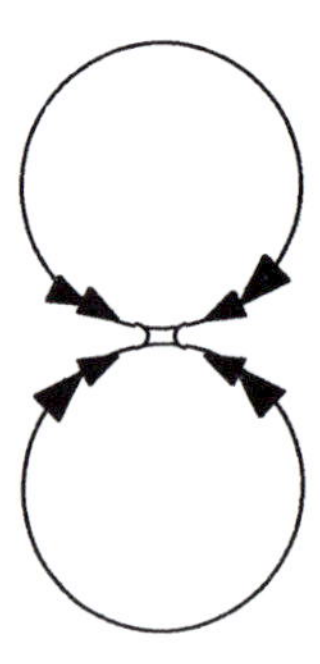

Fig. 7.24 The main driving force for sintering is the surface tension of the (solid) ceramic particles.

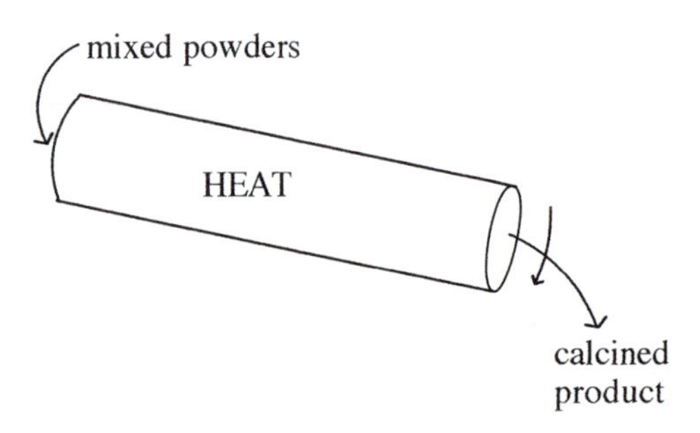

Fig. 7.25 Calcining is an early firing stage (like the pre-wash on your washing machine) designed to reduce shrinkage distortion.

7.7 Examples of ceramic artefacts in electrical and electronic engineering

Ceramics are used as (poor) conductors, as insulators and as capacitors. You will also find ceramics mentioned in Chapters 10 (on magnetic materials) and 11 (on optical materials and superconductors). Some examples of ceramics in use have already been referred to earlier in this chapter. The aim of this section is to give you a flavour of where ceramic components are used as conductors, insulators and capacitors and some examples of the manufacturing techniques described in the last section, in action.

Ceramics as conductors

Ceramics are used as conductors for the following applications:

(i) Heating elements (high temperature)
(ii) Resistors
(iii) Varistors
(iv) Thermistors
(v) Solid electrolytes
(vi) Gas sensors
(vii) Superconductors (see Chapter 11)

You will find further details concerning these and their manufacture, involving many of the methods outlined above, in the book by Moulson and Herbert (see Recommended Reading at the end of the chapter).

Ceramics as insulators

Typically $\epsilon_R < 15$. For the most part they are low-cost and mass produced. I shall describe two quite different examples:

- Electrical porcelains
- Alumina

(See also Fig. 7.1.)

Electrical porcelains
These are of two types: clay-based and talc based.
In the clay-based variety there are three main constituents:

Clay = kaolin (a type of clay) = $Al_2(Si_2O_5)(OH)_4$ (aluminosilicate)
Flux = feldspar = $KAlSi_3O_8$ (potassium aluminosilicate)
Filler = quartz (silica) = SiO_2

A common manufacturing process is jiggering (see Section 7.6):

1. Blend with water
2. Press out water → slip
3. Air at e.g. 50 °C (1 day)
4. Press/machine to shape (c.f. potter's wheel)
5. Glaze
6. Fire at 1200 °C (1 day)

Porcelain insulators are used for both high and low tension insulation. The porcelain consists of quartz crystals in a glass matrix (Fig. 7.26).

In talc-based electrical porcelains talc ($Mg_3Si_4O_{11}.H_2O$) replaces feldspar. The resulting ceramic is a mixture of Si, Al and Mg oxides (see Table 7.3).

Alumina is a good insulator with high thermal conductivity, which combination of properties makes it attractive as an electronic substrate. Its resistivity is very sensitive to purity. When pure it is totally crystalline; its manufacture requires sintering the powder at 1750 °C, which is very expensive. At 95% purity (5% Ca & Mg silicates → glass) the sintering temperature is down to 1350 °C which helps to make it much cheaper. This version is used for spark plug bodies and is already technically an aluminous porcelain. At the other end of the scale, aluminous porcelains can contain as little as 25% alumina.

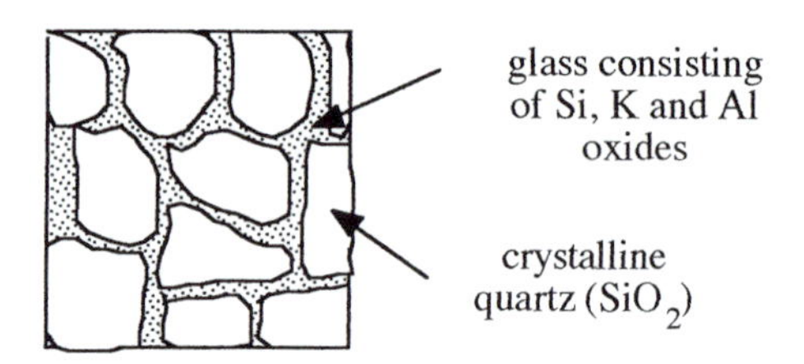

Fig. 7.26 The microstructure of electrical porcelain (clay type).

Table 7.3 Talc-based electrical porcelains.

Ceramic	Raw materials	Result	Applications
Steatite	Clay (15%), talc (85%)	$MgSiO_3$ in glass matrix	Low dielectric losses: capacitors, coil formers, substrates etc.
Cordierite	Clay (85%), talc (15%)	Mg alumino-silicate in glass	Good thermal shock resistance: high power electrical fuse holders, supports for high power resistors and fan-heater elements

(Forsterite Discontinued)

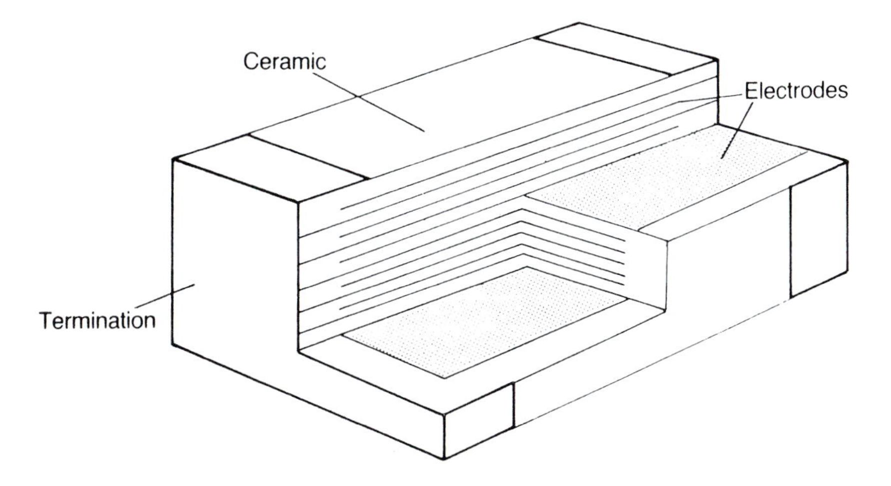

Fig. 7.27 A multilayer ceramic capacitor. (From *Electroceramics* by A.J. Moulson and J.M. Herbert by kind permission of Chapman and Hall.)

Ceramics as capacitors

Single capacitors are generally in the shape of discs or tubes. The shape is first formed by a combination of extruding, cutting and pressing a pre-calcined ceramic powder which is held together by an organic binder. Firing gets rid of the binder and sinters the powder particles together. Metal electrodes are attached by, for example, immersion in a solder bath or by electroless nickel plating (see Chapter 6). The capacitor is canned in epoxy resin (more exactly thermoset—see page 227).

In multilayer capacitors (Fig. 7.27) the 20 μm thick layers, whether of ceramic or electrode, are produced by screen printing. The ink can be 'wet' (i.e. aqueous) or 'dry' (i.e.with organic binder only). The final steps are the same, involving firing, attaching leads and encapsulation.

You can find more detail in, for example, Moulson and Herbert (see Recommended reading).

7.8 Summary

1. Insulators are used to prevent the flow of electricity and to store electric charge (capacitors).
2. A ceramic phase is a compound of a metal and a non-metal. Commercial ceramics consist usually of a mixture of phases.

3. The chemical bonding in a ceramic phase is a mixture of ionic and covalent.
4. Ceramic phases may be crystalline or amorphous (glassy). Crystalline phases tend towards ionic bonding while amorphous phases tend towards covalency.
5. The formation of a glass occurs on cooling past T_g and is encouraged by fast cooling.
6. Electricity is conducted via electrons and ions. Both processes are thermally activated and inflenced by doping.
7. Electrical breakdown occurs at voltage gradients of 10^7–10^8 Vm^{-1}. It is encouraged by microstructural inhomogeneities, which act as field raisers.
8. The four mechanisms of capacitance are, in order of response time, (a) electronic, (b) ion displacement, (c) ion jumping and (d) space charge. As the applied frequency increases past the relevant resonant frequency, the dielectric constant drops to a lower level and power losses go through a maximum.
9. Mechanical properties are dominated by fracture from pre-existing cracks. The statistics of fracture are described by the Weibull distribution.
10. Manufacturing methods for glass artefacts exploit a range of viscosity not available for materials which crystallise and include pressing, blowing and pulling.
11. Non-preponderantly glassy insulators are formed to shape as a powder (wet or dry) and then fired. They may also be glazed.

Recommended reading

Introduction to ceramics by W.D. Kingery, H.K. Bowen and D.R. Uhlmann (published by John Wiley, Inc.)

Physical ceramics by Y.-M. Chiang, D. Birnie III and W.D. Kingery (published by John Wiley, Inc.)

Electroceramics by A.J. Moulson and J.M. Herbert (published by Chapman and Hall)

Questions (Answers on pp. 323–324)

7.1 (I am writing this sitting in my kitchen.)
Which of the following are made from ceramic?:
(a) A wall tile.
(b) The refrigerator.
(c) The electrical plug connecting the fridge to the mains.

(d) The window glass.
(e) The window frame.
(f) The brick walls.
(g) The fluorescent light tube in the ceiling.

7.2 Assuming that the ions are touching spheres, what is the percentage of free space in $BaTiO_3$? (The ionic radii of Ba^{2+}, Ti^{4+} and O^{2-} are 0.135, 0.068 and 0.140 nm respectively.)

7.3 What are the electrical resistivities of copper, steel, soda glass and alumina? How would they vary if the temperature rose from room temperature to 300 °C?

7.4 What potential would you have to apply across 1 mm of porcelain before it explodes (i.e. breaks down and fragments)?

7.5 What is the capacitance of a 1 cm × 1 cm × 0.1 mm piece of steatite whose faces are completely coated with thin metal electrodes?
(The permittivity ϵ_0 of free space is 8.854×10^{-12} Fm^{-1}.)

7.6 The electrical conductivity of Mn_3O_4 is measured at various temperatures as follows:

T(K)	$(\Omega m)^{-1}$
200	3.63×10^{-11}
300	3.00×10^{-7}
400	1.66×10^{-5}
500	1.57×10^{-4}
600	1.37×10^{-3}
700	5.31×10^{-3}
800	1.00×10^{-2}
900	2.56×10^{-2}
1000	5.05×10^{-2}

By plotting the logarithm of the conductivity vs $1/T$, determine the activation energy of the conduction process.

8

Plastics

Chapter objectives

- What is a plastic? A polymer?
- The classification of polymers into thermoplastics, rubbers and thermosets
- How electricity is conducted through polymers (DC properties)
- Capacitance and power loss mechanisms in polymers (AC properties)
- The mechanical properties of polymers
  - instantaneous: Young's modulus and the tensile test
  - time dependent: viscosity/creep
  - fatigue
- Manufacturing methods

8.1 Introduction: plastics and ceramics as electrical insulators

This is the second of two chapters on electrical insulators. In the previous chapter I described ceramics. In this chapter I shall be concerned with plastics. Ceramics and plastics are very different, although they both involve non-metallic bonding. During this chapter I shall compare ceramics and plastics and explain when one is used and when the other, and why.

I will very largely follow the same scheme as for ceramics. To avoid repeating myself I shall assume some of the introductory material on electrical properties in Chapter 7, Sections 7.2 and 7.3. I shall warn you explicitly when you need to be familiar with previously presented material.

8.2 What *is* a plastic?

A plastic consists of a *polymer* plus various additives:

$$\text{plastic} \quad = \quad \text{polymer} \quad + \quad \left\{ \begin{array}{l} \text{dye} \\ \text{fire retardant} \\ \text{filler} \\ \text{etc.} \\ \text{etc.} \end{array} \right.$$

The science of plastics is therefore largely the science of polymers.

A polymer consists of giant molecules with carbon backbones. There are three basic types of polymer which are illustrated schematically in Fig. 8.1, where the lines represent strings of carbon atoms linked by *covalent* bonds. Thermoplastic molecules are linear (but coiled up) or branched molecules which are connected to each other by entanglement or by weak van der Waals bonds. In thermosets the covalent bonding

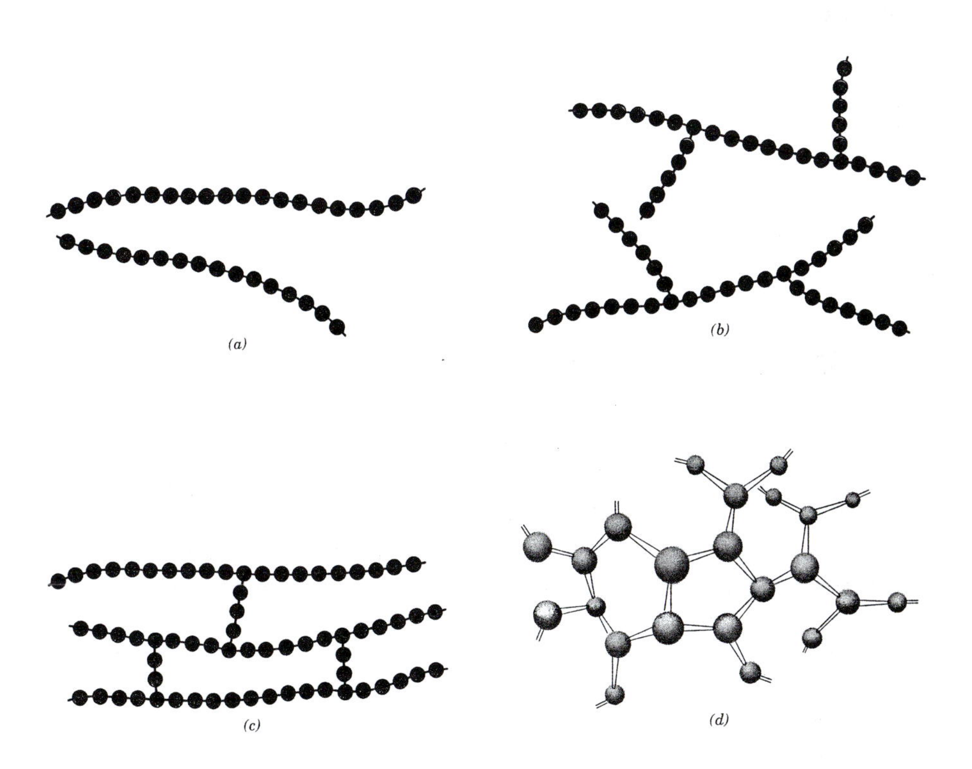

Fig. 8.1 The three basic types of polymer (a) linear and (b) branched *thermoplastic*, (c) lightly crosslinked *rubber* and (d) crosslinked *thermoset*. (From *Materials science and engineering: an introduction* by W.D. Callister, courtesy of John Wiley & Sons, Inc.)

makes a 3-D network. Rubbers, in this context, are a sort of halfway house between the two, with light cross-linking. Plastics take on the same name as the polymer which forms their major constituent. Most of the plastics we will encounter in this chapter, or with which you are already familiar, (or both) are *thermoplastics*.

Thermoplastics

Many thermoplastics have truly linear molecules which fit into the following scheme:

$$\text{etc.}\quad -\left[\begin{array}{ccc} | & & | \\ \mathrm{C} & - & \mathrm{C} \\ | & & | \end{array}\right]-\left[\begin{array}{ccc} | & & | \\ \mathrm{C} & - & \mathrm{C} \\ | & & | \end{array}\right]- \quad \text{etc.}$$

The simplest is polyethylene (polythene):

$$-\left[\begin{array}{ccc} \mathrm{H} & & \mathrm{H} \\ | & & | \\ \mathrm{C} & - & \mathrm{C} \\ | & & | \\ \mathrm{H} & & \mathrm{H} \end{array}\right]-$$

where I have bracketed one of the repeating $-\overset{|}{\underset{|}{\mathrm{C}}}-\overset{|}{\underset{|}{\mathrm{C}}}-$ groups. To make polyethylene, ethene (what used to be called ethylene) is reacted chemically in such a way that a large number of molecules add together:

$$\begin{array}{ccc} \mathrm{H} & & \mathrm{H} \\ | & & | \\ \mathrm{C} & = & \mathrm{C} \\ | & & | \\ \mathrm{H} & & \mathrm{H} \end{array} \xrightarrow[\text{chemistry}]{\text{clever}} \begin{array}{ccccc} & \mathrm{H} & & \mathrm{H} & \\ & | & & | & \\ - & \mathrm{C} & - & \mathrm{C} & - \\ & | & & | & \\ & \mathrm{H} & & \mathrm{H} & \end{array} \xrightarrow{\text{polymerise}} \begin{array}{ccccccccc} & \mathrm{H} & & \mathrm{H} & & \mathrm{H} & & \mathrm{H} & \\ & | & & | & & | & & | & \\ - & \mathrm{C} & - & \mathrm{C} & - & \mathrm{C} & - & \mathrm{C} & - \\ & | & & | & & | & & | & \\ & \mathrm{H} & & \mathrm{H} & & \mathrm{H} & & \mathrm{H} & \end{array}$$

ethene

monomer
=ethene
group

polymer
= polyethylene
(trade name
'polythene')

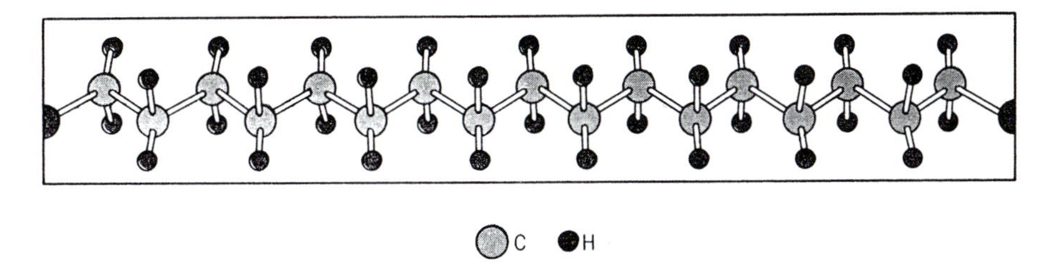

Fig. 8.2 3-D view of part of a thermoplastic polymer molecule. (From *Materials science and engineering: an introduction* by W.D. Callister, courtesy of John Wiley & Sons, Inc.)

The carbon covalent bonds are tetrahedrally disposed at 109° to each other (see Fig. 1.12) and so what I have depicted above as a flat molecule really looks as shown in Fig. 8.2. The hydrogen atoms in ethene may be substituted by other organic groups. Some common thermoplastics used in electrical and electronic engineering are listed in Table 8.1.

The exact length of each molecule cannot be controlled and so there is a *distribution* of lengths. The polymer chemist can however influence the *average* and the *spread* of the distribution and also more subtle aspects like the shape which the molecule adopts, how much it branches (if it does) etc., etc. A substantial part of polymer engineering lies in this type of *molecular* engineering which influences the most basic level of structure of the polymer. This is a totally different world from that of metals, where the immediate atomic neighbourhood of each atom is as it comes and is largely uncontrollable. Ceramics are in between, in that their rate of solidification controls whether we see a glass or a crystal (which is also true of many polymers). The other way of influencing the structure of a polymer will be described later on under processing.

Thermoplastics can be anything from partially crystalline to totally amorphous. They can never be totally crystalline because the molecules are just too large. Polymer crystals are rather different from those we have met before because the crystal is smaller than the molecule! What happens is depicted in Fig. 8.3. The crystallites are plate shaped, 10–20 nm thick and about 1 μm across. They are linked together by the polymer molecule itself. Several crystallites together form a little sphere, or *spherulite*, whose size can be anything from $\sim$1 μm up to 1 mm—see Fig. 8.4). Because the lengths of the molecules vary it is not surprising that polymer crystals melt over a range of temperatures, instead of at a single temperature as for metal and ceramic crystals.

The remainder of the thermoplastic (all of it in many cases) is occupied by an amorphous phase. At low temperatures this is a glass in the same sense as that which I described for ceramics. As the glass is warmed it goes through a 'glass transition' at a temperature T_g, which again is a function of the rate of heating. Values of T_g and the melting temperature, T_m, for some common thermoplastics are shown in Table 8.2.

At low temperatures Young's modulus is $\sim$2 GPa, as compared with 100 GPa for a metal or ceramic. Physically this elastic modulus comes via weak van der Waals bonds.

LIVERPOOL JOHN MOORES UNIVERSITY
LEARNING SERVICES

Table 8.1 Seven common thermoplastics used in electrical and electronic engineering.

Abbreviation	Full name	Repeating group
PE*	Polyethylene	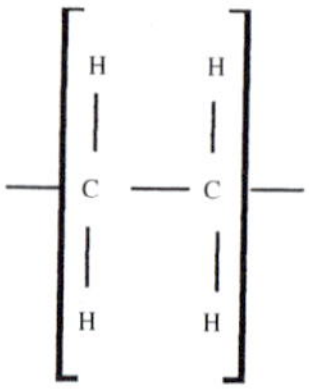 $-[CH_2-CH_2]-$ (H, H / C — C / H, H)
PVC	Poly(vinyl chloride)	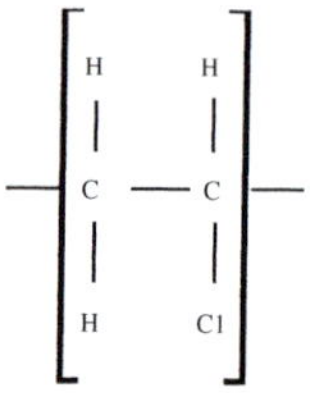 $-[CH_2-CHCl]-$ (H, H / C — C / H, Cl)
PP	Polypropylene	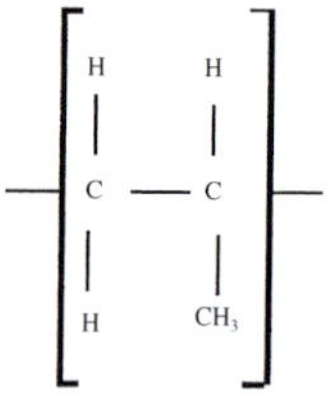 $-[CH_2-CH(CH_3)]-$ (H, H / C — C / H, CH_3)
PS	Polystyrene	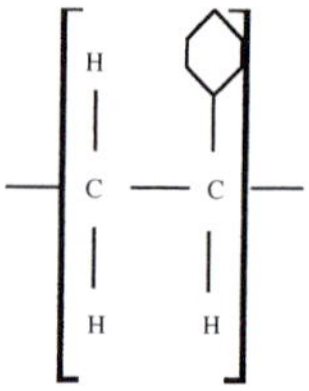 (H, phenyl / C — C / H, H)
ABS	Acrylonitrile-butadiene-styrene copolymer	(mixture of polymers)
SAN	Styrene-acrylonitrile copolymer	
PA†	Nylon	e.g. $-[(CH_2)_m-NH-CO-(CH_2)_n-CO-NH]-$ where m = 4, n = 6 ⇒ nylon 6,6

* LDPE, HDPE: Low Density PolyEthylene, High Density Polyethylene.
† PA6, PA66, PA610: Nylon 6, Nylon 66, Nylon 610.

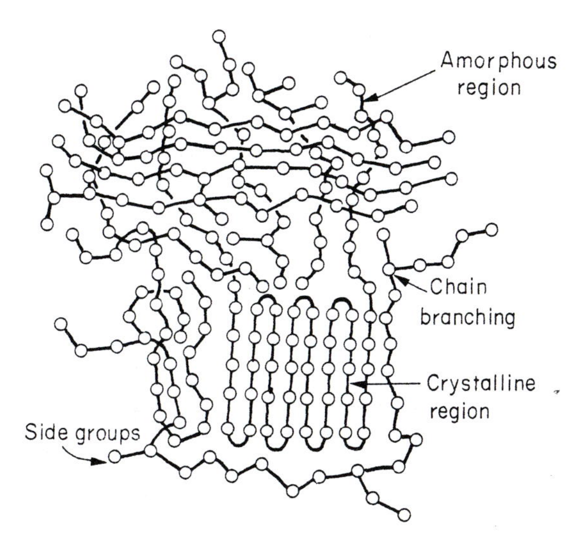

Fig. 8.3 Polymer crystals are smaller than polymer molecules. (From *Engineering materials I* by M.F. Ashby and D.R.H. Jones, courtesy of Elsevier Ltd.)

At higher temperatures, in the rubbery stage, the molecules are rubbing past each other. This region is called viscoelastic. It is characterised by viscous flow and large elastic displacements (hence *visco elastic*). The viscous flow is a little like a very fast version of creep in a metal, but unlike in a metal the creep is elastic, not plastic, and is mainly recoverable. In other words, if you hang a weight on a thermoplastic in its rubbery region it will immediately extend by several % (small Young's modulus → large elastic displacement) and then slowly grow by several more %. On removing the weight the instantaneous strain is recovered immediately and most of the remaining strain over roughly the same period as it took to accomplish.

The mathematics of describing this physical behaviour is quite similar to that for describing AC electricity in dielectrics. Polymer engineers often use a mechanical analogy to describe electrical behaviour in polymer dielectrics (see Section 8.4 below), so I would think it quite reasonable for an electrical engineer to use an electrical analogy in describing mechanical behaviour!

At higher temperatures still the polymer is a very viscous liquid and this is where polymer processing is carried out, in a viscosity range not accessible for crystalline materials (see below).

Semicrystalline polymers like PE show a mixture of crystal and glass behaviour.

Thermoplastics have the very useful property that they can be remelted and reprocessed, and so can be recycled.

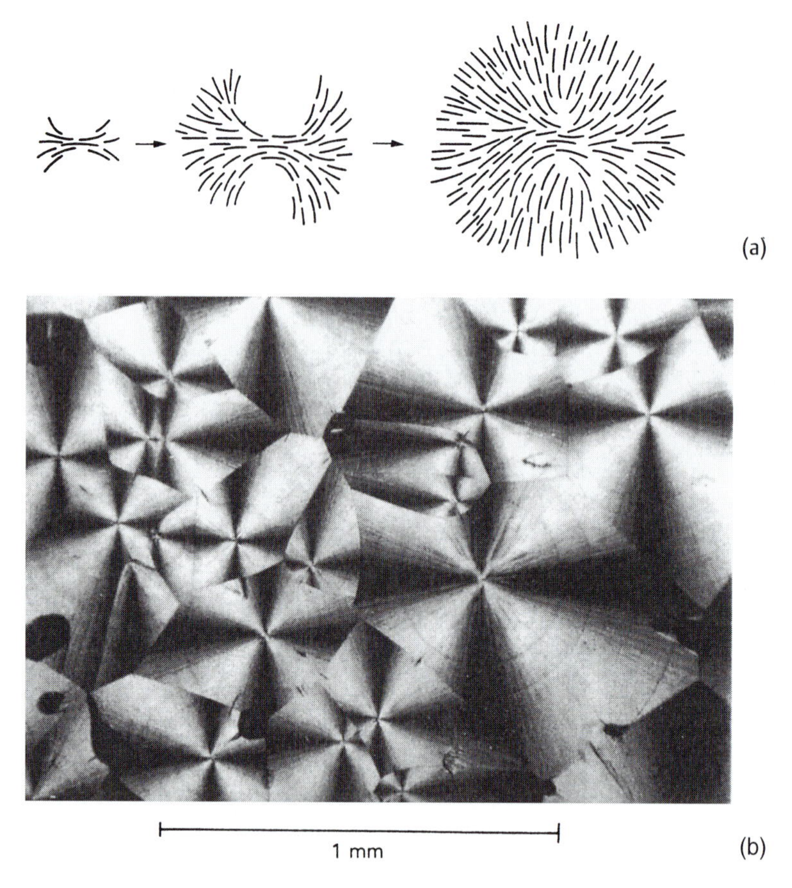

Fig. 8.4 (a) Schematic and (b) photograph of spherulites in PEO (polyethylene oxide). These are far larger than in a normal commercial polymer (μm's). (From *Plastics* by N.J. Mills, courtesy of Edward Arnold Ltd.)

Table 8.2 T_g (glass temperature) and T_m (melting temperature) for the thermoplastics of Table 8.1. (ABS and SAN are *copolymers* (mixtures of polymers).) Roughly, T_g in K is about two thirds T_m (also in K).

	T_g (°C)	T_m (°C)
PE	-100	125
PVC	80	210
PP	-20	175
PS	100	240
PA (66)	57	265

Rubbers

At the higher temperature end of the rubbery region Young's modulus becomes very small and enormous elongations of several 100% are possible. Unfortunately this is succeeded by rapid viscous flow. This viscous flow can be prevented and a very useful material, known simply as a *rubber*, can be produced by introducing light covalent cross-linking.[†] Because we have removed the viscous flow from a

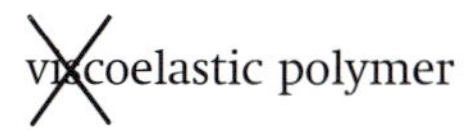

we call the result an

elastomer

On the same scheme as used earlier, natural rubber, for example, has a repeating chemical monomer group of

$$-\left[\begin{array}{ccccccc} H & & H & & H & & H \\ | & & | & & | & & | \\ C & - & C & = & C & - & C \\ | & & & & | & & | \\ H & & & & CH_3 & & H \end{array}\right]-$$

Silicone rubbers are unusual in that they have a silicon–oxygen backbone, rather than carbon.

Thermosets

Here the covalent bonding forms a 3-D network. This confers increased rigidity on thermosets[‡] as compared with thermoplastics; the disadvantage is that they cannot be remelted and reused. Once a thermoset has been made, that is that. Thermosets are typically used as matrices for composites (where they are reinforced by glass fibres) or as glues. Many thermosets are made by reacting a *resin* with a *hardener*. For example, in an *epoxy* thermoset, a 'prepolymer' resin with an epoxy group at either end is reacted with an amine hardener. Other resins include amines, polyesters, phenolics and

[†] First discovered by Goodyear in 1839 in the U.S.A. He *vulcanised* natural rubber using sulphur.

[‡] Do you remember Bakelite? Bakelite was the first thermoset, developed by Baekeland early in the 20th century.

urethanes, some of which names you probably recognise. Thermosets also make good glues, because they react chemically with other surfaces. 'Araldite' is an epoxy thermoset: the two tubes in the packet contain the resin and the hardener.

Thermosets are usually totally amorphous and glassy.

8.3 DC electrical properties

I described the DC electrical properties of ceramics in Section 7.3. It might be worth you glancing through that section first, before reading this one, particularly the early part which was quite general and referred to all types of insulator.

Polymers are insulators because, like ceramics, they are constructed using non-metallic bonds. Ceramics are dominated by ionic and covalent bonds. Polymers are dominated by covalent and van der Waals bonds. The electrical properties of some common polymers are shown in Table 8.3. If you compare these values with those given for ceramics in Table 7.2 and Fig. 7.8 you will see that the DC resistivity and breakdown voltage are quite similar for plastics and ceramics: the big difference is how resistant they are to heat and environment. Ceramics are *much* more robust in this sense. This is because the van der Waals bonds in plastics are easily broken. Even sunlight can cause a plastic to deteriorate (the energy of a photon $E = h\nu = 2$ eV > the van der Waals bond strength) as you will be aware, but you wouldn't expect a ceramic

Table 8.3 Average electrical properties of some common thermoplastics (see also Tables 8.1, 8.2, 8.4 and 8.5). (By kind permission of BASF AG.)

Polymer	Electrical resistivity (Ωm)	Dielectric strength[†] (V m^{-1})	Dielectric constant 50 Hz/1 MHz	tan δ 50 Hz/1 MHz
PE	$>10^{15}$	1.5×10^8	2.3/2.3	$2.0/2.0 \times 10^{-4}$[¶]
PVC[§]	10^{14}	0.5×10^8	4.0/3.0	$1.5/0.3 \times 10^{-2}$
PP	$>10^{14}$	10^8	2.3/2.3	$0.7/2.0 \times 10^{-4}$
PS	$>10^{14}$	1.35×10^8	2.5/2.5	$0.9/0.7 \times 10^{-4}$
ABS	$10^{12} - 10^{13}$	$1.2 - 1.6 \times 10^8$	3.2/2.9	$0.9/1.2 \times 10^{-2}$
SAN	10^{14}	0.95×10^8	3.0/2.8	$0.4/0.7 \times 10^{-2}$
PA	10^{13}[*]	1.2×10^8[*]	3.8/3.4	$0.5/2.5 \times 10^{-2}$

* Lower when plastic is moist.
† If glass fibre is added to increase Young's modulus (see Table 8.4), dielectric strength will be reduced.
§ Values much affected by plasticiser (see Fig. 8.8).
¶ LDPE.

cup to fall to pieces because it was left outside for a few months. Another good example with which you will immediately be familiar is how rubber bands perish after a year or so. This type of consideration controls when we use ceramics and when the generally cheaper and more convenient plastics. Anyway, back to Table 8.3. The electrical properties of polymers, as made, are just as good as those of ceramics. What electrical conduction there is takes place in the same ways as in ceramics: thus there is electronic conduction and ionic conduction, and so polymers have band gaps, just like ceramics. The exact mechanisms in polymers, however, are not as well understood as those in ceramics. This is partly because the science of polymers is younger than that of ceramics, and partly because the molecular structure of polymers is very variable indeed, as I tried to emphasise in Section 8.2. When a voltage is applied across a polymer, the initial current dies away as the charge difference (whether electrons or ions) builds up and no more charge carriers are available. If the current after, say, one second is plotted on a log-log plot against $\frac{1}{T}$ (an 'Arrhenius' plot—see Question 7.6, and Question 8.6 at the end of this chapter) the straight line shows that electrical conduction in polymers is thermally activated ($\propto e^{-\frac{Q}{kT}}$). Q is generally $\sim\frac{1}{2}$ eV per atom ($\sim$48 kJ per mole). Physically the thermally activated processes will correspond to electrons hopping from molecule to molecule or to ions making their way through the plastic.

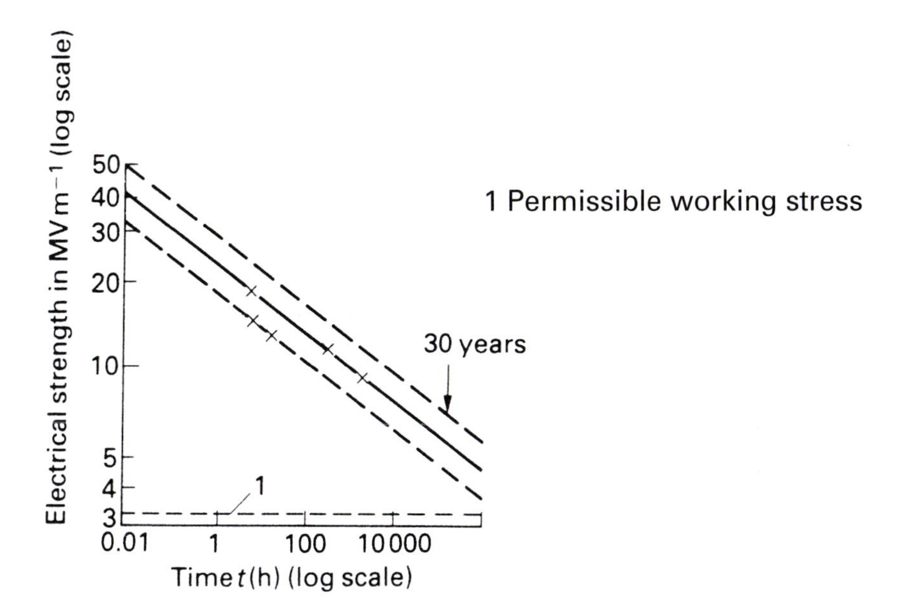

Fig. 8.5 How polythene deteriorates with both the time of exposure and with the time for which the high tension is applied. The upper and lower broken lines refer to set confidence levels. (From *Plastics* by N.J. Mills, courtesy of Edward Arnold Ltd.)

Fig. 8.6 Electrical 'tree' in polythene. (From *Plastics* by N.J. Mills, courtesy of Edward Arnold Ltd.)

The breakdown voltage again depends on heterogeneities (discontinuities) in the structure and so holes, chunks of inorganic filler, glass fibres etc. all bring down the breakdown voltage. In practice breakdown voltages for all electrical insulators are around 10^7–10^8 Vm^{-1}. The breakdown strengths of polymers deteriorate with age. (Fig. 8.5). When the polymer does break down, a *tree* is often formed (Fig. 8.6).

Like ceramics, plastics are generally intended as electrical insulators (including capacitors). Also like ceramics, however, sometimes we actually *want* the plastic to conduct. There are two ways of accomplishing this: *extrinsically* or *intrinsically*. In extrinsic plastic conductors, a conductor is mixed up with the conductor before it is made. The conductor may be metal fibres, or carbon black (i.e. soot). There needs to be enough conductor to form a continuous conducting path through the plastic. Typical uses are packaging for static-sensitive computer chips (see Chapter 9) and plastic bags where there must not be any chance of a spark—for example when they contain ammunition. Another way of reducing the chance of static electricity is to coat the plastic with a conducting wax. This is quite popular, for example, for large stretches of nylon carpet—have you ever walked across the floor of a department store and woken yourself up with an electric shock off the metal stair handrail?

In *intrinsic* plastic conductors it is the plastic itself which is made to conduct. At the moment this is a very fast moving and exciting field. The all-plastic battery is on the horizon. The basic problem is that plastics are made conducting by incorporating some ionic bonding. This also has the effect of increasing chemical reactivity—especially with water.

8.4 AC electrical properties

In Section 7.4 I introduced the concept of capacitance (reminded you of it, probably) and explained the four basic mechanisms which cause insulators to heat up when an alternating voltage is applied. For ceramics, it was ionic polarisation and jumping which made the major contribution to the dielectric constant in the GHz to MHz range. In polymers, with their strong covalent bonds and rather lower packing density, molecular rotation becomes relatively more important as compared with bond stretching.

i.e.

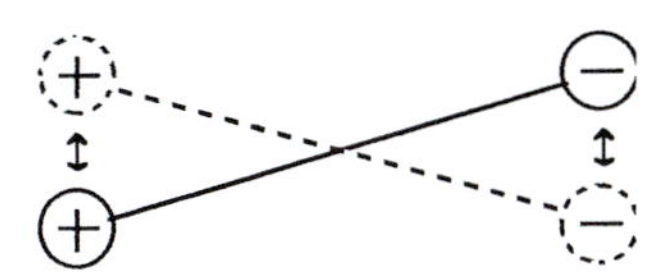

or, more statistically

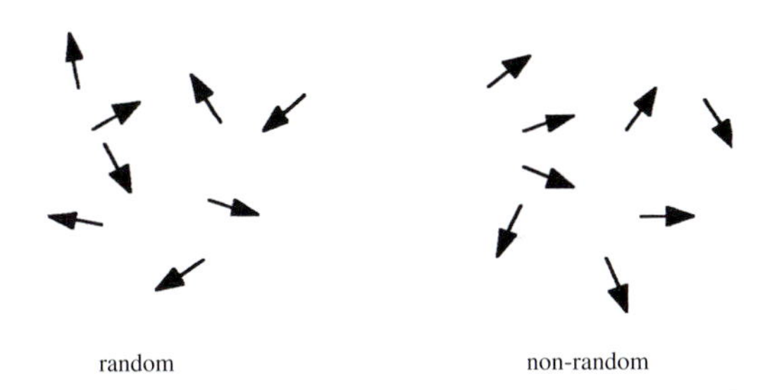

For this mechanism to contribute to polarisation and thus the dielectric constant, the relevant molecule or part of the molecule must be charged. For an atom covalently bonded to carbon, the *ionicity* of the bond will depend on how electronegative the atom is. For example, PVC, with its electron hungry chlorine atoms, heats up so much under AC fields that it is starting to become unfashionable for domestic wire insulation (poisonous chloro-hydrocarbons given off when it burns are an additional reason) and can actually be welded using an RF generator.

Because bond rotation is a heating ($\equiv I^2R$) mode, as with ceramics, this form of polarisation is expected to interact strongly with temperature. Figure 8.7 shows how the loss factor tan δ (see Section 7.4) varies with temperature for low and high density polyethylene. The various peaks are due to the different pairs of bonded atoms in the polymer. Figure 8.8 shows how dielectric constant (labelled in this (borrowed) figure as ε' rather than the ε_R used in Chapter 7) and ε'', the imaginary part of the dielectric constant ($\tan\delta = \varepsilon''/\varepsilon'$), vary with temperature for PVC. You can see why there is a pay-off between the improvement of the mechanical properties of PVC as plasticiser is added, and the resultant electrical properties.

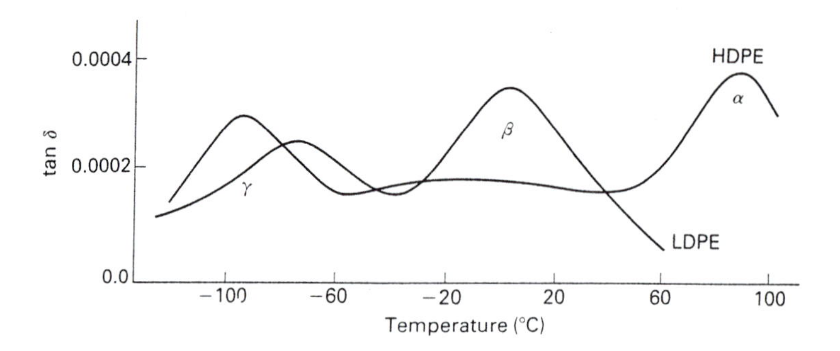

Fig. 8.7 How tan δ for polyethylene varies with temperature. 'HD' means 'High Density' and 'LD' means 'Low Density'. (From *Plastics* by N.J. Mills, courtesy of Edward Arnold Ltd.)

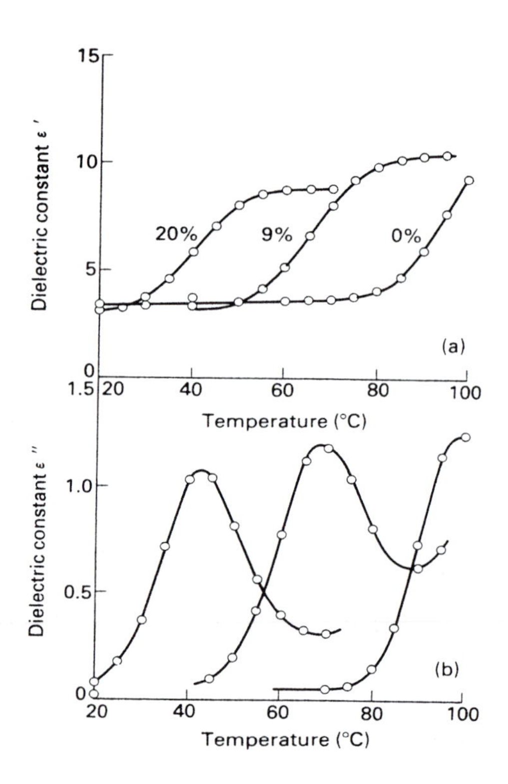

Fig. 8.8 The variation with temperature of dielectric constant (here ε' rather than k) and its imaginary component ε'' (tan $\delta = \varepsilon''/\varepsilon'$) at 60 Hz for three types of PVC. 0, 9, and 20% refer to plasticiser content (From *Plastics* by N.J. Mills, courtesy of Edward Arnold Ltd.)

Polymers are popular materials from which to make capacitors. On to a PS or PP film a few microns thick is evaporated a much thinner layer of aluminium. The polymer is then rolled up and sealed in epoxy resin (thermoset) or in a can. Paper impregnated with an insulating oil is also still used. Polymer capacitors are cheap and versatile, with capacitances ranging from 0.01 pF to 100 μF. For more information about capacitors, see *Electroceramics* by A.J. Moulson and J.M. Herbert (see Recommended Reading at the end of Chapter 7).

8.5 Mechanical properties

As was explained in Section 8.2, thermoplastics are viscoelastic materials. There is a viscous response to stress (like in a liquid) and at the same time an (instantaneous) elastic response. As the temperature rises away from T_g the balance between the two changes in favour of the viscous aspect. Plastics can be tensile tested just like metals (Section 3.5) (see Fig. 8.9). The resulting curve is similar to that for a metal, but involves much lower stresses. Mechanical properties for some common plastics are shown in

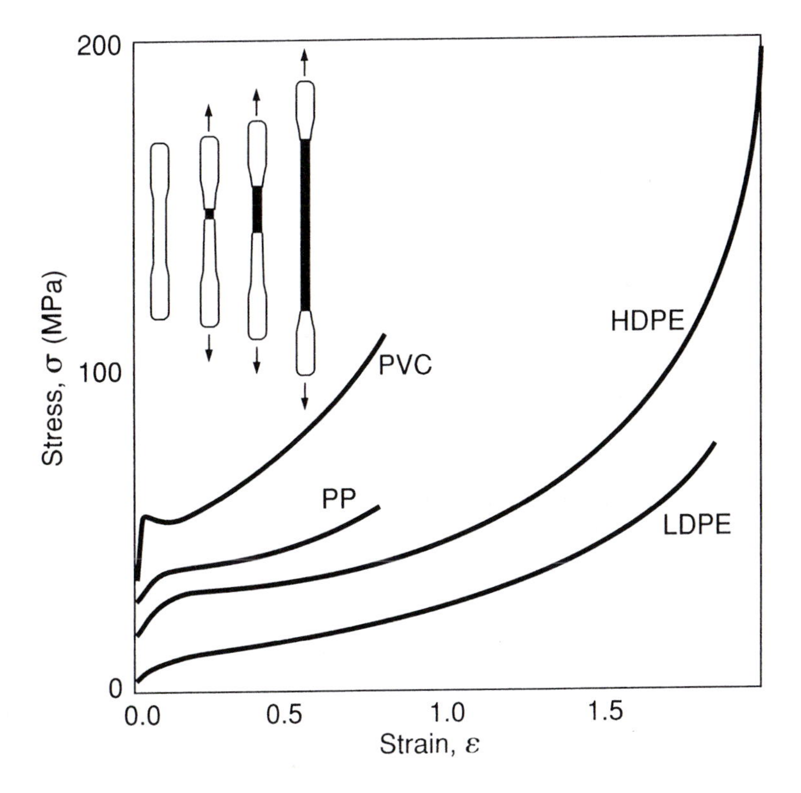

Fig. 8.9 Tensile test curves for a variety of polymers at room temperature. (From *Engineering materials science* by M. Ohring, courtesy of Academic Press, Inc.)

Table 8.4 Typical mechanical properties at room temperature for some common plastics used in electrical and electronic engineering. (Courtesy of BASF AG.)

Polymer	Young's modulus (Pa)*	Yield strength (Pa)	Tensile ductility (%)*
LDPE/HDPE	$0.2/1.0 \times 10^9$	$10/30 \times 10^6$	600
PVC†	3.0×10^9	55×10^6	30
PP	1.5×10^9	35×10^6	500
PS	3.5×10^9	56×10^6	3
ABS	2.3×10^9	48×10^6	20
SAN	3.8×10^9	80×10^6	3
PA	1.6×10^9	60×10^6	150

* The addition of, for example, 25% glass fibre can increase E by ~4 times and reduce ε_L by 30 times.
† Values much affected by plasticiser.

Table 8.4. In the elastic region the gradient of the curve, Young's modulus, E is about 100 times less than that for a metal ($\sim 10^9$ Pa c.f. 10^{11} Pa for a metal). This is because E depends on van der Waals bonds. For a rubber, E is even lower and $\sim 10^6$ Pa, which is caused by, and corresponds to, tangles between molecules.

You will note from Fig. 8.9 that there is a sort of 'yield point' after which the stress rises far less quickly. In metals this point is fairly distinct and the 'plastic' region of the curve corresponds to dislocation motion. The far more disorganised structure of a polymer renders dislocation motion† unnecessary and unlikely, in equal measure.

A defect unique to polymers, especially below T_g, is the *craze*. This is a crack-like cavity within the glassy part of the polymer. The material within the craze has locally adopted a fibrillar structure. Crazes start to form in the elastic region and are thought to be the origin of the final fracture. Fracture toughnesses in polymers rise from fairly low ceramic-type values below T_g to very high values above T_g (see Table 8.2). This is a ductile–brittle transition, like that described earlier for steels (Section 5.2). In this case, though, it is nothing to do with dislocations. Another mechanical property described for metals, but also displayed by polymers, is fatigue (Section 3.8). Clearly, like the ductile–brittle transition, fatigue also is not necessarily or particularly connected with dislocations, but is a general property of plasticity, accomplished by no matter what mechanism.

The important thing to note about plastic deformation of metals and plastics is that after a metal is plastically deformed, the structure is essentially unchanged (particularly after an anneal). This is not true of plastics. There, plastic deformation alters the structure, usually unacceptably. Plastic forming is *not* a permissible processing operation for plastics. We might say that metals are more truly plastic than plastics!

† Polymer crystals, large examples of which can, with difficulty, be made, do indeed deform via dislocations.

(a)

(b)

Fig. 8.10 Some common household plastic electrical components. (Courtesy Caradon MK Electric Ltd.) (N.B. Not the bulbs themselves!)

(c)

Fig. 8.10 (*continued*)

Table 8.5 Typical applications for some common plastics used in electrical and electronic engineering.

Polymer	Typical applications
PE	High voltage and/or high frequency cable insulation
PVC	Domestic cable insulation
PP	Housings for simple household appliances, e.g. coffee makers
PS	Alternative to ABS; high frequency insulation; lamp shades, diffusers
ABS	High quality housings for: radio, TV, video recorders/players, washing machines, telephones, vacuum cleaners, hairdriers
SAN	(Transparent) windows in washing machines; battery casings
PA	Plugs and sockets; terminal strips; switch bases; electric tool housings; coin box telephones

(i)

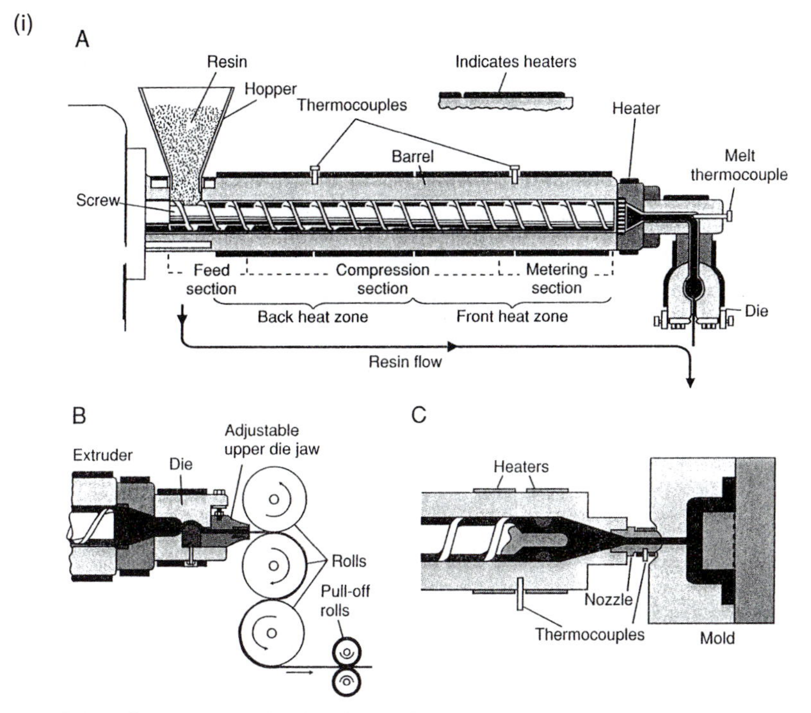

Polymer forming processes based on the use of a screw-type extruder: (A) Extrusion. (B) Extrusion of sheet. (C) Injection molding. (D) Blow moulding a bottle

Fig. 8.11 Common plastic forming processes. (i) based on extrusion, (ii) blow moulding and (iii) pressing and themoforming. (From *Engineering materials science* by M. Ohring, courtesy of Academic Press.)

8.6 Processing plastics

The processing of plastics is quite similar to the processing of glasses, because a similar range of viscosity is used, achieved by raising the temperature. The working range for a plastic corresponds to the viscosity, η, lying between 10^3 and 10^5 Pas. (As a rough comparator, 10^{12} Pas is effectively solid and a liquid metal will have $\eta = 10^{-4}$Pas.) Figure 8.10 illustrates a selection of common household electrical artefacts and Fig. 8.11 shows schematically some common processes. Table 8.5 lists typical applications for the polymers and copolymers which have appeared in previous tables. Other processing methods include pressing and calendering (see Section 7.6).

8.7 Summary

1. A *plastic* consists of a *polymer* plus various additives.
2. There are three types of polymer: thermoplastics, rubbers and thermosets.

(ii)

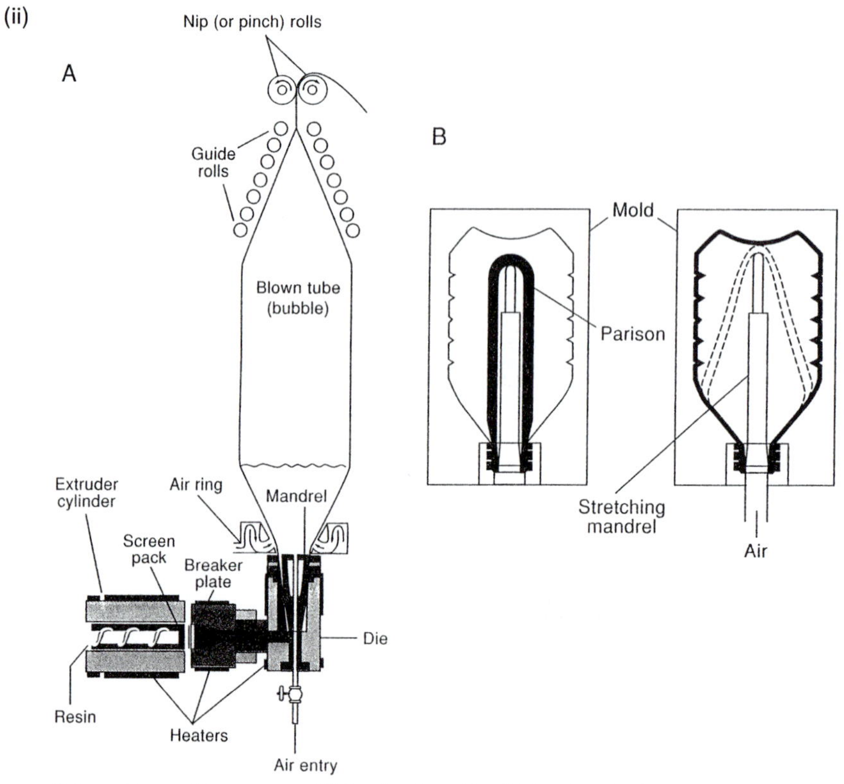

Polymer blow molding processes: (A) Forming blown film. (B) Forming a bottle.

(iii)

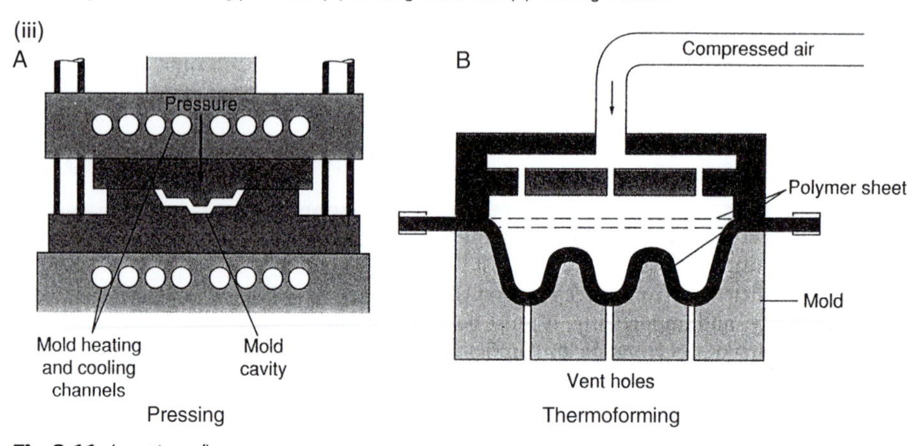

Fig 8.11 (*continued*)

Thermoplastics have linear or branched molecules, rubbers are lightly covalently cross-linked and thermosets heavily cross-linked.

3. Thermoplastics can be remelted and recycled.
4. Electricity is conducted through polymers by electrons and/or ions.

5. Electrical resistance and breakdown voltages are similar to those of ceramics, but polymers degrade more via temperature, sunlight and environment generally, because of the van der Waals bonding.
6. Polymers can be made to conduct electricity intrinsically and extrinsically.
7. Capacitance and power loss mechanisms include molecular rotation, as well as the electronic, bond stretching and space charge mechanisms characteristic of ceramics.
8. Polar polymers, like PVC, are very susceptible to AC heating.
9. The mechanical properties of polymers are time-dependent: polymers are *viscoelastic*.
10. Young's modulus due to van der Waals bonds (i.e. thermoplastics) is $\sim 10^9$ Pa and that due to molecular entanglements (i.e. rubbers) is $\sim 10^6$ Pa.
11. Polymers have ductile–brittle transitions and suffer from fatigue.
12. Manufacturing methods are similar to those for ceramic glasses, but are employed at lower temperatures.

Recommended reading

Many of the general materials texts listed under 'Recommended reading' for Chapter 1 have excellent sections on plastics. See, for example, *Engineering materials science* by M. Ohring (published by Academic Press), Sections 4.1–4.4, Section 7.5 and Sections 8.5–8.6.

More specialist texts:

Plastics by N.J. Mills (published by Arnold)
Introduction to polymers by R.J. Young and P.A. Lovell (published by Chapman and Hall)
Plastics for electronics by M.T. Goosey (published by Elsevier)

Questions (Answers on pp. 325–327)

8.1 (Another question written in my kitchen!—see Question 7.1)
Which of the following are plastics?
(a) Floor tiles
(b) Washing machine
(c) The plug for the washing machine
(d) The wall socket into which the plug fits
(e) The sink
(f) The table

8.2 What is the difference between a plastic and a polymer?

8.3 What is a typical number of mers (monomer units) in a polypropylene molecule? (You will need to look in other books for the answer to this.)

8.4 Are the following thermoplastics, thermosets or rubbers?
(a) polythene
(b) PVC
(c) wood
(d) araldite
(e) perspex
(f) paper
(g) neoprene
Are they crystalline, amorphous or a mixture?

8.5 If a 1 kg weight were hung from a 1 m long 1 mm^2 cross-section PVC wire, by how much would the wire instantaneously extend? (You may find Table 8.4 useful; g, the acceleration due to gravity, is 9.81 ms^{-2}.)

8.6 The DC resistivity of nylon is measured over a range of temperatures with the following results:

T (°C)	ρ (Ωm^{-1})
40	3.16×10^{13}
50	4.37×10^{12}
60	6.31×10^{11}
70	8.71×10^{10}
80	1.51×10^{10}
90	3.09×10^{9}
100	6.49×10^{8}
110	1.86×10^{8}

What is the activation energy of the conduction process?

8.7 A 1 cm $\times$ 1 cm $\times$ 0.1 mm sheet of PS has metal electrodes evaporated onto its square faces. What would the capacitance of the PS sheet be? A 1000 V potential difference is applied between the electrodes. What current would you expect to flow? If the voltage were increased, at what point would breakdown occur?
(The permittivity of free space, ϵ_0, $= 8.854 \times 10^{-12}$ Fm^{-1}.)

8.8 How and from what would you make
(a) a three pin plug
(b) a domestic electric socket
(c) the casing for
- a TV set
- an electric iron
- an electric drill
- a refrigerator
- a washing machine?

Part IV **Semiconductors and other materials**

9

Semiconductors and the electronics industry

Chapter objectives

- What is a semiconductor?
- Why dope semiconductors?
- Why silicon?
- The microelectronic engineer requires . . .
- The four basic interfaces
- Producing a clean single crystal wafer of silicon
- Microlithography
- Deposition and doping
- Packaging
- Problems with chips
- Spin-offs from the microelectronics industry

9.1 Introduction

In Chapter 1 (Section 1.10) I introduced the concept of a *semiconductor*, a covalently bonded solid, like silicon, whose bonding electrons are nearly, but not quite, free and which therefore has a conductivity halfway between that of an insulator and that of a conductor. The *band-gap* of a semiconductor is $<\sim 2$ eV. *Doping* the semiconductor with atoms of the wrong valency can produce violent changes in electrical conductivity, yielding n- or p-type materials (**n**egative or **p**ositive charge carriers). Semiconductors form the basis of the enormous present day microelectronics industry. The phenomena which are exploited go far beyond the simple concepts of doping and conductivity and how microelectronic devices work in electronic terms will occupy a major part of your time over the rest of your course. This is not the purpose of this book, however. What

I am interested in here is the *materials* which are used—how they are fabricated and some of the evil things which can happen to them during use. I shall leave it to someone else to dazzle you with the whole topic of microelectronics, where perhaps theoretical physics and engineering come most wonderfully and magically together.

Of course, I must say just a little about what we are aiming at. I shall describe in the simplest possible terms what the microelectronic engineer requires, and why, and then, in the main part of this chapter, I shall explain how it is currently delivered.

9.2 The four basic interfaces

Because, principally, of its band-gap (1.1 eV) and the sheer momentum of the expertise which has been developed around its processing, silicon dominates the microelectronics market.

Microelectronic devices are designed around four basic building blocks: in fact, four different interfaces (see Recommended Reading: *Materials for semiconductor devices* by C.R.M. Grovenor). These are illustrated in Fig. 9.1.

p and n refer to oppositely doped semiconductors. In a p-type semiconductor, it is *holes* in the valence band which conduct (see Section 1.10); in an n-type semiconductor it is electrons. The p–n junction is inherently rectifying, in that current can flow easily in one direction (involving both electrons and holes travelling in opposite directions, of course) but not in the other.

In an oxide–semiconductor junction the oxide exerts a field on the semiconductor, via its stored dipole moment (see Section 7.4). This can affect and control the conductive properties of the semiconductor.

Metal–semiconductor junctions are of two types. The common type has a barrier which the electrons must overcome. This type is called *Schottky*. Like the p–n junction it is a rectifier. The *ohmic* contact has no such barrier (or at least it is irrelevantly small) and shows normal resistive behaviour (hence the name).

Using these simple building blocks, microelectronic devices with a huge range of complexity can be built up. Figure 9.2 shows four very simple ones, whose names may be familiar to you.

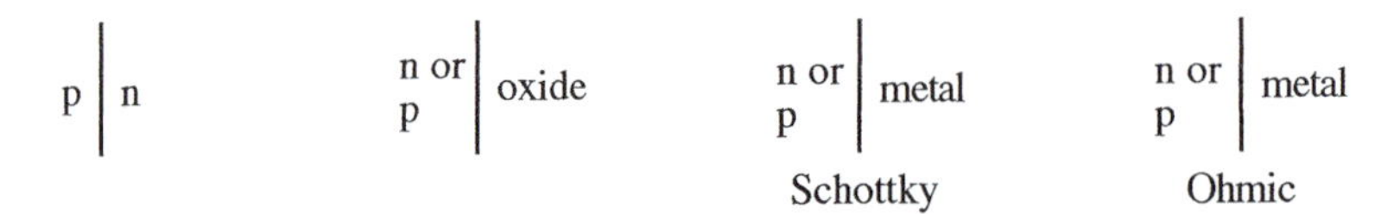

Fig. 9.1 The four basic interfaces of microelectrics (see Recommended Reading: *Materials for semiconductor devices* by C.R.M. Grovenor).

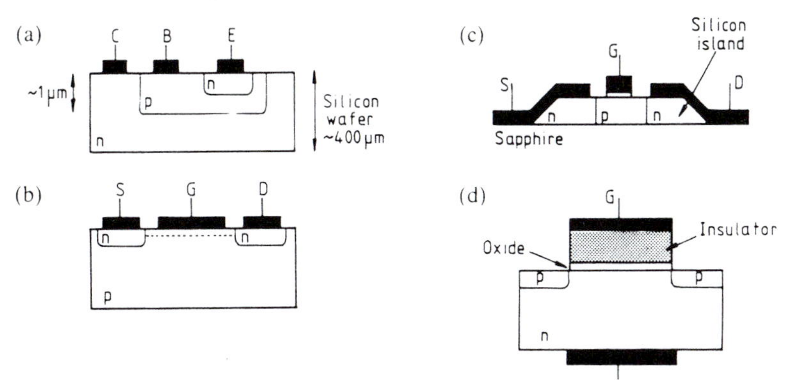

Fig. 9.2 Schematic illustrations of the structure of some simple semiconductor devices: (a) an npn bipolar transistor in a silicon wafer; (b) an n-channel field effect transistor; (c) a silicon-on-insulator field effect transistor; (d) an MIOS memory device (from *Microelectronic materials* by C.R.M. Grovenor, by kind permission of Adam Hilger).

What, then, is the microelectronic engineer asking for? The different regions in the devices of Fig. 9.2 must have clearly defined electrical properties and clearly defined boundaries (thus interfaces). They have to be connected electrically to the outside world. They have to be *packaged* in such a way that they can be handled easily and will not damage or deteriorate during use. Finally, and most notoriously, they should be small. They should be small because heat losses are smaller, because communication times between individual devices are smaller, but supremely because a centimetre^2 of silicon on which is a million transistors is an altogether more attractive engineering proposition than a houseful of small individual silicon blocks.

How does the materials engineer satisfy these requirements? It is evident from Fig. 9.2 that current technology is essentially a *surface* technology.† The reason for this will become evident in what follows. The starting material for fabricating devices such as those shown in Fig. 9.2 is a *wafer* of high purity silicon and my first step, therefore, is to describe how this is produced. I shall then go on to deal with the other steps in fabricating a microelectronic circuit.

9.3 Fabrication

Producing a silicon wafer

What does 'high purity' mean? Average doping levels are at about the *parts per million* (ppm) level and so impurity levels of electrically active elements must be considerably

† More three dimensional structures are gradually achieving a greater significance at the very top end of the market.

less than this. A silicon wafer is a thin disc of very pure silicon which has been cut from a large cylindrical single crystal. Silicon (Si) is mined in the form of *quartz*, which is a form of *silica* (SiO_2). (Silica is a very common everyday material: most glass is silica based (see Section 7.2; optic fibres are silica—see Section 11.4).) The quartz is reduced to silicon using coke:

$$SiO_2 + 2C \rightarrow Si + 2CO$$

The resulting *metallurgical grade* silicon is rather impure. It is dissolved in hydrochloric acid (HCl) to give $SiHCl_3$ (a low boiling point liquid) which is purified by distillation and then reconverted into very pure *electronic grade* silicon using hydrogen. 'EGS' contains ~5 ppm of the potentially harmful transition elements (Fe, Cr, Mn . . .). At this stage the silicon is polycrystalline and the next move is to convert it to a large perfect single crystal. This is because grain boundaries and dislocations are electrically active and would spoil the performance of any microelectronic device.

There are several methods for producing single crystals: I shall describe just one of them. This is the *Czochralski* method. It is shown both schematically and for real in Fig. 9.3, along with the final product. The *principle* of the Czochralski method is fairly straightforward. The silicon is melted in a large silica crucible. A *seed* crystal is lowered into the molten silicon and then slowly withdrawn. The silicon solidifies on the bottom of the seed crystal *epitaxially* (with the same orientation) and the final large crystal is therefore a magnified version of the original small seed crystal. Clearly the speed of withdrawal is important: too fast and the seed crystal and melt will lose contact with each other; too slow and the seed crystal will melt into the liquid silicon. The seed crystal and the crucible are slowly rotated (~1 rpm) in opposite directions. This is to distribute homogeneously the final remaining impurities and to ensure that they collect mainly at the end of the single crystal. The slow swirling action also has the effect of dissolving the walls of the crucible (silica, SiO_2) into the silicon and so Czochralski silicon is saturated in oxygen (10^{18} oxygen atoms/ cm^3) (see Fig. 9.4). Luckily the oxygen is electrically inactive and therefore harmless.

I said above that the gentle stirring action helped the more dangerous impurities, like iron, to the end of the bar. We can understand why this is by reference to the *phase diagrams* which I introduced in Section 4.5 and Fig. 4.10. Most of those for silicon and another element (e.g. Fe) look like the example in Fig. 9.4. If the overall concentration of the impurity X in the molten silicon is c_0, then the first solid to form has a concentration of $X = kc_0$, distinctly less than c_0. For very slow cooling, of course, as we pass through the two phase liquid + solid region, solid state diffusion would slowly bring the concentration of X in the solid silicon back to c_0. In the Czochralski process, though, the cooling rate is such that diffusion in the liquid (helped by convection from the stirring action) can occur, but not to any significant extent in the solid. The concentration of X (e.g. Fe) in the liquid slowly rises, but most of it is purged, as I mentioned above, to the end of the bar, resulting in a concentration profile like those shown in Fig. 9.5.

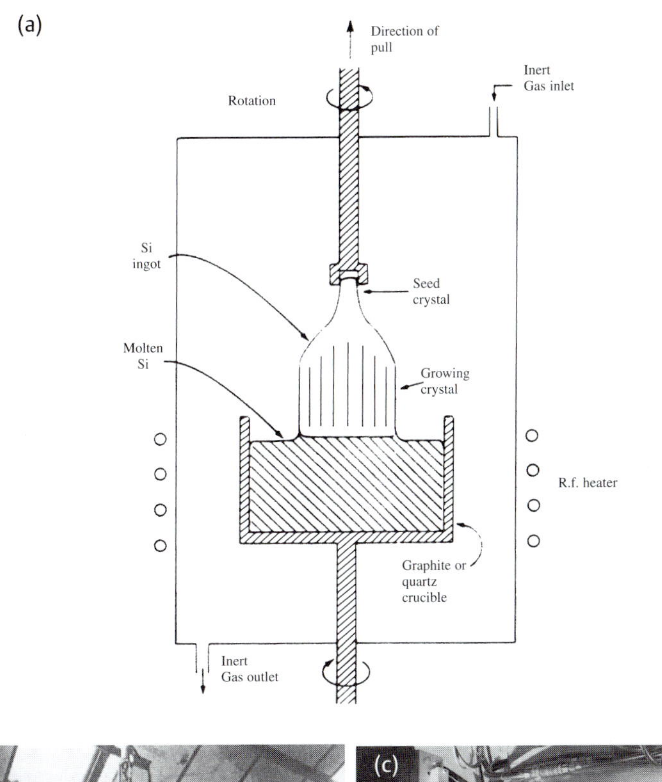

Fig. 9.3 The Czochralski process for growing single crystals of silicon (a) schematic (b) the real thing and (c) a single crystal of silicon, the main part of which contains no dislocations and parts per billion only of harmful impurities. (c) Courtesy of Wacker Siltronic AG, Burghausen, Germany)

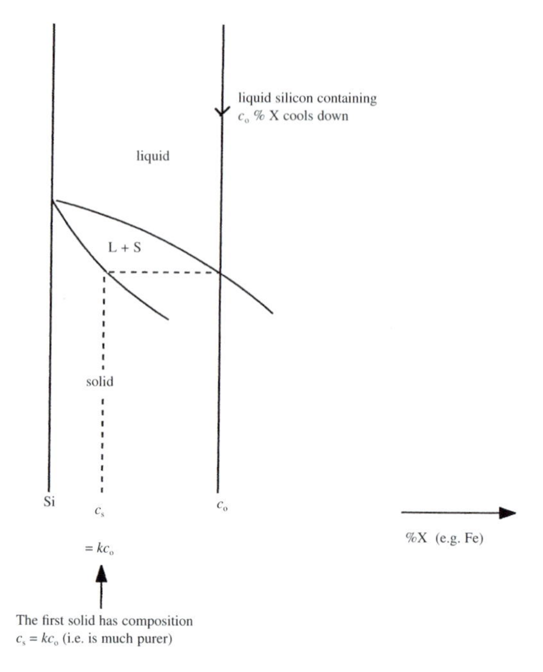

Fig. 9.4 The purification of silicon as it solidifies. X is some impurity, for example iron. This is the silicon end of the Si–X *phase diagram* (see Section 4.5).

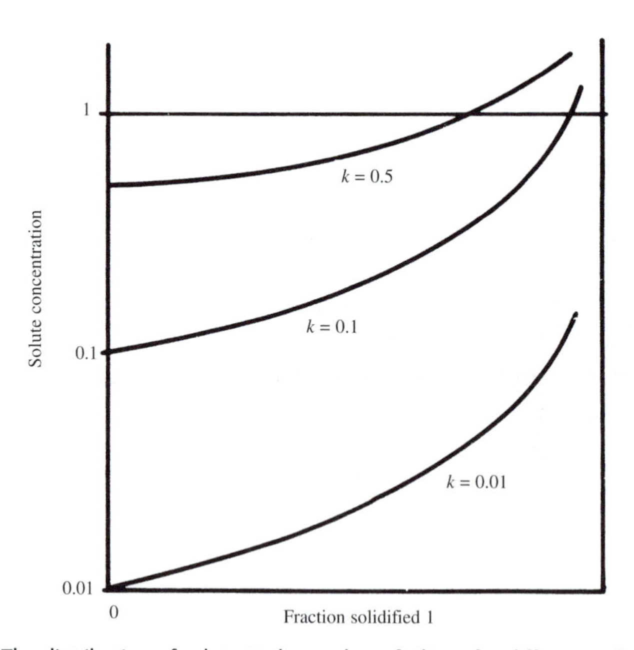

Fig. 9.5 The distribution of solute X along a bar of silicon for different values of the *partition coefficient k*. In fact $c_s = k\,c_0\,(1 - x)^{k-1}$ (note the log scale on the graph.) ($x = 0$: beginning; $x = 1$: end of bar.) (From *Materials for semiconductor devices* by C.R.M. Grovenor, courtesy of The Institute of Materials).

Typical values of k are

P 0.032   C 0.07   Al 0.02   Fe 10^{-5}

Oxygen, at $k = 1.25$ the odd man out, is actually concentrated slightly towards the beginning of the bar.

Figure 9.3(c) shows the final result. This single crystal is over a metre long and $\sim$30 cm across and, apart from the beginning and final few centimetres, it is entirely free of dislocations. As good a testament to 20th century technology as you will find and unimaginable to Bardeen and Shockley when they were performing their epoch-making experiments on germanium in the late 1940s in America.

The end of the bar containing most of the impurities is chopped off and the sides polished until the bar is cylindrical. A flat is ground along one side of the bar parallel to a certain crystal plane. The bar is then cut using a diamond saw into slices a fraction of a mm thick, called *wafers* (Fig. 9.6). One side of the wafer is polished to a mirror finish.

Fig. 9.6 A 300 mm wafer. The industry standard is currently (2000) changing from 200 mm to 300 mm. This example has already gone through the next few processing steps. (Courtesy Semiconductor Leading Edge Technologies (Selete) Inc.)

Lithography and deposition

We are now ready to produce the very small devices (transistors etc) which make up integrated microelectronic circuits. The smallness of these devices is achieved by *lithography*. The best way to explain this is via an example.

A *diode* (Fig. 9.7) is one of the simplest, most basic and most common building blocks which make up an electronic circuit. It uses two of the basic interfaces I introduced in Section 9.2, the p–n junction and the n-ohmic metal junction.

We start with a p-type wafer. The dopant (e.g. boron) will have been introduced during the Czochralski process. The wafer is oxidised by heating it in a mixture of oxygen and water vapour at around 1000 °C to produce a SiO_2 layer of the order of a micron thick. The surface of the wafer is then covered with a *photoresist*. This is a polymer which is sensitive to UV radiation. The wafer is spun and the polymer solution dripped on. This produces a thin uniform layer. The photoresist is baked to drive off the solvent and then irradiated by UV light through a *mask*. The mask defines the current (in the sense of 'now') stage of the circuit. The IC (integrated circuit) containing our p–n diode is originally drawn by computer, reduced photographically and then reduced again and at the same time stepped sequentially in two dimensions so that the final mask, which consists of chromium on a flat piece of glass, contains an array of many identical circuits. The glass plate is placed in contact with the photoresist and irradiated with UV light (Fig. 9.8). Alternatively, one circuit mask is used and the final projection and stepping to produce

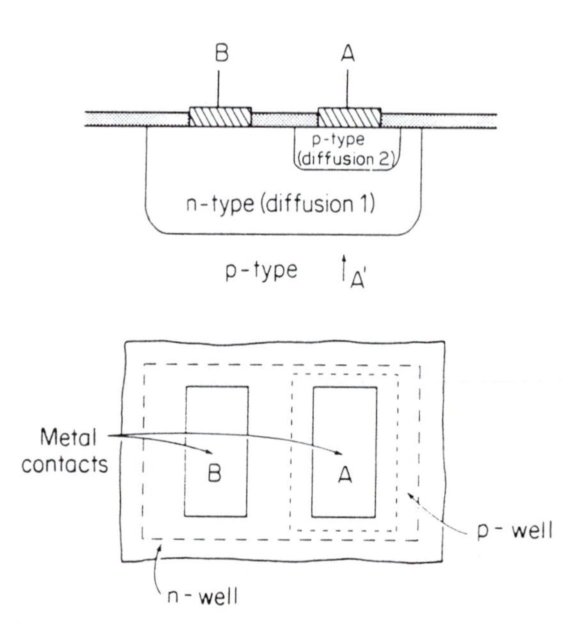

Fig. 9.7 A schematic diagram of a single p–n diode (from *An introduction to semiconductor microtechnology* by D.V. Morgan and K. Board, courtesy of John Wiley & Sons Inc.).

Fig. 9.8 PAS 5500/700 Step & Scan Lithography System (courtesy of ASML Inc.).

the multiple identical circuits are performed directly on to the photoresist. UV is used for the final patterning stage because the smaller wavelength of UV as compared with light permits greater spatial resolution (i.e. smaller devices) with details down to a fraction of a micron. Now comes the clever bit. The UV light causes the photoresist to *cross-link* and form a thermoset (see Chapter 8, Section 8.2). The photoresist is next dissolved in a solvent, but the cross-linked regions have become insoluble and are not affected.The exposed SiO_2 is etched away, for example with hydrofluoric acid. The temperature of the wafer is raised and the remaining photoresist burnt off. *Positive* photoresists, which work exactly the opposite way round, are also available.

The n-type region is created by *doping* the p-type silicon with, for example, arsenic. This can be done in two ways—either gaseously or by ion-implantation. In gaseous doping a gas containing the dopant (e.g. arsine, AsH_3 for arsenic) is allowed to contact the hot wafer. This decomposes the arsine and a layer of arsenic forms on the surface and then diffuses into the silicon. In ion implantation arsenic atoms are ionised and then fired into the surface of the silicon wafer. This can be done at a variety of voltages and so a fairly exact depth profile is built up (see Fig. 9.9). Ion implantation requires an anneal afterwards. This is because firing the ions in makes the silicon amorphous, or at

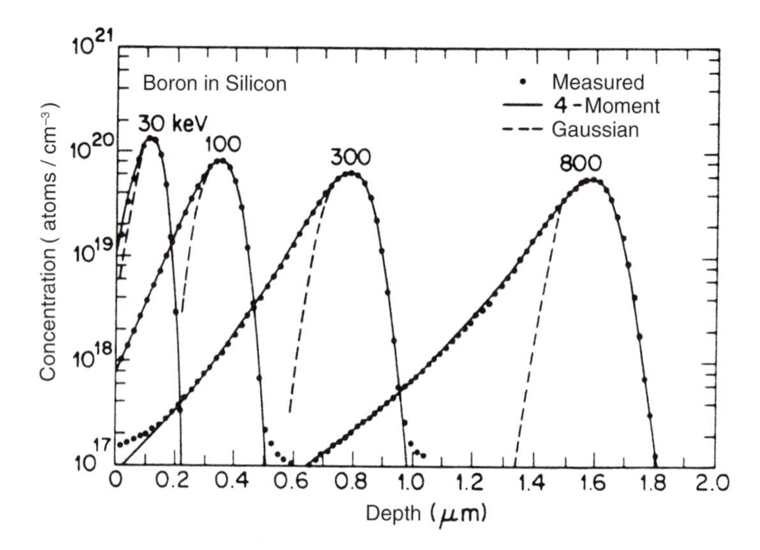

Fig. 9.9 Profiles of implanted boron atom distributions. The numbers above the curves refer to beam energies in keV. By combining different (sequential) implantations one with another, the desired profile can be built up. (From *VLSI technology* edited by S.M. Sze, courtesy of McGraw-Hill.)

the very least leaves the arsenic atoms in *interstitial* positions where they are not electrically active. Annealing carries with it the danger of smearing out carefully created and abrupt composition profiles: special *flash annealing* techniques (for example involving high intensity flash guns) which take a fraction of a second, have been developed to avoid this. The p-type region is now created using, for example, a boron implant. Since a separate mask is used it must be aligned very accurately with the preceding one. This is achieved via registration marks which form part of the mask.

The ohmic contacts can now be deposited. These might be aluminium or gold and they can simply be evaporated on to the wafer. In a reasonable vacuum the aluminium or gold source is heated and the wafer kept cool. Sometimes the metal is *sputtered* using Ar^+ ions. The aluminium or gold condenses on to the wafer. The SiO_2 and excess metal are dissolved off and we are now ready for the final step—the evaporation of aluminium–copper–silicon tracks which connect the different microcomponents one with another. Aluminium–4%Cu–1.5%Si has been a very common *interconnect* for a variety of reasons, not least because it bonds easily to SiO_2 and is corrosion resistant. This step can be performed directly on to a photoresist pattern: oxidation and doping, because they involve heat or radiation, cannot.†

† Al–4Cu–1.5Si is currently being replaced as an interconnect by copper. Copper, which is the obvious metal to choose (see Chapter 4) has not been used hitherto because it could not be prevented from diffusing into the silicon, thereby ruining its electrical properties. It has now been discovered, however, that a thin layer of tantalum (Ta: a heavy transition metal element) or tantalum nitride applied before the copper prevents it from diffusing into the silicon. The copper can be deposited by CVD (chemical vapour deposition), or, for thick layers (say >0.5 μm) by electroplating (see Chapter 6).

Finally a thin protective layer of SiO_2 is deposited by CVD (chemical vapour deposition) all over the IC surface, except for the metal contacts. Silane, SiH_4, is burnt in oxygen and the wafer is held at 300 °C or 400 °C.

There are two ways of joining the interconnect pads to the outside world. If the chip is to sit on a substrate as part of a hybrid circuit, ordinary lead–tin solder will probably be used (see Section 4.5). Solder will not *wet* the aluminium alloy pad and so thin intermediate layers of Cr and Cu are evaporated on first. The solder may be applied via the *flip-chip* method (see Fig. 9.10).

The other way of doing it, when the chip is to function in a stand-alone capacity, is to attach a gold or aluminium wire, which is heated and may be vibrated ultrasonically to scour away the aluminium oxide (Al_2O_3) film and establish a true metal–metal contact.

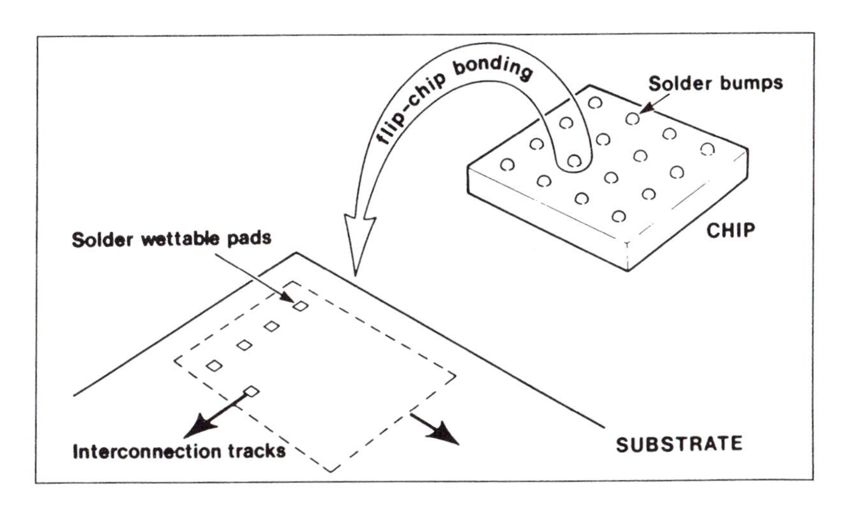

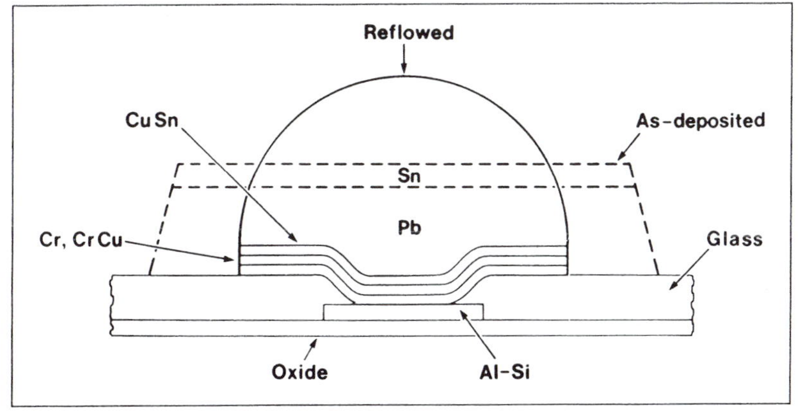

Fig. 9.10 The flip-chip solder bonding method. The chip is provided with an array of solder-wettable pads, on each of which a bead of solder is attached. The chip is 'flipped' and contact made with a similar array on the substrate. (Courtesy The Institute of Metals.)

Finally, the individual IC's are separated one from another by scribing the surface of the wafer and then cleaving it along the scribe by applying a bending force.

During this section we have encountered a number of very important unit operations:

- photolithography
- etching
- CVD[†]
- oxidation
- doping
  - gaseous
  - implantation
- evaporation
- sputtering

Although I have illustrated these by reference to fabricating a p–n diode, the same operations are involved in producing *any* microelectronic circuit. Furthermore, the production route I have chosen is a fairly arbitrary one. Many variations are possible.

The IC's must finally be *packaged*.

Packaging

The electronic devices as we left them at the end of Section 9.3.2 were hopelessly exposed and too sensitive to use. *Packaging* is the process whereby the microelectronic device or chip, whose manufacture we have been following, is made into a form in which it can be handled and used (and sold!). There are two basic forms: metal/ceramic and plastic. Consider a single chip package (see Fig. 9.11) which we first encountered in Chapter 6. The chip is mounted on a metal plate which itself is mounted in a glass base. The contact pads on the chip are connected to Kovar pins (Kovar is an iron–nickel–cobalt alloy with the same thermal expansion coefficient as the glass: less likelihood of changes of temperature causing cracks) via thin gold wires. These are welded to the chip pads via a thermomechanical method (heat and ultrasound). A metal can is sealed on to the glass under vacuum or in an inert gas atmosphere.

The Dual-in-line (DIL) we also met in Chapter 6. The chip is mounted on a plate (e.g. Kovar) which is mounted on a lead frame. It was the creation of the lead frame and contact pads which was described in Section 6.3, Example 4. The wires are

[†] You may also come across *PVD*—Physical Vapour Deposition—and *MOCVD*—Metal Organic Chemical Vapour Deposition.

Fig. 9.11 (a) Right: single mounted chip; left: with can removed. (b) Dual-in-line (DIL) next to 13A fuse.

thermomechanically bonded at either end and the whole package is *potted* in a thermosetting plastic (see Section 8.2). A thermoset is used because it provides a better barrier against moisture. As you can imagine, the chip and especially the wires are easily damaged. A gentle form of injection moulding (see Section 8.6) called *transfer moulding* is employed.

Fig. 9.12 Micromechanics, one of the spin-offs from the micrelectronics industry. This is a 10 μm long cantilever cut from a 500 nm thick silicon nitride layer using a focused ion beam (FIB). The contact at the end is FIB deposited tungsten $2 \times 2 \times 1$ μm. The cantilever has a resonant frequency in the MHz range and is being developed as a wetted microrelay for fast current switching. (Courtesy Professor P.D. Prewett.)

9.4 Problems

Most of the problems which microelectronic circuits suffer from we have met earlier in this book. For example, the continual heating and cooling of parts of the circuit lead to *thermal fatigue* (see Section 3.8), particularly of the soldered joints. These also suffer from the formation of *intermetallic compounds*—for example the exotically named 'purple plague'—a gold–aluminium compound formed where gold and aluminium are in contact (Section 4.5, p. 115). There is a multitude of *corrosion* problems (Section 6.4) because of the variety of metals in contact with each other. One problem which perhaps is more specific to microelectronics is *electromigration* where the electric currents cause chemical segregation (remember the currents themselves may be quite small, but the current *densities* can be quite large). Finally I should point out that the enormously precise and sophisticated and delicate techniques which have been developed over the last fifty years to create the vast microelectronics industry have provided spin-offs which themselves have nucleated new industries. Just to give one example, photolithography and etching techniques have spawned the new field of *micromechanics*, where tiny machines, whose total size may be a fraction of a millimetre, can be constructed: the idea of a small submarine which can investigate and repair the arteries round your heart is no longer science fiction (Fig. 9.12).

9.5 Summary

1. A semiconductor is an insulator with a band-gap of $<\sim 2$ eV.
2. Doping produces either n- or p-type semiconductor.
3. Silicon dominates the microelectronics industry.
4. The four interfaces basic to all microelectronic devices are (a) p:n, (b) p/n:oxide, (c) p/n:metal (Schottky) and (d) p/n:metal (Ohmic), where p and n refer to p- and n-type semiconductors.
5. Silicon is prepared by reducing silica, SiO_2, refining to electronic grade silicon and growing into a single crystal by, for example, the Czochralski process. Zone refining reduces impurity levels to parts per billion.
6. In microlithography, UV light shone through a mask crosslinks a photoresist to a thermoset. The remaining photoresist is dissolved. Device sizes down to $\frac{1}{4}\mu$m are possible.
7. SiO_2 is deposited by oxidation or CVD.
8. Metals (contacts or interconnects) may be evaporated directly onto the photoresist pattern.

9. Doping is achieved gaseously or by ion implantation through a SiO_2 pattern. Ion implantation must be followed by annealing.
10. The chip is connected either to a substrate by solder (e.g. flip-chip) or to a lead frame by Al or Au wires.
11. Chips are packaged in either a thermoset or a ceramic.
12. Among the many spin-offs from the microelectrics industry are micromechanical devices.

Recommended reading

Microelectronic materials by C.R.M. Grovenor (published by Adam Hilger)

Materials for semiconductor devices by C.R.M. Grovenor (published by The Institute of Materials)

An introduction to semiconductor microtechnology by D.V. Morgan and K. Board (published by John Wiley & Sons Inc.)

VLSI technology edited by S.M. Sze (published by McGraw-Hill, Inc)

Questions (Answers on pp. 327–328)

9.1 Name four semiconductors.

9.2 Which dopants are used for silicon?

9.3 If electronic grade silicon contains 1 ppm (part per million) Fe, how much would remain in the middle of a bar after one zone-refining pass?

9.4 What accelerating voltage is necessary to implant boron into silicon to a depth of 1 μm?

9.5 Why, until recently, was copper not used as an interconnect?

9.6 What are typical impurity and doping levels for silicon?

9.7 What is a negative photoresist?

9.8 Name a positive photoresist, a negative photoresist and a potting compound.

The materials described in Chapters 10 and 11 involve magnetic, superconducting and optical applications. They cover the whole range of material types which we have encountered in earlier chapters.

10

Magnetic materials

Chapter objectives

- The parameters of magnetism: field, induction, magnetisation, permeability and susceptibility
- Diamagnetism, paramagnetism and ferromagnetism
- B-H curves: measurement, origin and significance
- Soft, medium and hard magnets: examples and applications

10.1 Introduction

Magnetism is very important to electrical engineers. It is also tremendous fun—basically, I think, because something that can't be seen can make things suddenly jump around. Magnetism is electricity on the move, just as moving magnetic fields generate electricity. This symmetry is enshrined in the very beautiful Maxwell equations—perhaps the high point of Victorian science. Physically, magnetism is the effect of special relativity on electric charge. When an electric charge moves relative to an observer its space frame contracts, its charge doesn't and magnetism is the imbalance. That is why the speed of light keeps turning up like a magic genie all through the equations linking magnetism with electrostatics.

Magnetism is of enormous technical importance. Motors, generators, transformers—the very heart of electrical engineering—all depend on magnetism, and the choice of magnetic material is crucial. In this chapter I shall be describing how we choose between magnetic materials—which criteria we use—and why one material is suitable for a motor, but another for a transformer. In common with the other chapters in this book, I shall also explain a little about the manufacturing process—about how magnetic materials are processed. In order to pick the right material for the job, we need to be able to assess its magnetic properties and consider what we are asking it to do. How do we define and measure magnetic properties? That is what Section 10.2 is about.

10.2 Magnetic parameters

How do we describe magnetism and magnetic materials? We need some suitable magnetic parameters. The most important of these is

magnetic field H

A magnetic field is produced by a moving electric charge. It has the following effects:

- it exerts a force on a moving electric charge (including a current in a wire) perpendicular both to the direction of movement and to the direction of the field (10.1)
- it exerts a torque (a twisting force) on a magnetic dipole, such as a bar magnet, or a loop of electric current, or an electron orbit in an atom (10.2)
- when it changes in time, an electrical potential is produced (10.3)

All of these effects are described quantitatively by Maxwell's equations. By convention, the strength of a magnetic field is defined in terms of the way it is produced. A convenient method which provides a uniform magnetic field is via a long *solenoid* (see Fig. 10.1). This is a spiral of wire wrapped round a cylindrical former; the homogeneous field is produced at the centre of the solenoid. The magnitude of the field depends on the electric current and on how many turns of wire there are per unit length.

$$H = ni \tag{10.4}$$

where i is the current and there are n turns of wire per unit length. The units of H are therefore chosen to be amps m^{-1}.[†]

If, instead of defining magnetic field in terms of how we produce it, we define it in terms of what it does, we name the resulting parameter *magnetic induction*, B, because it is a result, not a cause. B is measured in *tesla* (T). A tesla is defined as that induction which will produce a force of 1 N per metre on a wire carrying a current of 1 amp

[†] Table 1.1 of *Introduction to magnetism and magnetic materials* by D. Jiles (see Recommended reading) gives some examples of typical field strengths in a variety of situations.

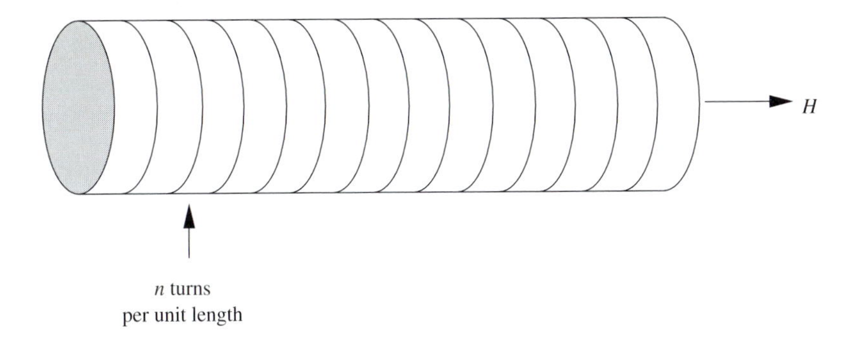

Fig. 10.1 A uniform field is produced using a *solenoid*: a spiral of wire, usually supported on a neutral former.

perpendicular to B (see effect (10.1) above). Equivalently, the tesla can be defined as that induction which, when it passes through a loop of conductor 1 m^2 in area and is reduced uniformly to zero in 1 s, produces (for that second) an electrical potential of 1 V (see effect (10.3) above). Similarly, a definition in terms of effect (10.2) could be devised.

Maxwell's equations incorporate the following relationship between B and H in a vacuum:

$$B = \mu_0 H \tag{10.5}$$

using B and H as defined above. μ_0 is called the *magnetic permeability* of free space. Its magnitude is $4\pi\ 10^{-7}\ \mathrm{VsA^{-1}m^{-1}}$ or Henries $\mathrm{m^{-1}}$ or $\mathrm{Hm^{-1}}$.

When a piece of solid material is introduced, B can be very much altered.

(Notice how similar all this is to my description of electrostatic polarisation, with

$$\begin{aligned} B &\equiv P \\ \mu &\equiv \varepsilon \\ \text{and} \quad H &\equiv E. \end{aligned}$$

The terminology is, in the main, merely transposed also.)

Then

$$B = \mu H \tag{10.6}$$

where μ is the permeability of the material concerned. The *relative permeability* is $\mu_r = \frac{\mu}{\mu_0}$. The reason for using induction, as defined above, is very much connected with eqn (10.5) and the effect of a material on a magnetic field (or *vice versa*, if you prefer). μ, B and H are actually all we need for the whole of this chapter, but another parameter, called *susceptibility* and written χ is sometimes more convenient to use than μ, and also you may come across it in your reading and need to know what it means.

Table 10.1 The major magnetic parameters expressed in SI units (used in this book) and the older cgs units and how to convert between them.

		SI (this book) $B = \mu_0(H + M)$	cgs $B = H + 4\pi M$	cgs → SI Multiply by:
Field	H	A m^{-1}	Oe	79.58
Induction	B	T	G	10^{-4}
Magnetisation	M	A m^{-1}	emu cc^{-1}	1000
Permeability	μ	H m^{-1}	—	
Susceptibility	χ	dimensionless	emu cm^{-3} Oe^{-1}	
Energy product	$(B\ H)_{max}$	kJ m^{-3}	MG Oe	

We write

$$\begin{aligned} B &= \mu H \\ &= \mu_0 (H + M) \end{aligned} \tag{10.7}$$

where M is the *magnetisation* of the material.
Then the susceptibility χ is defined by

$$M = \chi H \tag{10.8}$$

χ is the amount by which a material *amplifies* an applied field. The technically important materials to be discussed in this chapter have susceptibilities from 1–1 000 000.

Magnetism is a minefield of conventions which, as a practical engineer, you can regard with a healthy contempt. If your hand is trapped between two permanent magnets—this can happen now that magnets can be made so strong—then it doesn't really matter whether the bruise was caused by a Gauss, an Oersted or a Tesla! Whenever you feel yourself getting confused by magnetic units, jump back on to the dry ground of real effects which you can see and feel and measure.

Table 10.1 relates the SI system of units, which I have used, to the older cgs system, which you may still come across (see what I mean!).

10.3 The three types of magnetic material

There are three ways in which materials respond to magnetic fields. They are classified according to susceptibility χ:

	χ
Diamagnetism	small and negative
Paramagnetism	small and positive
Ferromagnetism	large and positive

Everything displays diamagnetism. Several materials display paramagnetism and a few materials are ferromagnetic. If a material is paramagnetic or ferromagnetic the diamagnetism is present but is hidden by the stronger para or ferro magnetism. All of these effects come from the electrons within the material. I described in Chapter 1 how the electrons orbit the nucleus in an atom. Their angular momentum about the nucleus is specified by the second of the four quantum numbers l (see Section 1.3). They also spin about their own axes as defined by the fourth quantum number s. Both of these rotations combine to give the atom a magnetic field and dipole moment.

In diamagnetic materials there is no magnetic field. The contributions of the orbital angular momenta of all the electrons in the atom cancel out, as do the spin contributions. When a magnetic field H is applied, the electron orbits tilt slightly and *precess*, producing a small magnetic field which opposes the applied field, but rather weakly. χ is therefore small and negative—typically 10^{-5}. The diamagnetic response is always present. When the atom *does* have a resultant magnetic field from either or both of l and s then when a field H is applied the atoms rotate and *augment* the field slightly.

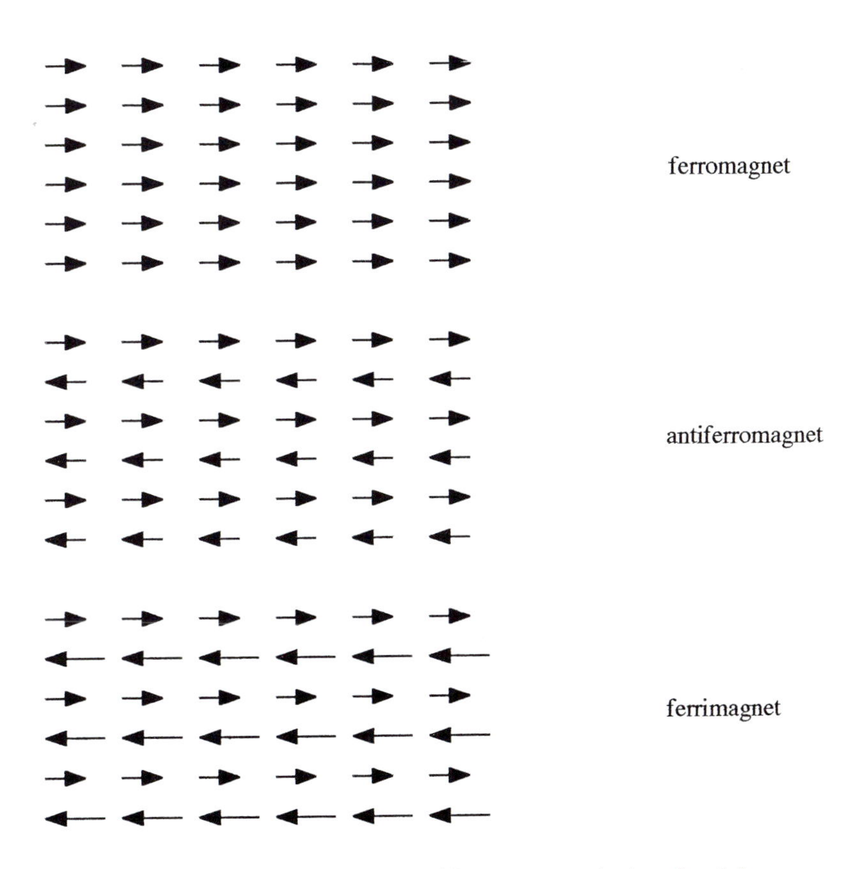

Fig. 10.2 Ferromagnets, antiferromagnets and ferrimagnets. The lengths of the arrows represent the strengths of the atomic magnets.

χ is therefore small and positive—typically 10^{-5}–10^{-3}. This is called *paramagnetism* (para: parallel; dia: oppose)[†].

For a paramagnetic solid, for particular diameters of the electron orbits and for particular separations between the atoms, it becomes energetically favourable[‡] for all the little atomic magnets to line up either parallel or antiparallel to each other. If parallel, this produces an enormous external magnetic field and we call this sort of material *ferromagnetic* ('ferrum' is Latin for iron, a common example—see also Section 5.2 and Question 5.4) (see Fig. 10.2). If the atomic magnets line up antiparallel to each other and cancel out exactly, the material is called *antiferromagnetic*. If the atomic magnets line up antiparallel to each other, but do not cancel out, because they have different magnetic moments—usually because there are different numbers of a given magnetic atom pointing in the two directions (parallel and antiparallel)—the material is *ferrimagnetic*.[*] All technologically important magnetic materials are ferromagnetic or ferrimagnetic. They have susceptibilities typically 10^2–10^4.

10.4 Measuring magnetic properties

A material's magnetic properties, which control its applications, are defined in terms of its B–H curve. Field H is applied and the resultant induction B is measured. This is typically performed in a permeameter (Fig. 10.3). A typical measured B–H curve might look as shown on the right of Fig. 10.4.

Ferromagnets as made tend to have no obvious external magnetic field—i.e. $B = 0$. This isn't because the little atomic magnets are not lining up: it is because they do so over small volumes called *domains*. These are oriented with the direction of magnetisation at various angles such that their effects cancel out—this reduces the overall energy. Figure 10.5 shows how magnetic domains can be imaged via a variety of techniques.

When an external field H is applied those domains which are oriented in the same direction as the field grow at the expense of others. The rate of change of B with H goes through a maximum and then decreases. Towards the top of the curve all the specimen is one domain, not magnetised exactly along H, but along a crystallographic direction of easy magnetisation. M then saturates as the direction of magnetisation rotates to H and the scatter of atomic magnet directions due to thermal fluctuations decreases. Finally B increases due to the increasing $\mu_0 H$ component only. Notice that μ changes all along the curve.

[†] more accurately dia = *across*, because diamagnets line up East-West in the Earth's field.

[‡] The interaction (*exchange* interaction) is actually electrostatic, although the result is magnetic.

[*] e.g. in strontium hexaferrite $Sr^{2+}O^{2-}.6Fe^{3+}{}_2O^{2-}{}_3$ the strontium and oxygen ions have no permanent magnetic moment. Eight of the Fe^{3+} align parallel to the field and four antiparallel.

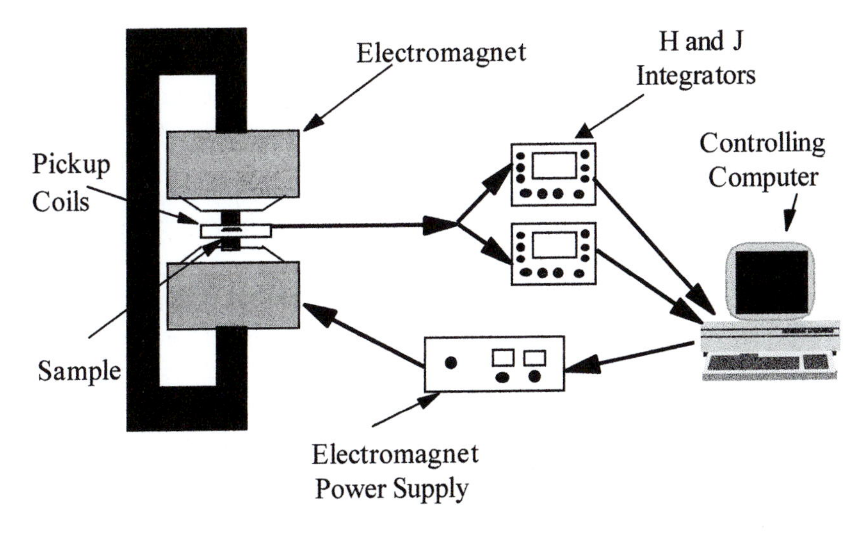

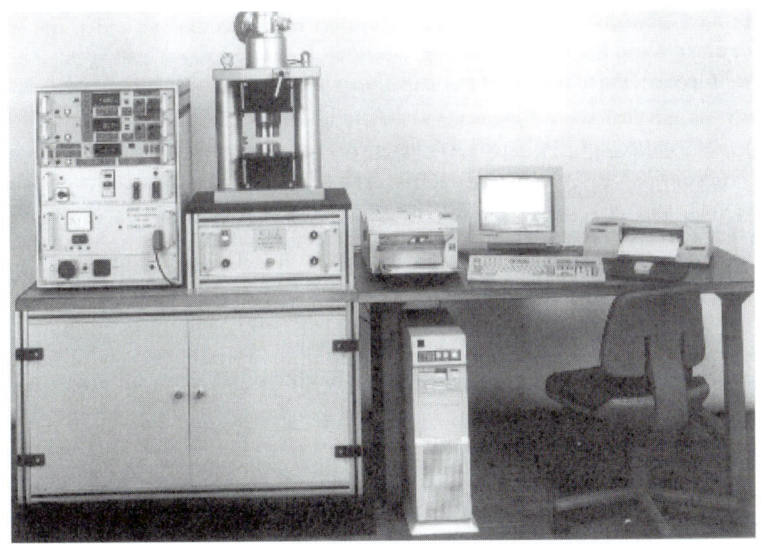

Fig. 10.3 (a) A schematic diagram of a permeameter (b) An example of the real thing. (Courtesy Dr Steingrover, GmbH)

As H is reduced and reversed and then increased again, B goes round a loop rather than up and down a single curve. This is called *hysteresis*. The area of the loop is the energy absorbed in a cycle. Physically this is accomplished by a mixture of new (reverse) domains nucleating and others rotating (magnetically!). If a *minor* loop is traced out (we haven't gone as far as saturation) then already existing reverse domains may grow.

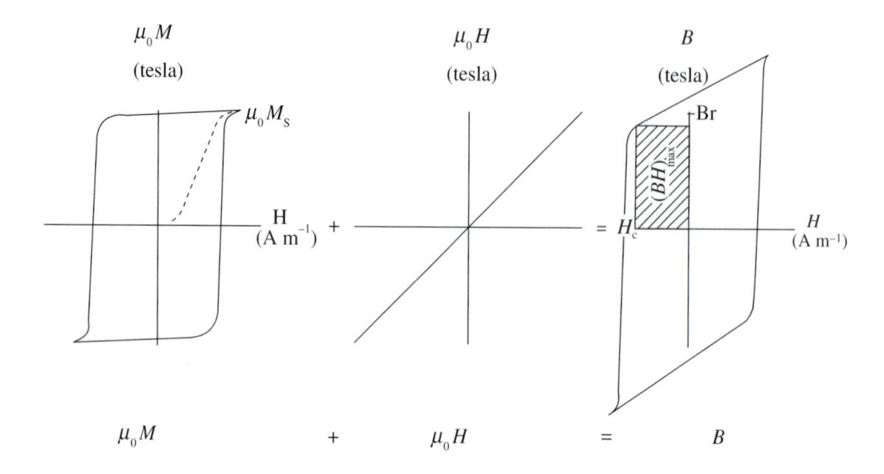

Fig. 10.4 A typical measured *B–H* curve. The induced magnetisation, *M*, plus the applied field, *H*, times μ_0, gives *B*. Notice the remanence (B_r) and coercivity (H_c).

When H is reduced to zero after saturation the value of B remaining is called the *remanence* B_r. H for $B = 0$—i.e. the reverse field necessary to demagnetise the specimen—is called the *coercivity* H_c. Two other important parameters are energy product, which I have shown, and saturation magnetisation, which may be inferred from eqn (10.7). All of these parameters will be important (under different circumstances) when we start to select materials for particular applications.

Why doesn't B just slide straight down the same curve like this: ?
Why does the curve show hysteresis? It is because the domain walls are *pinned* by defects in the structure. There are several mechanisms: the simplest just involves a particle replacing domain wall and thus saving energy:

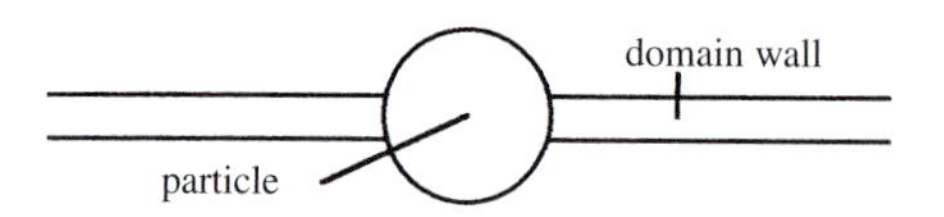

(Note that the domain wall has a finite width.)

Thus energy is required to nucleate reverse domains or to rotate those which are present. This is the *microstructure* controlling things again.

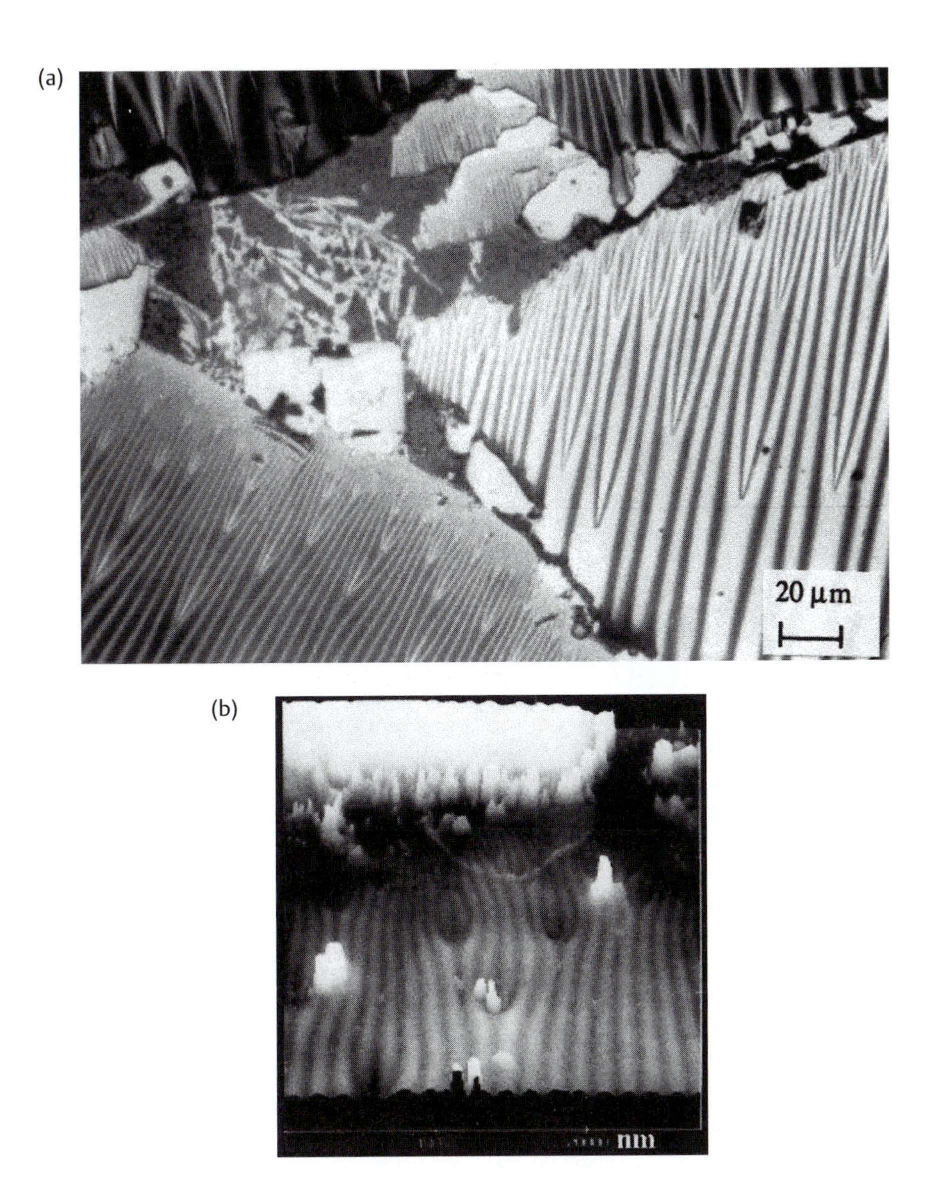

Fig. 10.5 Magnetic domains imaged via (a) Ferrofluid (colloidal equivalent of iron filings). (b) Ferrofluid + atomic force microscopy. (c) Kerr effect (polarised light). (d) Scanning electron microscopy.
In each case the specimen consists of an alloy based on $Nd_2Fe_{14}B$—see Section 10.7. (Courtesy Dr A.J. Williams.)

(c)

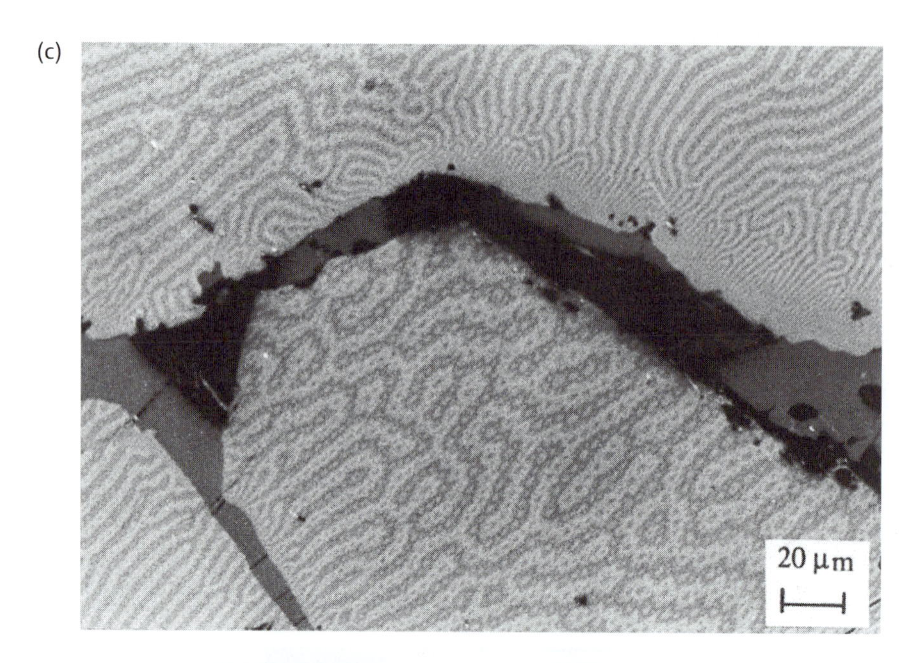

(d)

Fig. 10.5 (*continued*)

Thus the *saturation magnetisation* is controlled by *physics* but the *coercivity and remanence* are controlled by *microstructure*. This is very important and good news for engineers because it is a lot easier to control and alter the microstructure of a material than its basic physics.†

Ultimately one needs to look at the whole *B*–*H* curve, but most of the story can be gained by inspecting a few parameters which control the overall shape and size of the curve and which I have illustrated in Fig. 10.4. These parameters are as follows:

- saturation magnetisation
- remanence
- coercivity
- maximum energy product

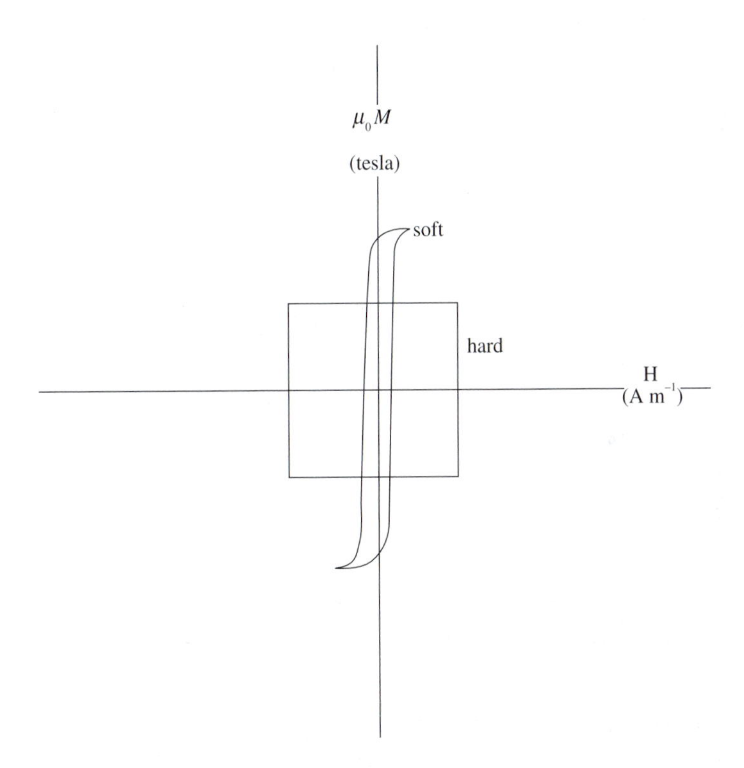

Fig. 10.6 Soft and hard magnets. Hard magnets are called 'hard' because within any single metallic magnetic material cold working increases both physical hardness and coercivity, both due to the increased dislocation density, which pins the domain walls.

† In fact if the material is cycled repeatedly around the curve it does eventually approach the version above: this is exploited for magnetic tape where an AC signal is superimposed to produce just this result—see Section 10.6.

Altogether, of course, they imply something about permeability and susceptibility. Also, they are not in practice independent—for example high μ often implies low H_c. So what do we want from a magnetic material in terms of its B–H curve? It all depends, of course! It all depends on exactly what we want to use it for. It is very helpful at this stage to classify magnets into *soft* and *hard* according to their coercivities, since these classes have TOTALLY different uses (see Fig. 10.6).

Roughly speaking, magnets may be classified on the basis of their coercivities† as suggested in Table 10.2. Tables 10.3 and 10.5 contain some examples of soft and hard magnets in common use.

Thus in electromagnets, transformers, etc. the primary requirements are high susceptibility and low coercivity. We want to be able to achieve high induction easily and remove it just as easily. With hard magnets we generally want a strong magnetic induction which is retained (high coercivity). In magnetic recording it must be reasonably easy to magnetise and demagnetise, but at the same time we don't want it to happen accidentally, so this is a halfway house.

For the remainder of this chapter I will take various applications in order, explain which material is used and why, and how the thing is made. I shall deal with applications in the overall order: soft (Section 10.5), medium (Section 10.6) and hard magnets (Section 10.7).

10.5 Applications of soft magnetic materials

1. Electromagnets and relays

In an electromagnet a coil of wire is wrapped around a core. The requirements are high permeability—to give high induction for minimum applied field—and low coercivity, so that demagnetisation is easy. The usual choice is soft iron of reasonable purity. This

Table 10.2 Applications of soft, medium and hard magnets. (Note that there is some overlap.)

Type	H_c (A m^{-1})	Applications
Soft	<1000	Electromagnets, transformers, motors and generators
Medium	$10\,000<H_c<100\,000$	Magnetic recording
Hard	$>50\,000$	Loudspeakers, headphones, videorecorders, watches, televisions, motors and generators (different from soft magnet versions)

† We could equally well use permeability—see *Introduction to magnetism and magnetic materials* by D. Jiles (2nd edn, Fig. 4.3).

typically would have a coercivity of $\sim$100 Am^{-1} and a maximum relative permeability of $\sim 10^4$, corresponding to a saturation induction of $\sim$2.15 T. Remember that for a soft magnetic material, what we are trying to do is *not* pin the domain boundaries. Therefore the purer the soft iron (for example), the better its properties. A reasonable compromise as far as cost *vs* properties goes is 0.1% (total) of C, Mn, Si etc. Unfortunately, as I explained above, 'soft' for a metallic magnet means not only low coercivity but low physical strength, because anything which pins domains, pins dislocations and *vice versa*. Thus soft iron is easily damaged and the resulting dislocations pin the domain walls, causing the magnetic properties to deteriorate. Soft iron is easily shaped, as required by the forming operations described in Chapter 4, but must be annealed subsequently to remove the dislocations and thus make it soft again.

Electromagnets and relays are the least demanding of soft magnet applications and soft iron is at the cheap end of the market.

2. AC transformers

AC transformers, generators and motors are the volume users of soft magnetic materials—in fact of all magnets considered as a class.

An AC transformer consists of two coils wrapped around a soft magnet yoke (see Fig. 10.7). Again, we require high permeability and low coercivity with a high saturation induction. The principal reason for low coercivity here is because the hysteresis loss is proportional to the area of the B–H curve. This loss appears in the form of heat. The hysteresis loss is not, however, the only loss mechanism. There is also power loss due to eddy currents

$$W_{ec} = \frac{\pi^2 B_{max}^2 d^2 v^2}{\rho\beta} \qquad (10.9)$$

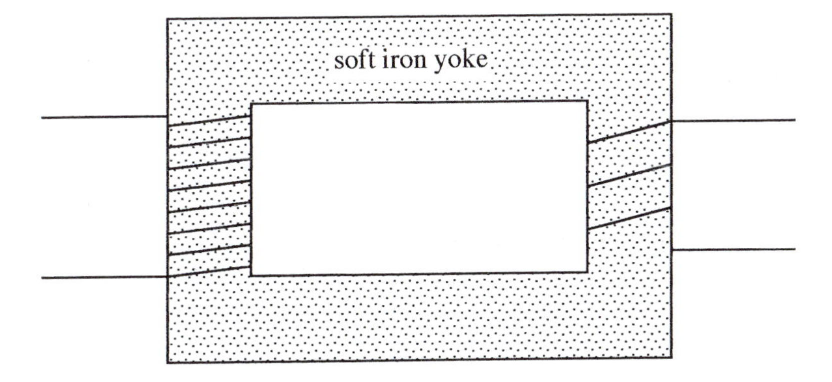

Fig. 10.7 An AC transformer. The voltage is changed according to the ratio of the numbers of windings.

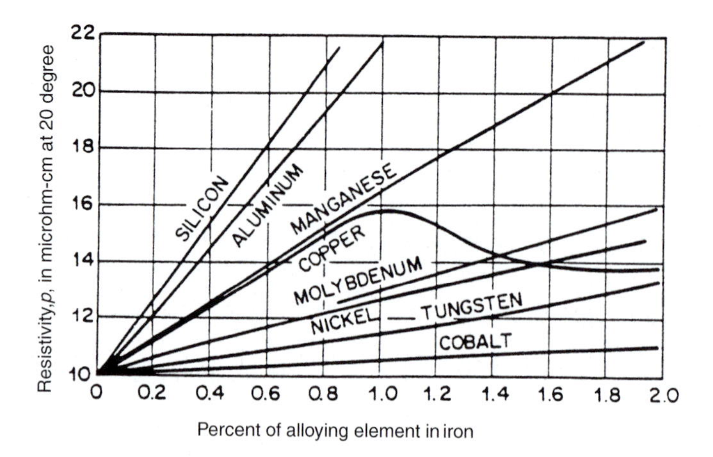

Fig. 10.8 The effect of various alloying elements on the electrical resistivity of iron (compare Fig. 4.22). (From *Introduction to magnetism and magnetic materials* by D. Jiles, courtesy of Chapman and Hall.)

where B_{max} is the maximum value of B reached in the cycle, d is the thickness of the magnet, ν is the AC frequency, ρ the electrical resistivity and β a geometrical factor. There are also losses due to magnetostriction (strain or stress caused by magnetic fields). Thus (material aside) transformer cores are laminated to reduce d and a material of high electrical resistivity is employed. The usual choice is iron–silicon: iron with 3–4% Si (by weight) added. Figure 10.8 shows how various solutes affect the electrical resistivity of iron. Silicon has the greatest effect because its valency is so different from that of iron. Compare Fig. 10.8 with Fig. 4.22 for copper. The same principles obtain (see Section 4.6, p. 122, subsection 3), but there we were trying *not* to increase the resistivity of copper: here it is just the opposite. 3% Si raises ρ for iron by a factor of 4! The laminations are chosen to have a thickness of $\sim\frac{1}{2}$ mm, which is roughly the AC magnetic field penetration depth at 50–60 Hz. Power losses are usually quoted in W kg^{-1}. The magnetic properties of the silicon–iron are further improved by rolling and annealing the sheets before lamination. This makes most of the grains (remember a metal consists of small crystals or 'grains') adopt an orientation such that the little cubic unit cells (recall iron is 'b.c.c.'—see Section 2.3) lie parallel to the laminations. (This is called by metallurgists the 'texture' of the material.) This orientation is the softest, magnetically speaking. Figure 10.9 shows domains in rolled Fe–3wt%Si and how their regularity can be improved by scribing.

For higher specification applications and higher frequencies nickel–iron alloys can be used because they can be tailored to have no anisotropy and no magnetostriction. Table 10.3 compares the relevant properties with those of silicon–iron. Nickel iron alloys can also be alloyed with other elements to improve even further their magnetic properties—for example see supermalloy and mumetal in Table 10.3. This family of materials is used for more purposes than just transformers—for example thin sheets of

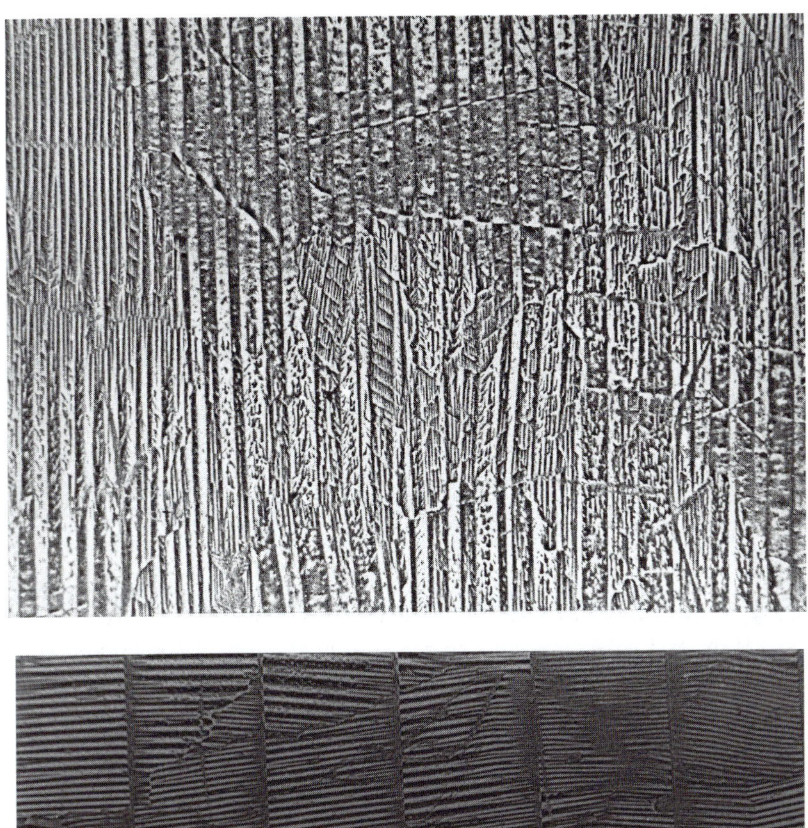

Fig. 10.9 (a) Domains in rolled and annealed Fe–3wt%Si. (b) This iron silicon has been mechanically scribed with a tungsten carbide wheel. The cutting of grooves in the surface of the iron pins the domains in a regular pattern and improves the magnetic properties. (Courtesy Mr R. Taylor, Corus Ltd.)

Table 10.3 Various magnetic parameters for a selection of soft magnets. μ_i is initial relative permeability and μ_{max} is maximum relative permeability. (From *Introduction to magnetism and magnetic materials* by D. Jiles by kind permission of Chapman and Hall.)

	Composition	Relative permeability μ_i	μ_{max}	Coercivity H_c (A m^{-1})	Saturation induction B_s (T)
Iron	100% Fe	150	5000	80	2.15
Silicon–iron (nonoriented)	96% Fe 4% Si	500	7000	40	1.97
Silicon–iron (grain-oriented)	97% Fe 3% Si	1500	40 000	8	2.0
78 Permalloy	78% Ni 22% Fe	8000	100 000	4	1.08
Hipernik	50% Ni 50% Fe	4000	70 000	4	1.60
Supermalloy	79% Ni 16% Fe, 5% Mo	100 000	1 000 000	0.16	0.79
Mumetal	77% Ni, 16% Fe 5% Cu, 2% Cr	20 000	100 000	4	0.65
Permendur	50% Fe 50% Co	800	5000	160	2.45
Hiperco	64% Fe 35% Co, 0.5% Cr	650	10 000	80	2.42
Supermendur	49% Fe 49% Co, 2% V		60 000	16	2.40

mumetal are the traditional way of screening components from magnetic fields. Remember that nickel is horrendously expensive and that all of this cleverness therefore comes at a (sometimes considerable) price. Also, a material with excellent magnetic properties can be unusable because it is so brittle.

For low power transformers, metallic glasses can be used. This is the first and only time in this book where we will come across a metal which is not crystalline. This is achieved by alloying iron with the metalloids B, P and Si and then solidifying the alloy incredibly quickly by shooting a stream of molten metal at a rapidly rotating water cooled copper wheel (Fig. 10.10). The resulting tape is annealed lightly to remove stress (which would pin domains), sprayed with an insulating coating and then used directly in the transformer core.

Table 10.4 shows how low the coercivity can be made and how high the permeability. Unfortunately these metallic glasses also have low saturation induction and core losses (hysteresis, eddy currents and magnetostriction) start to shoot up for high fields. They have found some application in the USA, however, where the electricity distribution often includes a small transformer at the domestic end of the chain.

(a)

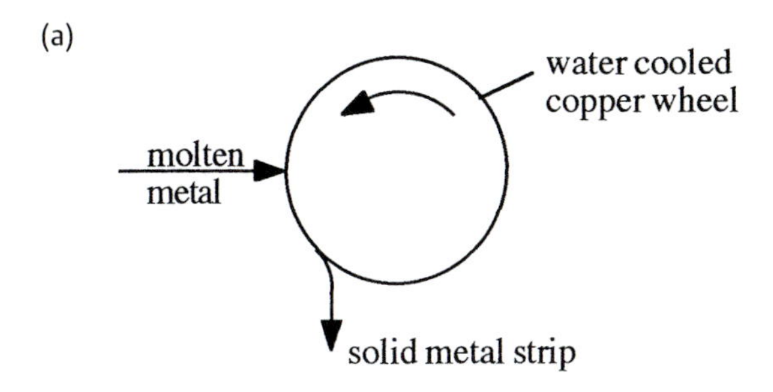

(b)

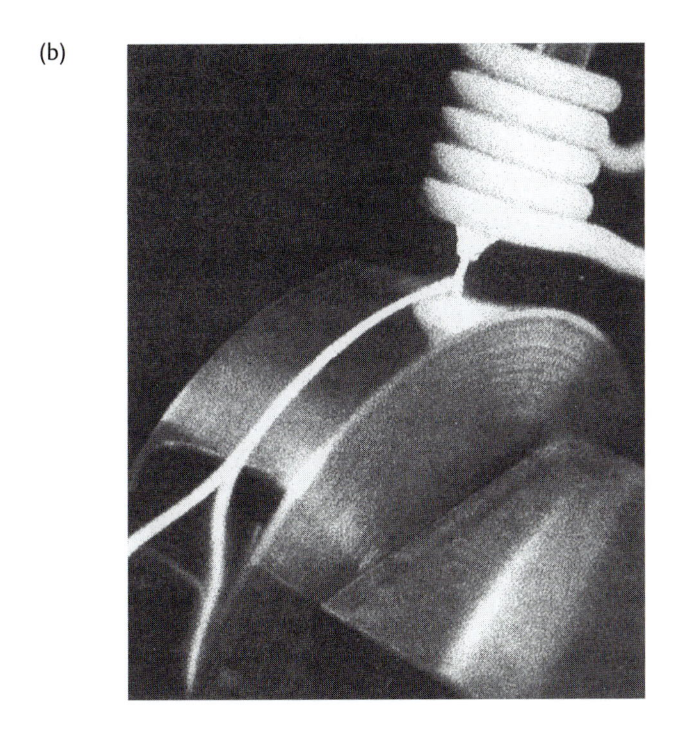

Fig. 10.10 Producing a metal glass by rapid quenching (a) Schematic (b) A real example. Liquid iron silicon boron hits the rotating water cooled copper cylinder and is led off as solid glass tape on the left. Cooling rates of $\sim 10^6$ °C s^{-1} are achieved. (Courtesy Professor H.A. Davies.)

Finally, for high frequency applications (see eqn (10.9)) it is crucial to raise ρ even further. This is achieved by turning to a ceramic magnet, which is actually an electrical insulator. By far the most common ceramic magnets are the ferrites. These have the general formula

$$\mathrm{MO}.x\mathrm{Fe_2O_3}$$

Table 10.4 The magnetic properties of metallic glasses. (From *Introduction to magnetism and magnetic materials* by D. Jiles, courtesy of Chapman and Hall.)

		As cast			Annealed		
Alloy	***Shape***	H_c ($A\,m^{-1}$)	M_r/M_s	μ_{max} (10^3)	H_c ($A\,m^{-1}$)	M_r/M_s	μ_{max} (10^3)
Metglas 2605 $Fe_{80}B_{20}$	Toroid	6.4	0.51	100	3.2	0.77	300
Metglas 2826 $Fe_{40}Ni_{40}P_{14}B_6$	Toroid	4.8	0.45	58	1.6	0.71	275
Metglas 2826 $Fe_{29}Ni_{44}P_{14}B_6Si_2$	Toroid	4.6	0.54	46	0.88	0.70	310
$Fe_{4.7}Co_{70.3}Si_{15}B_{10}$	Strip	1.04	0.36	190	0.48	0.63	700
$(Fe_{.8}Ni_{.2})_{78}Si_8B_{14}$	Strip	1.44	0.41	300	0.48	0.95	2000
Metglas 2615 $Fe_{80}P_{16}C_3B$	Toroid	4.96	0.4	96	4.0	0.42	130

where for soft magnets, as here, M is Ni, Mn, Mg, Zn or Fe itself. Because they are ceramics they must be fabricated using ceramic processing technology (see Chapter 7). Typically they are sintered to shape or alternatively dispersed as a fine powder in a plastic. Ferrites constitute the largest volume of magnet material used. Applications range far more widely than just transformers. Your AM radio almost certainly contains a ferrite antenna. They are also used as waveguides. (In the microwave region somewhat more complicated ferrites are used, such as yttrium iron garnet.)

3. Generators and motors

There are generally less demanding applications than power transformers. Silicon iron is a common choice, but without the grain orienting rolling and heat treatment schedule (to reduce the cost). At the cheap end of the market for generators, motors and transformers, simple unadorned steel is still used—remember from Chapter 5 that unless there is a good reason not to, we *always* use steel.

10.6 Magnetic recording media

You will recall these are in the medium coercivity range of 10 000–100 000 Am^{-1}. Magnetic recording media are tapes and discs (both floppy and hard). Most tapes are currently analogue and used for audio or video recording. Discs are for audio or computer storage and are digital. In all cases there is a substrate of plastic (tapes: polyethylene terephthalate

(PET); discs: polycarbonate (PC)—see Chapter 8) and on this is supported a large number of very small magnetic needles. The most common material is haematite, γ-Fe_2O_3. The needles are ~0.5 μm long by 0.1 μm wide. They have a coercivity of ~30 000 Am^{-1} and a saturation induction of ~0.5 T. Alternative materials are chromite (CrO_2) ($H_c \sim 60\,000$ Am^{-1}, $B_{sat} \sim 0.6$ T: a smaller particle size of 0.4×0.05 μm $\Rightarrow$ higher recording density) and cobalt-modified γ-Fe_2O_3 ($H_c \sim 48\,000$ Am^{-1}) which is especially used for videotape.

The particles are suspended in an organic liquid and applied to the tape. A magnetic field is applied which rotates all the (single domain) particles parallel to it. The liquid is then heated to evaporate it and the tape or disc is rolled or pressed to consolidate the coating (Fig. 10.11). Figure 10.12 shows the writing and reading processes for a magnetic tape (or disc). The little electromagnet has a pole gap of 0.3 μm and is made from, for example, a soft ferrite or a nickel–iron alloy. Whether the recording is digital or analogue makes no difference in principle to the actual recording process, but in fact for analogue signals an AC bias is used to remove hysteresis (see Section 10.4) and improve signal noise.

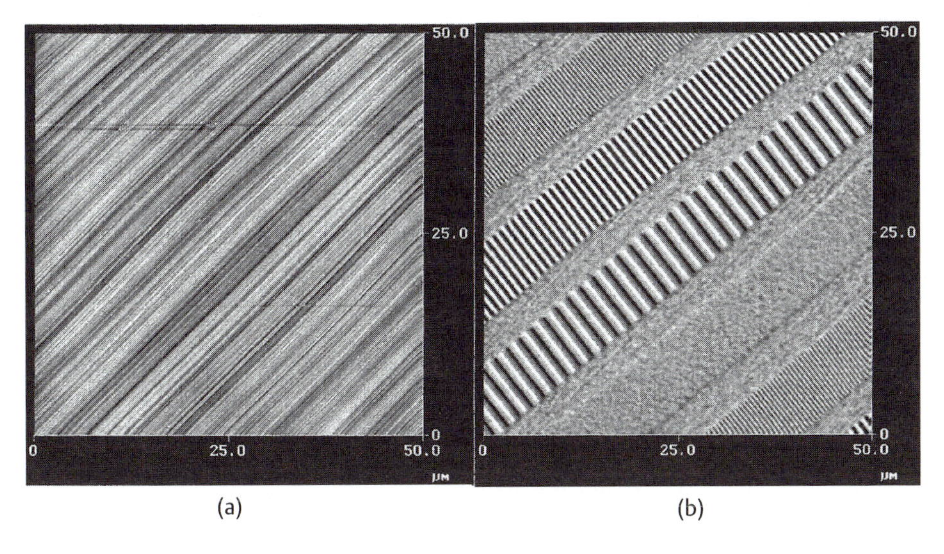

Fig. 10.11 (a) An Atomic Force Microscope (AFM) image of the surface of a magnetic hard disc showing the mechanical texture detail (i.e. the shape-grooves). This texture helps the recording head 'fly' properly over the surface of the disc. (b) A magnetic force microscope (MFM) image of the same area showing recorded tracks on the magnetic layer on the disc. The 'bits' are written at different frequencies, between 10^3 flux reversals mm^{-1} and 250 flux reversals mm^{-1}. AFM and MFN are examples of *Scanning Probe Microscopies*, where different signals are taken from an atomically sharp needle scanned across the specimen surface. (Courtesy Prof P.J. Grundy.)

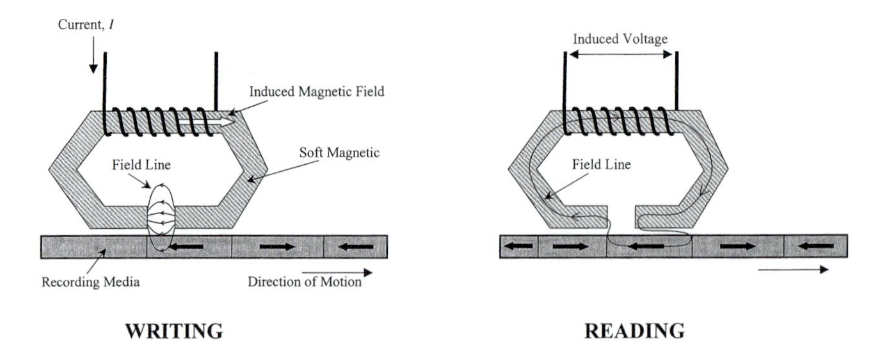

Fig. 10.12 Writing and reading a magnetic tape or disc. (Courtesy Dr A.J. Williams.)

10.7 Hard (permanent) magnets

These magnets are made to be used without any supporting magnetic field. They are magnetised at the beginnings of their lives and thereafter they are on their own. Thus the remanence (and saturation magnetisation) must be large. It is important that the magnet is not demagnetised by stray magnetic fields and thus the coercivity must be large also. In fact the single parameter most used to assess a permanent magnet is $(BH)_{max}$—the maximum value of the energy product BH (see Fig. 10.4). $(BH)_{max}$ is the maximum amount of useful work which can be done by the magnet. In addition and not entirely independently, the more square the M-H loop, the better.

Permanent magnets are less wedded to specific applications than soft magnets and so I shall organise this section by material. Table 10.5 shows some common permanent magnetic materials with corresponding values of the three parameters mentioned above. Figure 10.13 shows how permanent magnet materials have improved over the last hundred years or so. Table 10.5 shows that it is mainly the coercivity which has been altered.

Steels

An ordinary high carbon steel has an energy product of $\sim$1.5 kJm^{-3}. This can be raised (at a cost) by a factor of 5 by alloying with 36%Co, 4%W and 6%Cr.

Alnicos

These are also iron-based, but with substantial additions of Al, Ni and Co—as the name suggests. Like many of the other metal alloys described in this book (see Chapters 4 and 5) they are solution treated at high temperature, quenched and aged. This produces highly magnetic FeCo needles in a less strongly magnetic Ni–Al matrix. The alloys are so brittle that they either have to be cast to shape or processed using ceramic-type technology.

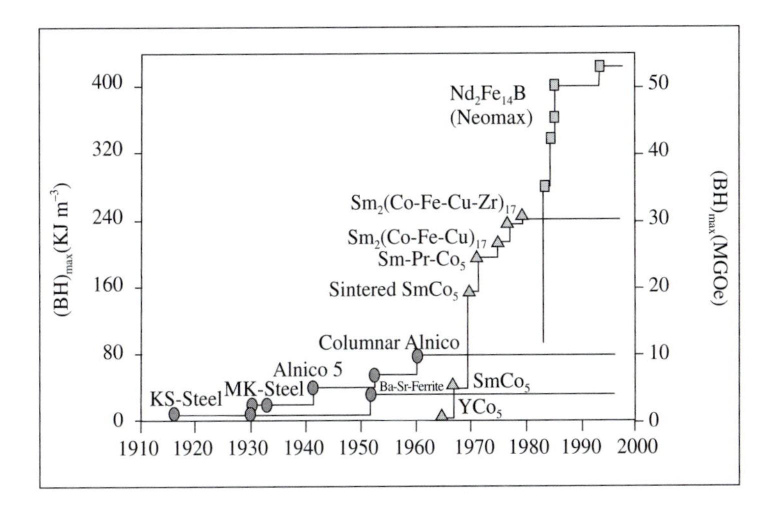

Fig. 10.13 How permanent magnets have improved over the last 100 years. $(BH)_{max}$ is a figure of merit (see text). (Courtesy Professor I.R. Harris.)

Table 10.5 Various magnetic parameters for a selection of hard (i.e. permanent) magnetic materials. (From *Introduction to magnetism and magnetic materials* by D. Jiles by kind permission of Chapman and Hall.)

Material	***Composition***		***Remanence* (T)**	***Coercivity* kA m^{-1}**	**$(BH)_{max}$ kJ m^{-1}**
Steel	99% Fe,	1%C	0.9	4	1.59
36Co Steel	36%Co, 5.75%Cr,	3.75%W 0.8%C	0.96	18.25	7.42
Alnico 2	12%Al, 3%Cu,	26%Ni 63%Fe	0.7	52	13.5
Alnico 5	8%Al, 24%Co,	15%Ni 3%Cu, 50%Fe	1.2	57.6	40
Alnico DG	8%Al, 24%Co,	15%Ni 3%Cu 50%Fe	1.31	56	52
Ba Ferrite	$BaO \cdot 6Fe_2O_3$		0.395	192	28
PtCo	77%Pt,	23%Co	0.645	344	76
Remalloy	12%Co, 71%Fe	17%Mo	1.0	18.4	9
Vicalloy	13%V, 35%Fe	52%Co	1.0	36	24
Samarium-cobalt	$SmCo_5$		0.9	696	160
Neodymium-iron-boron	$Nd_2Fe_{14}B$		1.3	1120	320

Table 10.6 Developing applications for strong permanent magnets. (From *'The attractions of rare earth magnets'* by I.R. Harris and A.J. Williams in *Materials World*, August 1999, p. 478, by kind permission of The Institute of Materials.)

Automotive	Starter motors, anti-lock braking systems, motor drives for wipers, injection pumps, fans and controls for windows, seats etc, loudspeakers, eddy current brakes, alternators, speedometers.
Telecommunications	Loudspeakers, microphones, telephone ringers, electro-acoustic pick-ups, switches and relays.
Data processing	Disc drives and actuators, stepping motors, printers.
Consumer electronics	DC motors for showers, washing machines, drills, citrus presses, knife sharpeners, food mixers, can openers, hair trimmers etc, low voltage DC drives for cordless appliances eg drills, hedgecutters, chainsaws, magnetic locks for cupboards or doors, loudspeakers for TV and audio, TV beam correction and focusing devices, CD drives, video recorders, computers, electric clocks, analogue watches.
Industrial	DC motors for magnetic tools, robotics, magnetic separators for extracting metals and ores, magnetic bearings, servo-motor drives, lifting apparatus, brakes and clutches, meters and measuring equipment.
Electronic and instrumentation	Sensors, contactless switches, NMR spectrometers, energy meter discs, electro-mechanical transducers, crossed field tubes, flux-transfer trip devices, dampers.
Astro and aerospace	Frictionless bearings, stepping motors, couplings, instrumentation, travelling wave tubes, auto-compass.
Biosurgical	Dentures, orthopaedics, wound closures, stomach seals, repulsion collars, ferromagnetic probes, cancer cell separators, magnetomotive artificial hearts, MRI scanners.

Ferrites

We encountered the ferrites ($MO.xFe_2O_3$) in the soft magnet section. There are also 'hard' ferrites where M is Ba or Sr. They are cheap, with quite adequate properties (see Table 10.5). They are either pressed and sintered, or alternatively supported in a plastic—which dilutes the magnetic properties but makes them much easier to handle and use.

Samarium-cobalt alloys

Samarium is the first example in this book of a *rare earth*† element. If you look back to Chapter 1, you may remind yourself that the rare earths are transition elements where

† a misnomer, really, because they are not in any sense 'earth' and they are not particularly 'rare'.

the 4f electron shell is partially empty. Because this is so deep within the atom (relative to the outside) the outer electrons, which dictate the chemical properties of the atom, are unchanged across the series: the rare earths are chemically almost identical and are very difficult to separate. Because of their electronic structures they have unique magnetic and optical properties and so find their applications in magnets and devices like lasers. There are two common Sm-Co based magnetic materials: Sm_2Co_{17} and $SmCo_5$. Both are intermetallic compounds (compounds between metals, which are themselves metallic) and their brittleness means that powder processing must be used. The high costs of cobalt and samarium were the main factors driving the search for other rare earth-transition metal strong magnetic compounds. This search led to the discovery of:

Neodymium-iron-boron

$Nd_2Fe_{14}B$ is another intermetallic compound (except for the boron) with a complicated crystal structure. It has to be produced with a very specific microstructure to exploit its unique magnetic properties. One form is as tiny particles of $Nd_2Fe_{14}B$, where each particle is a single domain, which confers great coercivity. The dependence of ease of

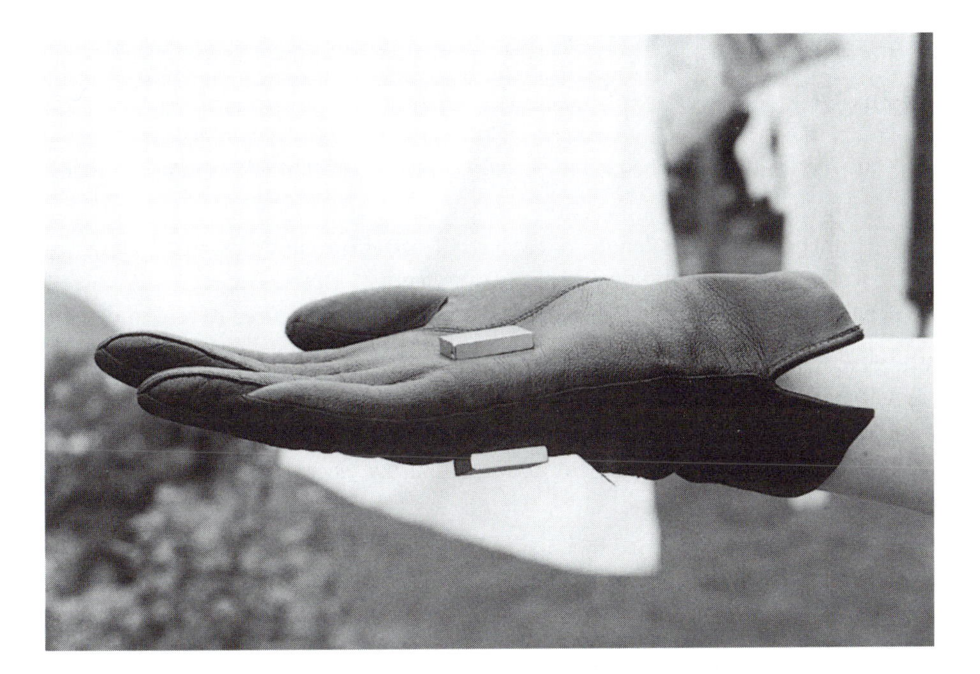

Fig. 10.14 The new strong magnets. The upper magnet in the palm of the hand is supporting the lower magnet against the force of gravity, despite their separation by several centimetres.

magnetisation on direction (magnetic anisotropy) is large, which rules out domain reversal by rotation and necessitates reverse domain nucleation. The small particle size is achieved by using rapid solidification (see metal glasses above) and then carefully controlled crystallisation. Another variant of $Nd_2Fe_{14}B$ uses powder metallurgy. A composition away from $Nd_2Fe_{14}B$ itself is chosen and the extra phase insulates the $Nd_2Fe_{14}B$ grains from each other. Each grain, however, contains several magnetic domains.

$Nd_2Fe_{14}B$ has wonderful magnetic properties—as you can see from Table 10.5—but it has two major disadvantages:

1. It has a low Curie temperature
2. It corrodes easily

At some temperature ferromagnets lose their ferromagnetism. This temperature is called the Curie temperature (T_c).† For iron it is 767 °C. For $Nd_2Fe_{14}B$ it is around 310 °C. At temperatures greater than $\frac{3}{4}T_c$ (in K) the magnetic properties start to deteriorate, so manufacturers must be very careful where and how they employ $Nd_2Fe_{14}B$.

I should not, however, end this section on a downbeat. Figure 10.13 shows how permanent magnets have improved recently. As a magnetic layman I find the strength and capabilities of some of the new magnets almost unreal (Fig. 10.14). New applications are continually being developed for these new materials (Table 10.6).

10.8 Summary

1. Magnetic field is defined via its *production* by a long solenoid as current × turns / unit length ($A\ m^{-1}$).
2. Magnetic induction is defined via its *effect*. 1 Tesla will produce a force of 1 N on a wire perpendicular to B and carrying 1 A. Equivalently, 1 T produces, for 1 s 1V when it passes through a loop of conductor 1 m^2 in area and is reduced to zero in that 1 s.
3. If $B = \mu_0 H$ in vacuo, $B = \mu H$ when a material is present. μ_0 is the magnetic permeability of free space ($H\ m^{-1}$) and $\frac{\mu}{\mu_0} = \mu_r$, the relative permeability of the material. $B = \mu_0\ (H + M)$ and χ, the susceptibility, $= \frac{M}{H} = \mu_r - 1$.
4. Diamagnetism is omnipresent and is caused by electron orbit precession. χ is small and negative.
5. Paramagnetism is caused by unpaired electron spins. χ is small and positive.
6. Ferromagnetism is cooperative paramagnetism, where exchange interaction between neighbouring atoms causes their magnetic dipoles to line up. χ is large and positive.

†Note that there is another T_c connected with superconductivity which appears in Chapter 11.

7. In antiferromagnets the dipoles line up in opposite directions and cancel out. In ferrimagnets the dipoles line up in opposite directions but do not cancel out.
8. In ferromagnets the atoms organise themselves into domains (c.f. ferroelectrics) such that the magnetic fields of the domains cancel.
9. B is measured in a permeameter.
10. M–H and B–H curves define a material's magnetic properties.
11. The area of the B–H curve represents the energy lost when H is cycled once. This is called *hysteresis*.
12. M_s is a physically *intrinsic* property; B_r and H_c are *extrinsic* properties, which may be engineered.
13. Soft magnetic materials have $H_c < 1000\ \mathrm{Am^{-1}}$, hard magnets have $H_c > 10^5\ \mathrm{Am^{-1}}$ and medium magnets are in between.
14. Soft magnetic materials are used for electromagnets and relays, transformers, generators and motors. Typical examples are soft iron, silicon iron and ferrites.
15. Medium hardness magnets are used for magnetic recording media. Typical examples are γ-Fe_2O_3 and CrO_2.
16. Hard magnets are used particularly for miniature components (e.g. in a personal computer). Typical examples of hard magnetic materials include tungsten steels, alnicos, Sr and Ba ferrites, SmCo compounds and FeNdB.

Recommended reading

Introduction to magnetism and magnetic materials by D. Jiles (published by Chapman and Hall)

For a more lighthearted read:
Driving force—the natural magic of magnets by J.D. Livingston (published by Harvard University Press)

Questions (Answers on pp. 328–330)

10.1 A long solenoid is wound with 10 turns/cm and a current of 10 A is passed through it. What is the field at its centre, away from the ends? What is the induction? What force would be exerted on another wire, perpendicular to the axis of the solenoid and also carrying 10 A? (The permeability of free space $\mu_0 = 4\pi \times 10^{-7} = 1.257 \times 10^{-6}\ \mathrm{H\ m^{-1}}$.)

10.2 A bar of material with relative permeability $\mu_r = 1000$ is placed within the coil of Question 10.1. What is the induction within the material? What is the susceptibility of the material?

10.3 The material referred to in Questions 10.1 and 2 has $H_c = 15\ \mathrm{A\,m^{-1}}$ and saturation induction $B_s = 2.15$ T. Would you describe it as magnetically soft or hard? The current in the coil is increased until the magnet is saturated, reversed to the same (negative) value and then returned to zero. Roughly how much energy is absorbed (order of magnitude estimate)?

10.4 Classify the following materials as *diamagnets, paramagnets* and *ferro/ferrimagnets*.

- copper
- polythene
- silicon
- iron
- aluminium
- magnetite (Fe_3O_4)

10.5 A piece of soft iron is dropped on the floor. Its coercivity doubles. Why is this?

10.6 Why does dissolving silicon into iron affect its electrical resistivity more than dissolving cobalt into it?

10.7 Classify the following as *soft* and *hard* magnetic materials:

- iron
- steel
- iron–silicon
- iron neodymium boron
- iron silicon boron glass
- alnico
- barium ferrite
- samarium–cobalt
- magnetite (Fe_3O_4)

11

Superconductors and optical fibres

Chapter objectives

- What is superconductivity?
- How and when does it occur?
- The two types of superconductivity
- Applications of superconductors
- How do optical fibres work?
- Manufacturing optical fibres

11.1 Superconductors

Chapter 4 was concerned with *metallic conductors*, whose use in electrical and electronic engineering derives from their very low electrical resistivity, many many orders of magnitude less than that of ceramics or polymers. Even though metals, generally, have such low electrical resistivities, the aim of the electrical engineer, notwithstanding this, is usually to minimise the electrical resistivity—hence the overwhelming position of copper. This is because power losses ($\equiv$ money lost) are proportional to electrical resistance (for a given current).

Some materials, not all of them metals, lose their electrical resistance entirely† below a certain temperature, called T_c. They are called *superconductors*. T_c varies from material to material and is always $\leq\sim 150$ K, usually much less. Why then, you will be asking yourself, do we not always use superconductors instead of, for instance, copper? The answer is threefold:

- the cost of cooling the superconductor
- the difficulty of fabricating and using the component (e.g. as wire)
- there is a maximum current density, above which superconductivity disappears.

† or at least the electrical resistance drops below what it is possible to measure—e.g. ρ_{Cu} at room temperature $\approx 1.7 \times 10^{-8}\,\Omega$m; $\rho_{\text{superconductor}} < 10^{-23}\,\Omega$m.

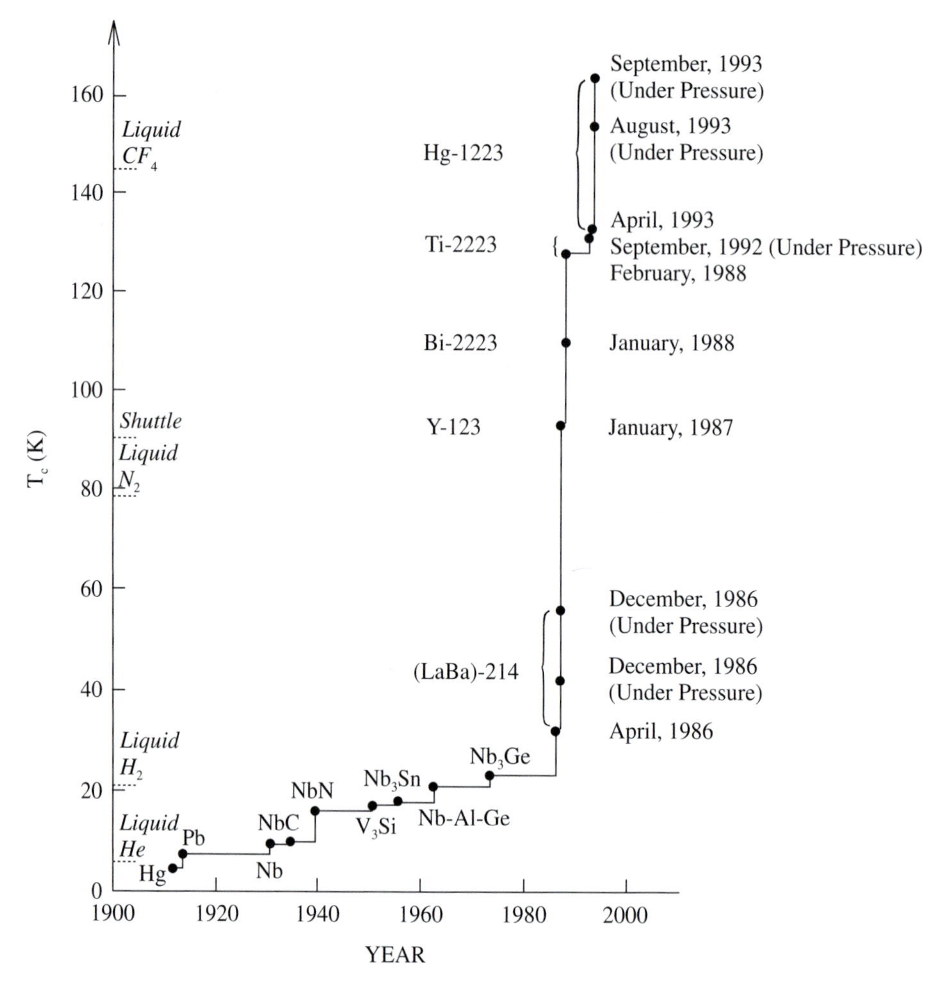

Fig. 11.1 The evolution with time of T_c for superconductors. (From C.W. Chu (1997). High temperature superconducting materials: a decade of impressive advancement of T_c. IEEE *Transactions on Applied Superconductivity*, **7**, 80. Courtesy of IEEE.)

The market share of superconductors is thus tiny. The recent rise in known T_c's from ~20K in the mid-1980s to ~150K today (Fig. 11.1) has meant a resurgence of interest in superconductors for engineers and is why they appear in this book.

11.2 How does superconductivity work?

Recall that electrical resistance in metallic conductors comes from the scattering of the conduction electrons by irregularities in the lattice (thermal vibrations, solute atoms ...). In a superconductor below T_c, the conduction electrons are loosely bound together in pairs. This means that when one electron of the pair is scattered, the 'spring' holding the two electrons together is strong enough to keep the would-be scattered electron on track. Thus the electrical resistance disappears. In band theory terms (see Section 1.10) a small band gap appears and the thermal energy available is not sufficient to cause scattering across it. There are two sorts of superconductor: metallic and ceramic. In metal superconductors, the spring or interaction holding the two electrons together has to do with phonons—the waves which travel through the lattice and which constitute its heat capacity. The first electron attracts the positively charged ions of the lattice towards it, giving a region of slightly net positive charge. If the second electron of the pair can fit into this region sufficiently far away from the first electron that its drop in energy, as a result of occupying a positively charged region, more than counterbalances the electrostatic repulsion between the two electrons, then the two electrons will be bound together. Not surprisingly, this very delicate balance is easily disrupted by raising the temperature and thus T_c is limited to a few K (see Fig. 11.1). The strength of the interaction controls not only T_c, but J_c, the maximum current density which the superconductor can support before reverting to an ordinary conductor. J_c is as important to an engineer as T_c, because it controls how much current can be passed through a superconducting wire.

In the other type of superconductor, the ceramic superconductors, the mechanism holding together the two conduction electrons is not yet understood, but is thought to be electronic in nature. Ceramic superconductors (discovered in 1986, as compared with metallic superconductors in 1911) can have much higher T_c's (currently up to $\sim$150K is known—see Fig. 11.1) but have much lower J_c's and are much more difficult to process.

11.3 Engineering applications of superconductors

I mentioned above that the market share of superconductors is currently tiny: *low temperature* (metallic) superconductors are used to make strong magnets which themselves are used in Magnetic Resonance Imaging (MRI) (body scanners) and in nuclear physics to make particle colliders. It is crucial for these applications that J_c be as large as possible, so that the corresponding critical allowed field, H_c, is also large and strong fields may thus be produced. All practical engineering superconductors are of the so-called *second type*, where the material divides into two regions: superconducting and non-superconducting. The non-superconducting regions consist of a series of tubes called *fluxoids* and the supercurrent travels down the walls of these. If these fluxoids can be pinned by the microstructure of the material (the dislocations, precipitates, etc.) they are prevented from coalescing under the influence of a field and H_c raised considerably. Thus H_c is not an intrinsic physical property of the material, it is something which can be *engineered*. There is a very close analogy here with the pinning of magnetic domains to make a hard magnetic material (see Section 10.4).

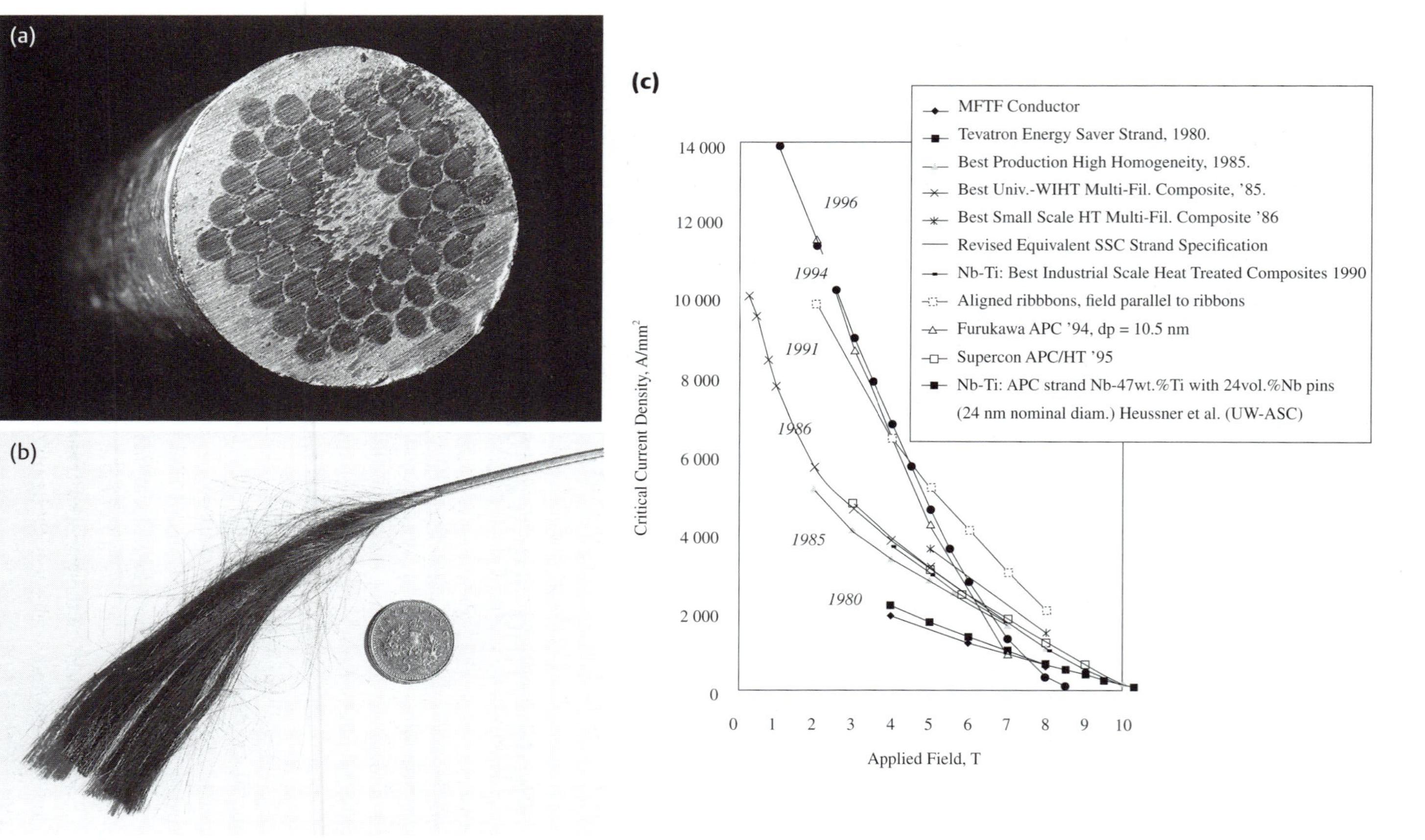

Fig. 11.2 Nb–Ti superconductor. (a) Nb–Ti filaments supported in a copper matrix. (b) After extrusion and wire drawing: very fine Nb–Ti braid in copper tube. (c) The critical current density of Nb–Ti has been improved by *engineering* the microstructure. (From D.C. Larbelestier (1997). The road to high temperature superconductors: 10 years do make a difference!, *IEEE Transactions on Superconductivity*, **7**, 90. Courtesy of IEEE.)

The most common material used is Nb–46wt%Ti with a T_c of ~9K. It is encased in copper (see Fig. 11.2) and drawn down until it consists of very fine filaments. It is then annealed at 400 °C to produce precipitates (see Section 4.6, p. 127, subsection 4) which help to pin the fluxoids. (This doesn't remove the dislocations, which also help to pin the fluxoids.) The copper is there, amongst other reasons, to protect the Nb–Ti and stop it from cracking.

The prospective market for *high temperature* (ceramic) superconductors is enormous, provided the considerable engineering challenges which they currently present can successfully be overcome. These challenges are of two sorts: materials processing and improving J_c. The latter involves understanding better than we do at the moment what exactly limits J_c. The grain boundaries in the ceramic and its 'texture' (see Section 10.5) are certainly both important. As far as the processing is concerned, imagine having to make a wire out of a ceramic! Various clever and innovative solutions are currently being investigated. Probably the first important applications of ceramic superconductors will be in the form of thin or thick films supported on substrates (like electronic materials). Suitable fabrication methods have been described in Chapter 7 (Section 7.6) for thick films and Chapter 9 (Section 9.3.2) for thin films.

11.4 Optical fibres

Why should a book on materials for electrical and electronic engineers contain a section on optics? Communications Engineering on the one hand and Electrical and Electronic Engineering on the other are very intimately connected—energy flow and information flow are first cousins—and some of you are probably doing joint degrees in Electrical and Communications Engineering, or some variation on this. Optical fibres are central to communications at the beginning of the third millennium. How they work is quite simple: making them is not.

In an optical fibre the light bounces along the inside of the fibre by total internal reflection (Fig. 11.3). There are three types of fibre: step-index multimode, graded-index multimode and step index monomode. These are illustrated in Fig. 11.4. In the fatter

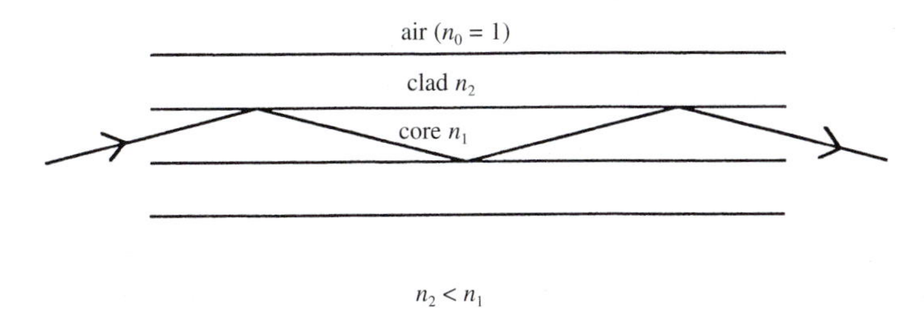

Fig. 11.3 In an optical fibre the infra-red radiation is confined to the core region of the fibre by *total internal reflection* off the inside of the clad. The refractive index of the clad, n_2, must be less than that of the core (n_1).

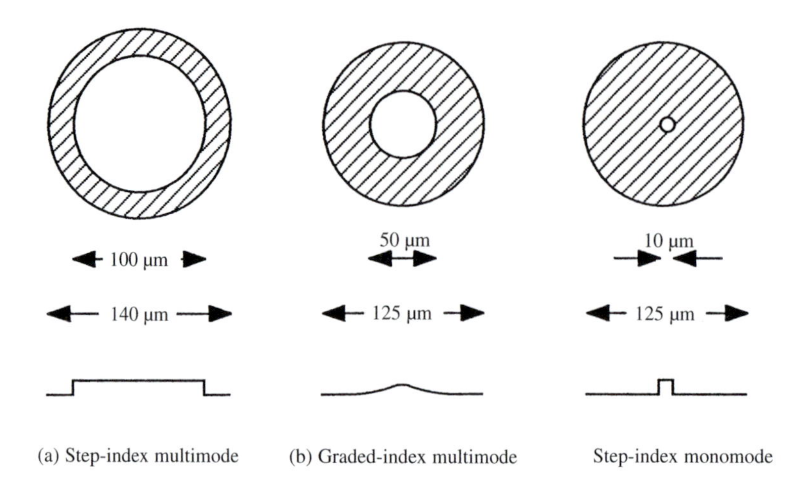

Fig. 11.4 The three main types of optical fibre. The profiles refer to the refractive index.

multimode fibres, more than one mode can be transmitted down the fibre. In the thinner *monomode* fibres only one mode is transmitted, which simplifies and improves the performance of the fibre. It is monomode fibres which are used for long distance telecommunications. Typically one monomode optical fibre can take ~2000 simultaneous telephone conversations requiring data transfer of ~140 Mbits^{-1}.

Just as conduction electrons are scattered by inhomogeneities in the lattice—hence electrical resistivity—so light is scattered by inhomogeneities in the glass of an optical fibre. The light is also absorbed and converted into heat. These two factors limit the length of fibre which can be used before the light becomes too weak to detect. The engineering challenge is to make the optical fibre sufficiently pure and perfect that practicable lengths of it can be used before the signal has to be detected and amplified back to its original level (currently every ~50 km). Here is how the problem was solved in the early 1970s and how optical fibres are made nowadays.

The fibre is made from very pure *silica* glass (SiO_2). The absorption mechanism I mentioned above is mainly connected with *hydroxyl* (OH) bonds. The infrared light is converted into vibrations of the O-H bonds (i.e. heat). The other attenuation mechanism, the scattering, decreases as the wavelength increases. The two factors taken together mean that there are three *windows* of transmission at 850, 1300 and 1550 nm (see Fig. 11.5). The infrared light at 850 nm is provided by a GaAs/GaAlAs laser and the two longer wavelengths by InGaAsP/InP lasers. The hydroxyl impurity has to be maintained <10 parts per billion.

The optical fibre consists of amorphous silica (see Section 7.2) which is made by burning $SiCl_4$ in oxygen on the inside of a cylinder to make a *preform*. The refractive index of the fibre is controlled by *doping* the $SiCl_4$ gas with, for example, $GeCl_4$ to increase it and, again for example, BCl_3 to decrease it (see Fig. 11.6). The preform is drawn

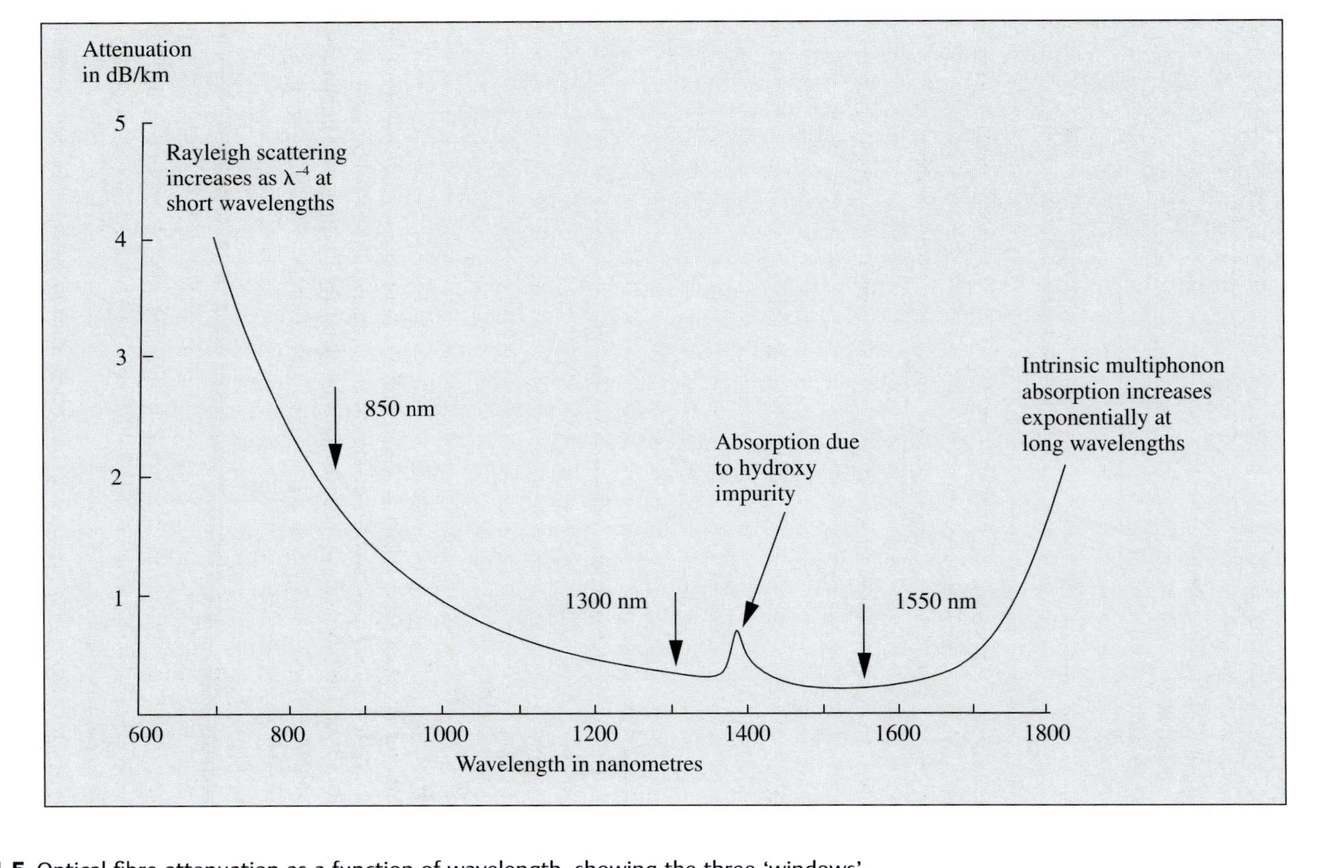

Fig. 11.5 Optical fibre attenuation as a function of wavelength, showing the three 'windows'.

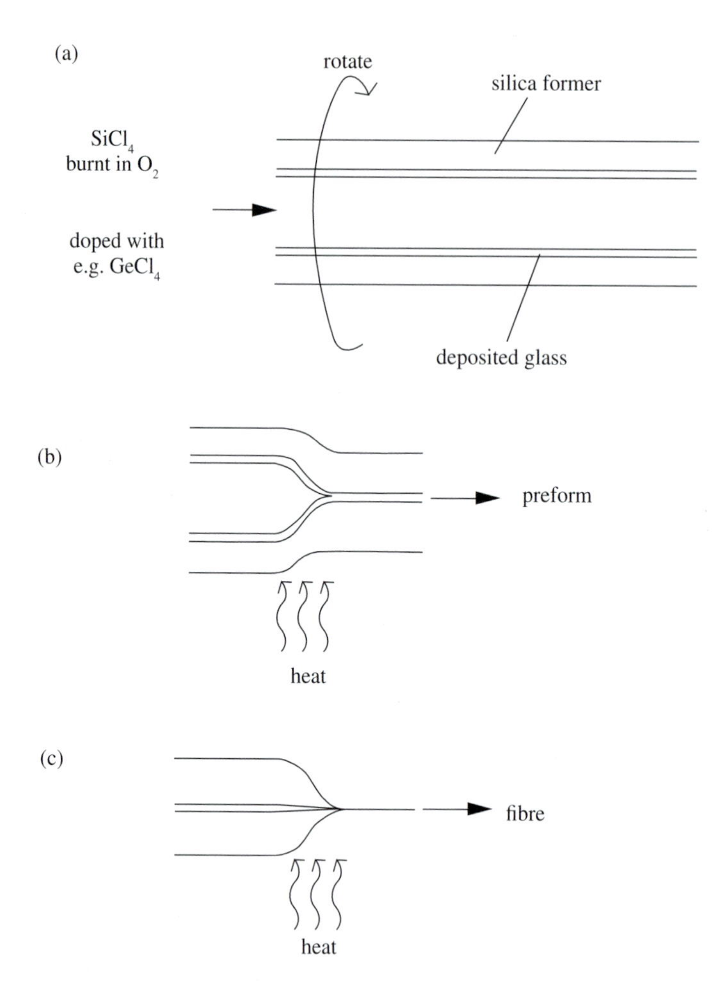

Fig. 11.6 Making an optical fibre. (a) Silica, doped with e.g. Ge is deposited on the inside of a hollow tube. The refractive index profile is defined at this stage via the partial pressure of the $GeCl_4$ gas. (b) The hollow tube is drawn down to a solid *preform*. (c) The preform is drawn down to an optical fibre.

out to its final diameter and the dopant/refractive index profile is thereby reproduced, but on a much smaller scale. The fibre is immediately protected by coating it in plastic, drawing it through a bath. The individual fibres are assembled into, for example, a trunk cable as depicted in Fig. 11.7.

Note that the technology for producing the doped silica is the same as that described in Section 9.3, p. 250. The vast optical fibre market is yet another spin-off from the electronics industry. It is possible to make plastic optical fibres, but the large polymer molecules (see Section 8.2) themselves act as inhomogeneity light scatterers and this

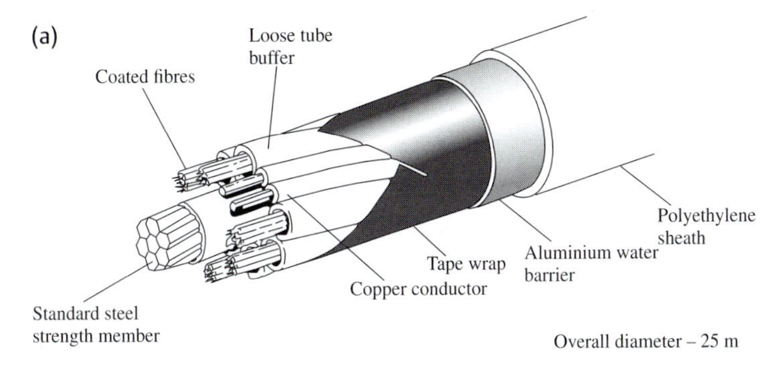

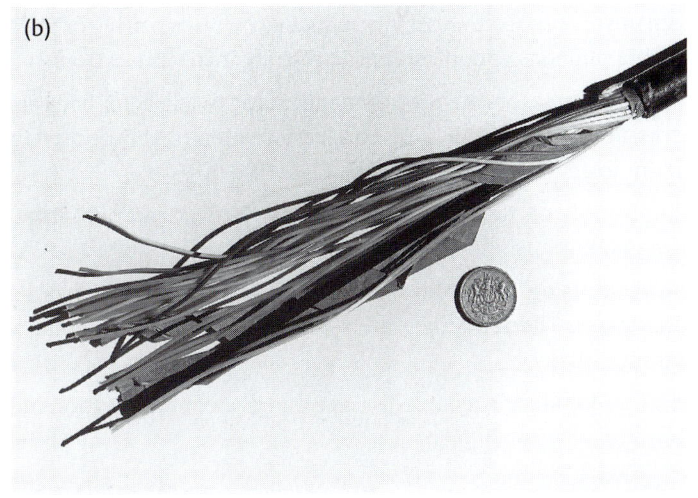

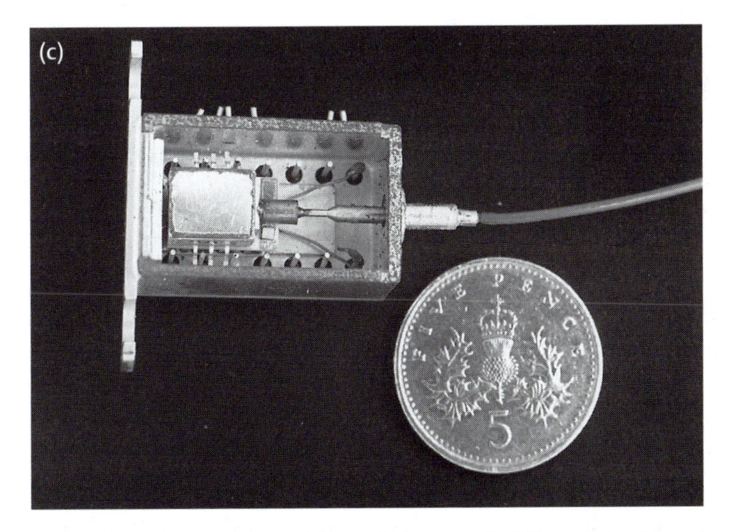

Fig. 11.7 Trunk telecommunications cable made from optical fibres: (a) schematic, (b) a real example, (c) the laser.

constitutes a very fundamental limitation of their performance. Plastic optical fibres still find many uses, however—for example in medicine.

11.5 Summary

1. Superconductivity is the reduction of electrical resistance below what can be measured.
2. It occurs in some metals at a temperature of a few K and in some ceramics at temperatures up to 150K.
3. The mechanism of superconductivity in metals involves pairing of the conduction electrons via phonons and in ceramics probably involves electronic effects.
4. Metallic superconductors are used to make strong magnets for body scanners and nuclear accelerators. Ceramic superconductors will probably be used first in thick or thin film form.
5. Optical fibres rely on total internal reflection from an interior change (drop) in refractive index.
6. They are made, using CVD, from silica, SiO_2, which is very pure in order to reduce scattering losses. The refractive index profile is created by doping with other elements e.g. Ge, B.
7. Optical fibres dominate the long distance and fast communication industries.

Recommended reading

Superconductivity

(Physics) ***Introduction to superconductivity*** by A.C. Rose-Innes and E.H. Rhoderick (published by Elsevier)

(Metallurgy) ***Metallurgy of superconducting materials*** edited by T. Luhmann and D. Dew-Hughes (published by Academic Press)

(Ceramic superconductors) High temperature superconductors: synthesis, processing and large scale applications edited by U. Balachandran, P.J. McGinn and J.S. Abell (published by Minerals, Metals and Materials Society)

Optical fibres

Understanding fibre optics by J. Hecht (2nd edn, published by SAMS)

Fibre optic cables by G. Mahlke and P. Gössing (published by Wiley)

Questions (Answers on pp. 330–331)

(Superconductivity)

11.1 If a 1 A current flows around a 1 m diameter circle of superconducting wire with 1 mm^2 cross-section for one year without diminishing detectably (e.g. loss < 1%), what is the maximum resistance of the wire?

(The current $i(t)$ flowing around a loop of wire decays according to

$$i(t) = i(0)e\frac{Rt}{L}$$

where t = time, R = electrical resistance and L = self-inductance.)

11.2 If a metal needs to be cooled to 2K to superconduct, estimate the size of the superconducting band-gap.

(Boltzmann's constant, k, is 8.62×10^{-5} eV K^{-1})

The next two questions will require some research (outside this book).

11.3 (See Fig. 11.1) What are the ceramic superconductors Y–123 and Bi–2223?

(Optical fibres)

11.4 (See Fig. 11.3) In, for example, a monomode fibre, what are n_1 and n_2? What is the angle for total internal reflection? What % of e.g. Ge would be required to produce this change?

Answers to questions

Chapter 1

1.1 Are the following solid materials electrical conductors, semiconductors or insulators?

(a) wood, (b) steel, (c) stone, (d) silicon, (e) copper, (f) polythene, (g) gold, (h) porcelain, (i) salt, (j) aluminium, (k) glass, (l) silver, (m) polystyrene.

	Conductor (metal)	**Semiconductor**	**Insulator**
Wood			natural polymer
Steel	√		
Stone			ceramic
Silicon		√	
Copper	√		
Polythene			polymer
Gold	√		
Porcelain			ceramic
Salt			ceramic
Aluminium	√		
Glass			ceramic
Silver	√		
Polystyrene			polymer

Notice that I have done a little more than just answer the question: I have classified the insulators into ceramics and polymers.

1.2 Which is smaller, the ratio of the radius of the sun to that of the solar system, or the ratio of the nucleus of a hydrogen atom to its atomic radius?

Radius of sun $= 7 \times 10^5$ km

Radius of solar system (= radius of Pluto's orbit?) $= 5.9 \times 10^9$ km

} ratio $= 10^{-4}$

Radius of hydrogen nucleus $= 0.5$ fm (10^{-15} m)

Radius of hydrogen atom $= 53$ pm (53×10^{-12} m)

} ratio $= 10^{-5}$

Thus the nucleus: atom ratio is ten times smaller than the sun: solar system ratio. A nucleus is a *very* small beastie.

1.3 Put the following elements in order of their constitution (by weight) of the earth's crust: Copper; oxygen; aluminium; silver; hydrogen.

Oxygen, aluminium, hydrogen, copper, silver.

1.4 The isotopes of iron, their relative atomic masses and their relative abundances in the earth's crust are as follows:

	RAM	Relative abundance
Fe^{54}	53.94	5.80%
Fe^{56}	55.93	91.72%
Fe^{57}	56.94	2.20%
Fe^{58}	57.93	0.28%

What is the terrestrial relative atomic mass of iron? Would you expect this to be the same on Mars? In the Andomeda galaxy?

The terrestrial relative atomic mass of iron

$$= 53.94 \times .058 + 55.93 \times .9172 + 56.94 \times 0.22 + 57.93 \times .0028$$
$$= 55.84$$

It would be the same on Mars (same original material), but might be a few per cent different in Andromeda.

1.5 Using the rules I gave you in Section 1.3, work out the maximum number of electrons which could occupy the third principal quantum shell in an element.

$n = 3$. This means that l can be 0, 1 and 2. (This corresponds to s, p and d.)

For $l = 0$, $m = 0$

$l = 1$, $m = -1, 0, 1$

$l = 2$, $m = -2, -1, 0, 1, 2$

For each m value, s can be $\pm \frac{1}{2}$

Then $l = 0$ 2 electrons

$l = 1$ 6 electrons

$l = 2$ 10 electrons

18 electrons in all

1.6 Using the rules in Section 1.3, derive the formula $2n^2$ for the maximum number of electrons in the nth principal shell.

The maximum number of electrons is

$$2\sum_{0} m = 2\sum_{l=0}^{n-1} (2l + 1)$$
$$= 2n^2$$

1.7 Does the 3d level have a higher energy than the 4s level in sodium? In silver? Are the 3d and 4s levels occupied in these two cases?

See Fig. 1.5.
3d has a higher energy than 4s in sodium ($Z = 11$) but not in silver ($Z = 47$). Both levels are occupied in silver, but not in sodium.

1.8 Sometimes energies are written as eV atom^{-1} and sometimes as kJ mole^{-1}, where mole $\equiv$ Avogadro's number of atoms. Work out the relationship between eV atom^{-1} and kJ mole^{-1}.

[The charge on an electron $e = 1.602 \times 10^{-19}$C; Avogadro's number, $N_A = 6.022 \times 10^{23}$ atoms mole^{-1}.]

Volts are SI units and so

$$1 \text{ eV} = 1.602 \times 10^{-19} \text{ J}$$

The number of atoms in a mole is $N_A = 6.022 \times 10^{23}$.

$$\therefore 1 \text{ kJ mole}^{-1} = \frac{10^3}{1.602 \times 10^{-19}} \times \frac{1}{6.022 \times 10^{23}} = 0.01037 \text{ eV atom}^{-1}$$

or, alternatively, 1 eV atom^{-1} = 96.47 kJ mole^{-1}

1.9 Write down the electronic structures of potassium (K) and bromine (Br) in solid potassium bromide (KBr), of Al and O in Al_2O_3 (alumina) and of Ca and O in CaO (calcium oxide, or 'quicklime').

Solid potassium bromide KBr is ionically bonded and is better written $K{+}Br^-$.

$K^+ = 1s^2\ 2s^2\ 2p^6\ 3s^2\ 3p^6$ (≡ argon)
and $Br^- = 1s^2\ 2s^2\ 2p^6\ 3s^2\ 3p^6\ 3d^{10}\ 4s^2\ 4p^6$ (≡ krypton)

Similarly, $Al_2O_3 = Al_2^{3+}\ O_3^{2-}$ and $CaO = Ca^{2+}O^{2-}$:

$Al^{3+} = 1s^2\ 2s^2\ 2p^6$ (≡ neon)
and $O^{2-} = 1s^2\ 2s^2\ 2p^6$ (≡ neon also)

In fact Al_2O_3 (alumina) (see Chapter 7) is quite covalent and so the bonding electrons are partially between the ions rather than unambiguously on the oxygen anion.

$Ca^{2+} = 1s^2\ 2s^2\ 2p^6\ 3s^2\ 3p^6$ (≡ argon)
and $O^{2-} = 1s^2\ 2s^2\ 2p^6$ (≡ neon again)

1.10 If the first ionisation energy of K is 4.34 eV and the electron affinity of Br is 3.36 eV, how much energy does it require to form a K^+Br^- ion pair? If the final equilibrium separation d_0 of the two ions is 0.334 nm, and the repulsive force between the two ions follows the law $\frac{B}{r^{10}}$, estimate the sublimation energy per ion pair. (The sublimation energy is the energy to convert something from solid to gas. The permittivity of free space, ε_0, is 8.854×10^{-12} Fm^{-1}.)

Ionisation energy required to form K^+Br^- pair = 4.34 - 3.36 = 0.98 eV.
There are two forces on the pair of ions: an attractive electrostatic force and a repulsive force. Thus at equilibrium:

$$\frac{e^2}{4\pi\varepsilon_0 d_0^2} = \frac{B}{d_0^{10}}$$

Substituting for d_0, $B = 3.573 \times 10^{-104}$ Nm^{10}
Then the electrostatic potential energy on bringing together the two ions from infinity to a separation of d_0 is $-\dfrac{e^2}{4\pi\varepsilon_0 d_0^2} = -4.31$ eV

$$\text{The repulsive potential energy} = \int_\infty^{d_0} -\frac{B}{r^{10}} \text{dr} = \frac{B}{9\, d_0^9} = 0.48 \text{ eV}$$

(Note in both cases I have converted from J to eV by dividing by $e = 1.602 \times 10^{-19}$ C.)
The total sublimation energy is therefore $-0.98 + 4.31 - 0.48 = 2.85$ eV per ion pair.

1.11 Work out the Madelung constant for the following linear chain of ions:

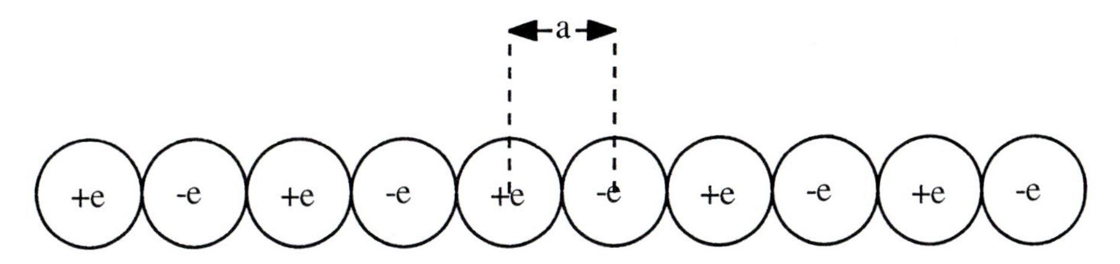

The potential energy between any two neighbouring ions $= \dfrac{e^2}{4\pi\varepsilon_0 a}$

The potential energy per ion for the whole infinite array $= -\left(\dfrac{e^2}{4\pi\varepsilon_0 a}\right)\mathbf{M}$

Consider any cation. The nearest neighbour anions contribute $= -\left(\dfrac{e^2}{4\pi\varepsilon_0 a}\right)2$ to the potential energy of the cation when brought from infinity.

The next-nearest neighbour cations contribute $\left(\dfrac{e^2}{4\pi\varepsilon_0 2a}\right)2 = -\left(\dfrac{e^2}{4\pi\varepsilon_0 a}\right)\left(\dfrac{-2}{2}\right)$

Thus the whole sequence of neighbours contributes

$$-\left(\frac{e^2}{4\pi\varepsilon_0 a}\right)\left[2\left(1-\frac{1}{2}+\frac{1}{3}-\frac{1}{4}+\frac{1}{5}\cdots\right)\right]$$

to the potential energy of the cation when brought from infinity, where the expression in square brackets $= \mathbf{M}$.

Thus $\mathbf{M} = 2\ \ln 2 = 1.386$

Note that we would arrive at the same answer if we started on an anion.

1.12 If the density of solid copper is 8960 kgm^{-3} and its relative atomic mass is 63.55, what are the weight and volume of a mole of copper atoms in the solid state?

By definition, a mole is chosen such that 1 mole of hydrogen weighs 1 g (give or take). 1 mole of copper atoms therefore weighs 63.55 g and occupies $\dfrac{63.55}{8.96 \times 10^6} = 7.09 \times 10^{-6}\text{m}^3$.

1.13 The latent heat of copper (the energy required to make it melt) is 13 kJ mole^{-1}. How far would 13 kJ raise 1 mole of copper in the earth's gravitational field? (See also preceding question.) (The acceleration due to gravity, g, is 9.807 ms^{-2}.)

The weight of 1 mole of copper is 63.55 g = 0.06355 kg
If the mole of copper is raised by a distance h, then

$0.06355 \times 9.807 \times \text{h} = 13000$ and $h = 20859$ m.

Given that the energy released in most chemical reactions is considerably greater than this, you can see why a litre of fuel can take your car so far.

1.14 Nitrogen forms covalently bonded N_2 molecules in the gaseous, liquid and solid states. How many electrons are shared between each bound pair of nitrogen atoms?

Nitrogen has electronic structure

$$1s^2\ 2s^2\ 2p^3$$

It lacks three electrons to make the inert gas structure $1s^2\ 2s^2\ 2p^6$ corresponding to neon. Thus two nitrogen atoms bond covalently together to form N≡N where each covalent bond contains two electrons and in all 6 electrons are shared between the two nitrogen atoms.

1.15 What is the electronic structure of an isolated germanium atom? Solid germanium adopts the same structure as silicon. How many bonding electrons does each atom contribute? What is the electronic structure of a germanium atom (a) disregarding all the bonding electrons and (b) assuming all the bonding electrons seen by an atom are possessed entirely by that atom?

An isolated germanium atom has electronic structure

$$1s^2\ 2s^2\ 2p^6\ 3s^2\ 3p^6\ 3d^{10}\ 4s^2\ 4p^2$$

By forming four covalent bonds, each germanium atom can achieve the electronic structure

$$1s^2\ 2s^2\ 2p^6\ 3s^2\ 3p^6\ 3d^{10}\ 4s^2\ 4p^6 \qquad (\equiv \text{krypton})$$

provided each atom considers that it 'owns', itself, all the bonding electrons it sees. If these are disregarded (i.e. the metallic situation), the electronic structure of each germanium atom reduces to

$$1s^2\ 2s^2\ 2p^6\ 3s^2\ 3p^6\ 3d^{10} \qquad (\text{not that of argon})$$

Each germanium atom contributes 4 bonding electrons.
In fact, as solid germanium forms, the outermost s and p orbitals *hybridise* to form 4 sp^3 orbitals. Each orbital interacts with that belonging to a neighbouring germanium atom, forming a bonding orbital and an antibonding orbital. The two electrons go into the bonding orbital. The empty antibonding orbitals overlap also (in space) and provide the *conduction band* for the germanium.

1.16 How many conduction electrons per atom would you expect the following metal elements to have?: (a) potassium (K), (b) magnesium (Mg), (c) copper (Cu), (d) aluminium (Al).

(a) Potassium has electronic structure $1s^2\ 2s^2\ 2p^6\ 3s^2\ 3p^6\ 4s^1$.
Each potassium atom will lose one electron to the conduction band, leaving the argon electronic structure $1s^2\ 2s^2\ 2p^6\ 3s^2\ 3p^6$.

(b) Magnesium has electronic structure $1s^2\ 2s^2\ 2p^6\ 3s^2$. Each magnesium atom will lose two electrons to the conduction band, leaving the neon electronic structure $1s^2\ 2s^2\ 2p^6$.

(c) Copper has electronic structure $1s^2\ 2s^2\ 2p^6\ 3s^2\ 3p^6\ 3d^{10}\ 4s^1$. Each copper atom will lose one electron to the conduction band, leaving the fairly stable electronic structure $1s^2\ 2s^2\ 2p^6\ 3s^2\ 3p^6\ 3d^{10}$, which does not correspond to any inert gas.

(d) Aluminium has electronic structure $1s^2\ 2s^2\ 2p^6\ 3s^2\ 3p^1$. Each aluminium atom will lose three electrons to the conduction band, leaving the neon electronic structure $1s^2\ 2s^2\ 2p^6$.

Chapter 2

2.1 If f.c.c., c.p.h. and b.c.c. crystals are made up from solid spheres which touch, what fraction of space is occupied by matter in each case?

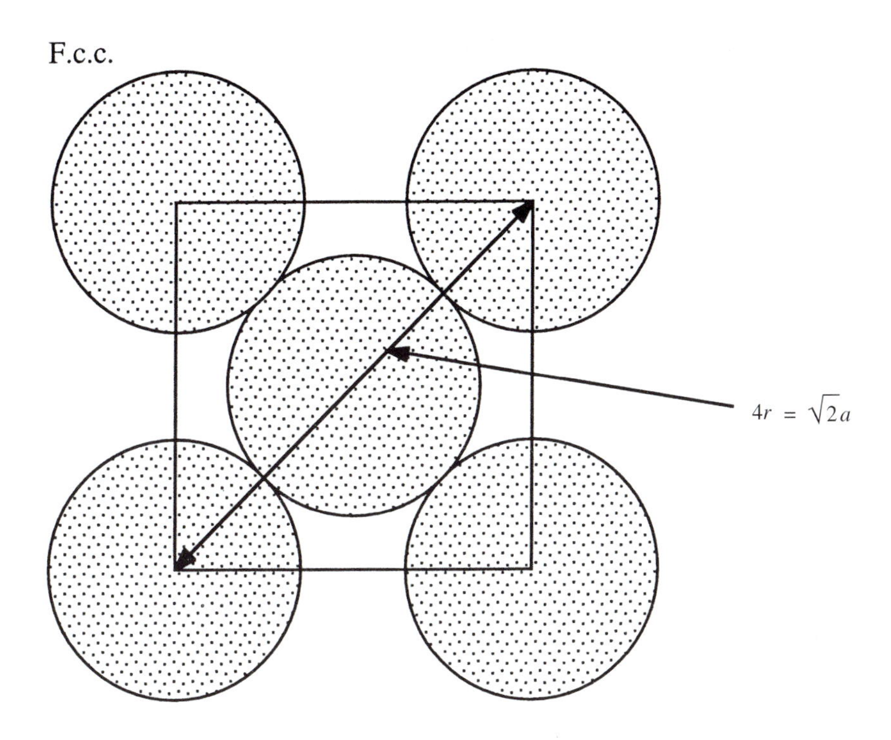

$$\text{Volume of unit cell} = a^3$$

$$\text{Volume of 4 atoms} = 4 \times \tfrac{4}{3}\pi r^3$$

$$\text{Fraction of space occupied by matter}$$

$$= \frac{\frac{16}{3}\pi r^3}{a^3}$$

$$= \frac{\pi}{3\sqrt{2}}$$

$$= 0.74$$

The answer has to be the same for c.p.h..

B.c.c.

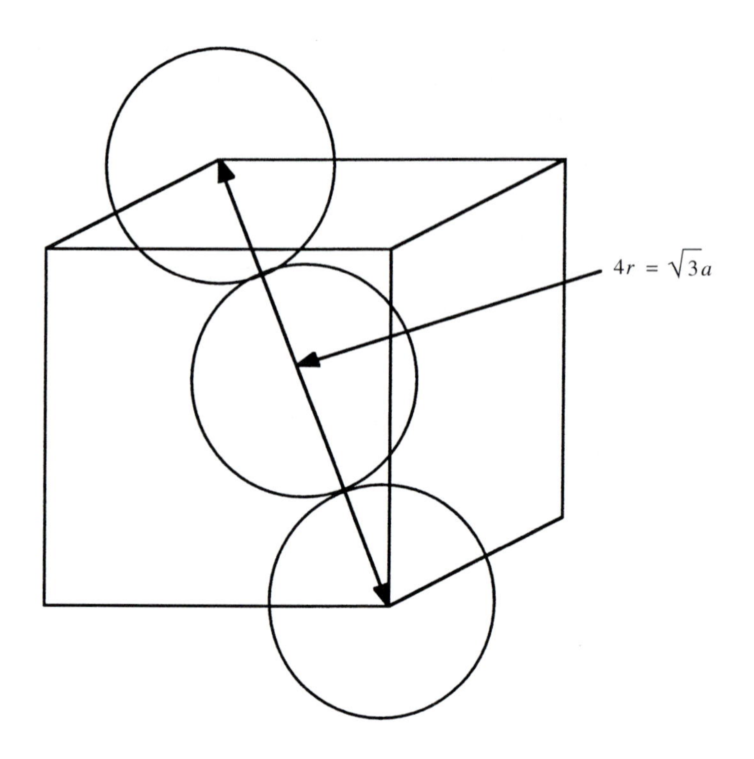

$$\text{Volume of unit cell} = a^3 \text{ again}$$

$$\text{Volume of 2 atoms} = 2 \times \frac{4}{3}\pi r^3$$

Fraction of space occupied by matter

$$= \frac{2 \times \frac{4}{3}\pi r^3}{a^3}$$

$$= \frac{\pi\sqrt{3}}{8}$$

$$= 0.68$$

2.2 How many atoms are there in the f.c.c., c.p.h. and b.c.c. unit cells?

The trick here is that all the atoms are shared between more than one unit cell and we have to recognise this fact in working out the answer. Referring to Fig. 2.8(b), each of the corner atoms is shared between 8 unit cells. There are eight of them and so they contribute $8 \times \frac{1}{8} = 1$ atom altogether. The face centre atoms are each shared between 2 unit cells; there are 6 of them and $6 \times \frac{1}{2} = 3$ atoms. So in total we have $1 + 3 = 4$ atoms in the f.c.c. unit cell.

Using the same logic, there are 2 atoms in the b.c.c. unit cell (Fig. 2.11).

The c.p.h. cell shown in Fig. 2.8(a) is, strictly speaking, not a unit cell because by definition unit cells should have 6 parallelogram sides, but ignoring this for the moment the cell shown in Fig. 2.8(a) contains 6 atoms. (The true unit cell is a third of this hexagonal cell, contains two atoms, but does not display the hexagonal symmetry of the whole structure.)

2.3 Is there any distinguishable difference between the surroundings of the four atoms in the f.c.c. unit cell?

If you work out the vectors from a face centred atom to its nearest neighbours (the touching atoms) and then do the same for a corner atom, you will arrive at the same answer. In an infinite crystal, there is no difference between the two types of atom. Remember that the unit cell is just a set of atoms cut from a larger assembly. The corner atom could be chosen to be *any* atom in the crystal.

What about the c.p.h. unit cell?

Yes and no! Although you could start the unit cell on any atom, once you have made that (arbitrary) choice, the vectors to the nearest atoms are different in the two cases—they are rotated through 60°. So the two types of atom (corner and interior) *are* different, strictly speaking.

And the two atoms in the b.c.c. unit cell?

As for f.c.c.—no difference.

2.4 An f.c.c. crystal is made up from solid spheres which touch. Where in the unit cell is the largest interstice and what is the size of the largest sphere which could be inserted into it?

Interstice means 'hole'. By inspection, the largest interstices in f.c.c. are at [$\frac{1}{2}$, 0, 0] etc. (Note that [$\frac{1}{2}$, $\frac{1}{2}$, $\frac{1}{2}$] is one of the set.)

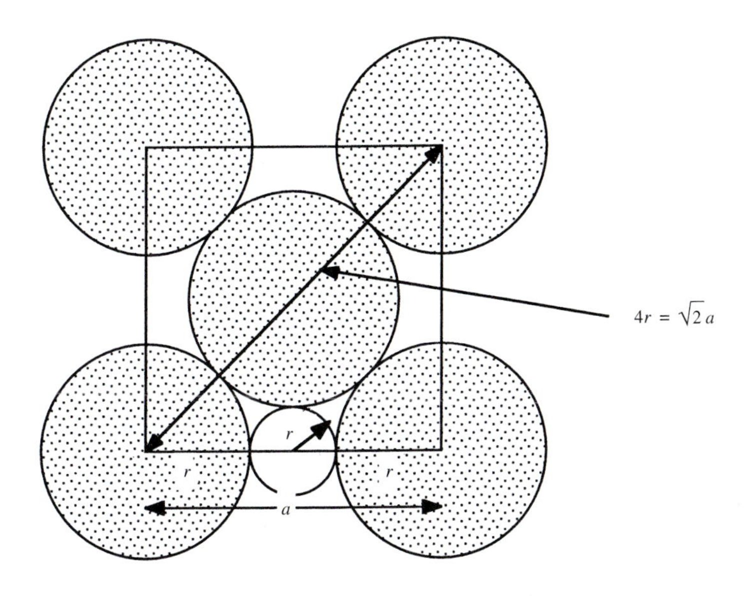

$$r' = a - 2r$$
$$= 0.41r = 0.15a$$

How many of these interstices are there per atom?

1 interstice per atom.

2.5 Answer Question 2.4 for c.p.h. and b.c.c. structures.

The answer for c.p.h. is very similar to that for f.c.c. The largest interstices are midway between pairs of next nearest neighbours and their size is the same as for f.c.c. There is one per atom, as there is for f.c.c.

For b.c.c. the largest interstices are at $\frac{1}{4}, \frac{1}{4}, \frac{1}{4}$ and similar sites. The size of this interstice is $0.29r$ (see Question 2.3) and there are four such interstices per atom.

The fact that there are fewer, but larger, interstices in f.c.c. as compared with b.c.c. turns out to be crucial to steel technology (see Chapter 5, Section 5.2 and Question 5.6).

2.6 How many nearest neighbours does an atom in an f.c.c. crystal possess?

12 nearest neighbours. (Think about the raft model of f.c.c.—for example see Fig. 2.7.)

How many second nearest neighbours are there?

6 second nearest neighbours along the cube directions.

2.7 Dislocation densities are measured in lines per unit area (like lines of magnetic flux cutting a surface). A well annealed metal sample will contain about 10^8 dislocation lines per m^2 whereas a heavily cold worked sample may contain 10^{16} lines per m^2. What are the average separations of the dislocations in the two cases?

For the annealed metal, 1 m will contain roughly $\sqrt{10^8} = 10^4$ dislocations and therefore the separation of the dislocations will be $\sim$100 μm. For the cold worked metal the rough separation will be 10 nm. (An atom is $\sim$0.25 nm.)

2.8 A good rule-of-thumb states that the concentration of vacancies in a metal is about 10^{-4} at the melting point T_m. Express the energy of formation of a vacancy, E_f^v, in terms of T_m. What is the concentration of vacancies at $\frac{T_m}{2}$? $\frac{T_m}{4}$? If the melting point of copper is 1085 °C, what is E_f^v? (Boltzmann's constant $k = 8.6\times 10^{-5}$ eV K^{-1}.)

Referring to Section 2.5 and eqn (2.1), if

$$10^{-4} = e^{-\frac{E_f^v}{kT_m}}$$

then

$$E_f^v = 4kT_m \ln(10) = \frac{T_m}{1262} \text{ eV}$$

At $\frac{T_m}{2}$, the concentration of vacancies, $c_v = (10^{-4})^2 = 10^{-8}$. At $\frac{T_m}{4}$, $c_v = 10^{-16}$.
If T_m for copper = 1085 °C = 1358 K, then $E_f^v \sim \frac{1358}{1262} = 1.1$ eV
(Experiment shows E_f^v for copper is indeed $\sim$ 1 eV.)

2.9 The electrical resistivity of copper at 160 K is 0.83×10^{-8} Ωm and that at 280 K is 1.69×10^{-8} Ωm. The resistivity of copper–1% nickel at 280 K is 3.12×10^{-8} Ωm. What is the resistivity of copper–3% nickel at 220 K?

$\Delta\rho$ for 1% nickel $= 1.43 \times 10^{-8}\ \Omega\text{m}$

ρ for copper at 220 K $= \frac{0.83+1.69}{2} \times 10^{-8} = 1.26 \times 10^{-8}\ \Omega\text{m}$

$\therefore\ \rho$ for copper - 3% nickel at 220 K $= 5.55 \times 10^{-8}\ \Omega\text{m}$

2.10 If the diameter of a copper wire is 1 mm, what would be the diameter of an aluminium wire with the same resistance per unit length? What would be the ratio of the two masses per unit length? (The density of copper is 8.9 Mg m^{-3} and that of aluminium is 2.7 Mg m^{-3}.) (Use Table 2.3.)

Referring to Table 2.3 and eqn (2.3), the aluminium wire would need a cross-sectional area $\frac{2.74}{1.70}$ times larger. Thus the diameter of the aluminium wire would need to be $\sqrt{\frac{2.74}{1.70}} = 1.27$ mm. The ratio of the mass of the aluminium wire to that of the same length of copper wire would be $\frac{2.74}{1.70} \times \frac{2.7}{8.9} = 0.49$. This is one of the reasons why overhead power cables are made from aluminium and not copper.

2.11 What would be the ratios of costs of silver, copper, gold and aluminium wires of the same resistance per unit length? (The densities are 10.5, 8.9, 19.3 and 2.7 Mg m^{-3} respectively.) (Use Table 2.5.)

From Table 2.5, and ratioing (arbitrarily) to copper:

Metal element	**Resistivity (& ∴ area)**	**Density**	**Cost per unit mass**	**Cost (same electrical resistance)**
Ag	0.95	1.18	37	35
Cu	1.00	1.00	1	1
Au	1.29	2.17	4593	5925
Al	1.61	0.30	0.69	1.11

Chapter 3

3.1 A metal cube with an edge length of 2 cm is placed on a smooth flat table. A mass of 100 kg is balanced on top of it. What stress acts on the cube? If Young's modulus for the cube is 10^{11} Pa, what is the resulting decrease in height of the cube?

Assuming that friction between the cube and the smooth table can be neglected, the cube will change shape as follows:

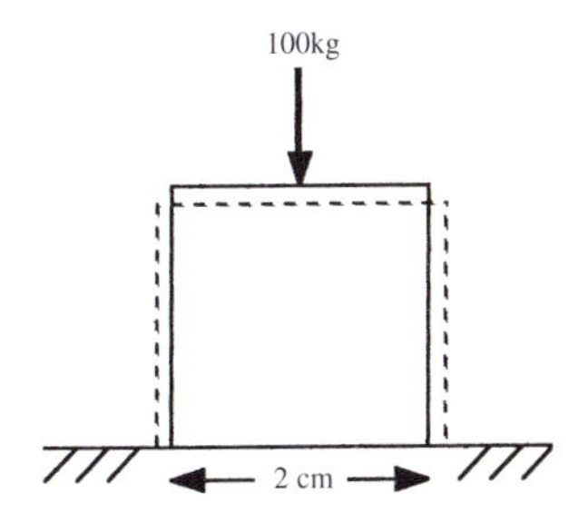

The stress acting on the upper surface of the cube is the force exerted divided by the area:

$$\sigma = \frac{100 \times 9.81}{(2 \times 0^{-2})^2} = 2.45 \times 10^6 \text{ Pa}$$

This is a uniaxial compressive stress. The resulting uniaxial compressive strain ε may be calculated from Young's modulus (E):

$$E = \frac{\text{uniaxial stress}}{\text{uniaxial strain}}$$

$$\text{strain } \varepsilon = \frac{2.45 \times 10^6}{10^{11}} = 2.45 \times 10^{-5}$$

The decrease in height of the cube is therefore $20 \times 2.45 \times 10^{-5} = 0.0005$ mm—not very much! Metals are very *stiff* (the elastic modulus is high).

3.2 The cube of Q3.1 is glued to the table and a rod glued to the top side

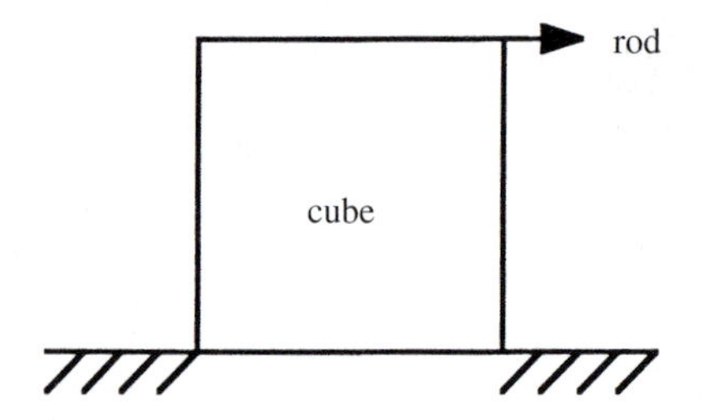

Via the rod a force of 3 kN is applied. If the shear modulus is 5×10^{10} Pa, describe the final shape of the cube and the strain within it.

This is a *shear* stress and the cube will turn into a parallelogram:

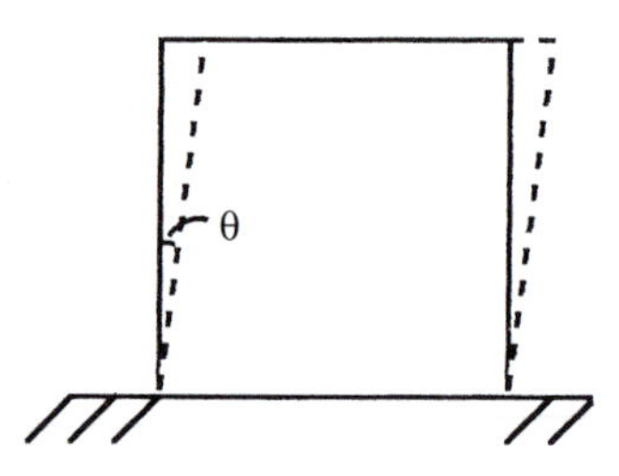

The angle θ in radians is the shear strain and

$$\theta = \frac{\text{stress}}{\text{elastic modulus}} = \frac{3 \times 10^3}{(2 \times 10^{-2})^2} \times \frac{1}{5 \times 10^{10}}$$
$$= 1.5 \times 10^{-4} \text{ radians}$$

3.3 In a tensile test, the gauge length of the cylindrical mild steel specimen was 5 cm and its diameter 13 mm. Yielding occurred at a load of 36.6 kN when the extension of the gauge length was 0.067 mm. The maximum load was 60.4 kN at an extension of 15.4 mm and the specimen finally fractured at an extension of 17.0 mm. Using these figures, work out

(a) the yield stress,

(b) Young's modulus,

(c) the (ultimate) tensile stress,

(d) the true strain at the onset of necking

and (e) the tensile ductility.

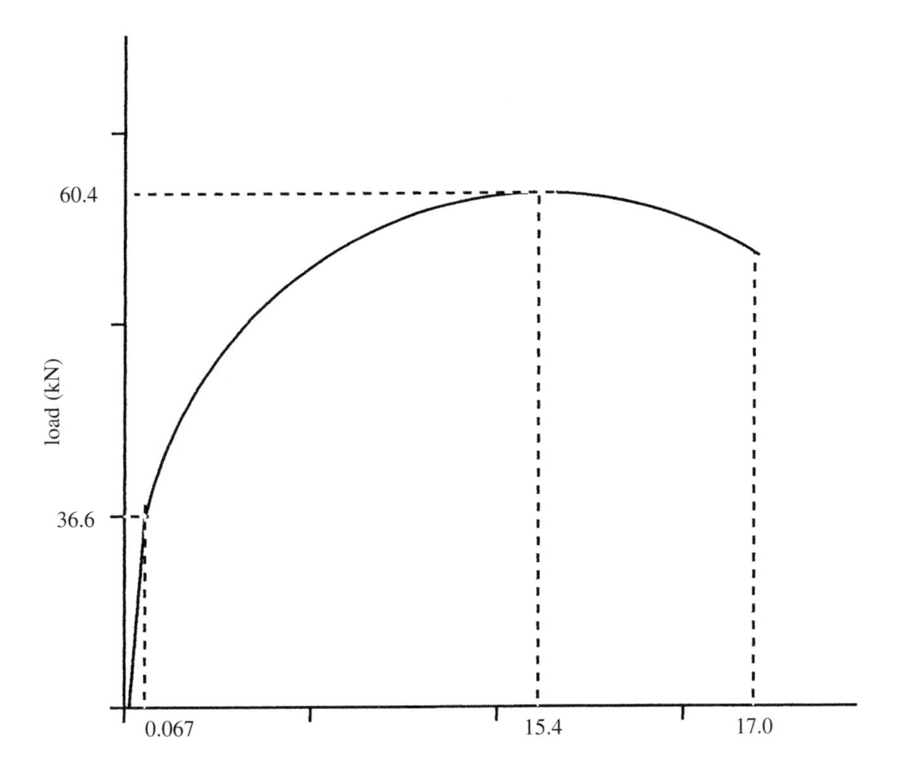

(Graph not to scale.)

(a) Yield stress $= \dfrac{\text{force}}{\text{area}} = \dfrac{36.6 \times 10^3}{\pi \times (6.5 \times 10^{-3})^2} = 276$ MPa

(b) Young's modulus $= \dfrac{\text{yield stress}}{\text{extension at yield}} = \dfrac{276 \times 10^6}{\left(\frac{0.067}{50}\right)} = 2.06 \times 10^{11}$ Pa

(c) UTS $= \dfrac{60.4 \times 10^3}{\pi \times (6.5 \times 10^{-3})^2} = 455$ MPa

(d) To work out the true strain ε_T at UTS prior to necking, note that

$$\varepsilon_T = \int_{l_0}^{l_1} \frac{dl}{l} = \ln\left(\frac{l_1}{l_0}\right) = \ln(1 + \varepsilon_N) \text{ where } \varepsilon_N \text{ is the nominal or engineering strain.}$$

$$\therefore \varepsilon_T \text{ at UTS} = \ln\left(1 + \frac{15.4}{50}\right) = \ln(1 + 0.308) = 27\%$$

i.e. the true strain is 27% when the nominal strain is 31% for uniform elongation (prior to necking).

(e) The tensile ductility $\varepsilon_L = \dfrac{17}{50} = 34\%$

3.4 A piece of machinery is suspended using two vertical solid cylindrical bars. The bars are 1 cm in diameter and 30 cm long. The material from which the bars are made had previously been assessed using a tensile test, from which the following parameters had been derived:

Yield stress	=	300MPa
Young's modulus	=	200 GPa
U.T.S.	=	450 MPa
Tensile ductility	=	20 %

(a) What weight of machinery would cause plastic yielding of the supporting bars?

(b) What would be the length of the bars when plastic yielding commenced?

(c) What weight of machinery would break the bars?

(d) What would be the length of the bars just before they broke?

(a) The yield stress is 300 MPa.
The cross-sectional area of each bar is $\pi\,(5 \times 10^{-3})^2\ \text{m}^2$
Thus force at yield $= 300 \times 10^6 \times 2 \times \pi\,(5 \times 10^{-3})^2$
$= 47.1$ kN

$$\therefore \text{ Weight of machinery} = \frac{47.1 \times 10^3}{9.81} = 4.8 \text{ tonnes}$$

(b) Strain $= \dfrac{\text{Yield stress}}{\text{Young's modulus}} = 0.15\%$

$\therefore$ The length of the bars would be $300(1 + 0.15 \times 10^{-2}) = 300.45$ mm.

(c) Similarly to (a), the weight of machinery which would break the bars

$$= \frac{450 \times 10^6 \times 2 \times \pi(5 \times 10^{-3})^2}{9.81} = 7.2 \text{ tonnes}$$

(d) Tensile ductility = 20% and so length of bars at fracture = 36 cm.

3.5 The data on p. 99 are taken from a Copper Development Association brochure.

Using the data in the table, work out at what load a tensile specimen of solution heat treated, cold worked and aged CuCr1 would break. The gauge of the testpiece is circular with a radius of 2 mm.

This question and the next one are designed to give you some practice in understanding the data in technical brochures, of which this is a genuine example. CuCr1, for example, is Cu–1wt%Cr. The meanings of 'solution heat treated' etc are explained in Chapter 4.
The tensile strength is 510 N mm^{-2} (i.e. MPa). The breaking load is therefore $510 \times 10^6 \times \pi(10^{-3})^2 = 1.6$ kN.

3.6 Again using the table on p. 99, estimate the flow stress of aged CuBe2 from its hardness and compare with the yield stress and UTS.

The hardness is 370. The units are kg mm^{-2}. The flow stress is therefore $\frac{370}{3} \times g = 1210$ MPa. The actual values of yield stress and UTS from the table are 930 and 1160 MPa. The estimate of $3 \times$ the hardness should lie between σ_y and UTS, which it doesn't, but it isn't far off.

3.7 The plastic part of the tensile true stress-true strain curve of copper is well approximated by the expression $\sigma = 320\sqrt{\varepsilon}$ MPa. Assuming that the volume of a copper tensile specimen does not change during plastic deformation, at what strain does necking begin?

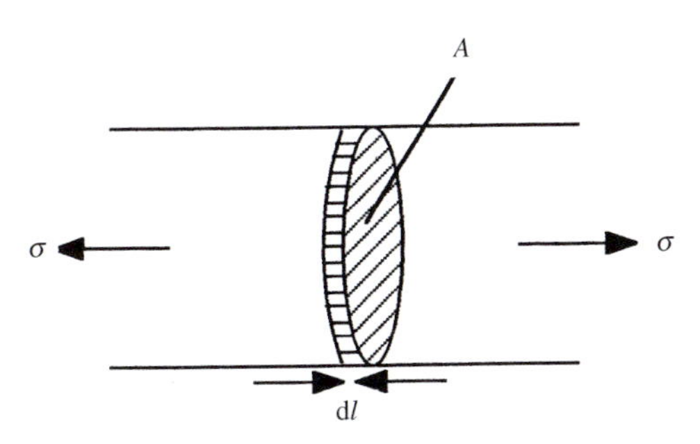

Necking will begin when the work hardening increment $A\mathrm{d}\sigma$ ceases to balance the increase in stress $-\sigma\mathrm{d}A$ ($\mathrm{d}A$ is negative), i.e.when $A\mathrm{d}\sigma = -\sigma\mathrm{d}A$

or $$\frac{\mathrm{d}\sigma}{\sigma} = -\frac{\mathrm{d}A}{A}$$

For constant volume, $$-\frac{\mathrm{d}A}{A} = \frac{\mathrm{d}l}{l} = \mathrm{d}\varepsilon$$

and $$\frac{\mathrm{d}\sigma}{\sigma} = \mathrm{d}\varepsilon$$

or $$\frac{\mathrm{d}\sigma}{\mathrm{d}\varepsilon} = \sigma \qquad \text{for necking}$$

If $$\sigma = 320\sqrt{\varepsilon}\,\text{MPa}$$

$$\frac{\mathrm{d}\sigma}{\mathrm{d}\varepsilon} = 320 \times \frac{1}{2} \times \varepsilon^{\frac{1}{2}} = 320 \times \varepsilon^{\frac{1}{2}}$$

and $$\varepsilon = 0.5 \text{ or } 50\%$$

3.8 In an impact test a pendulum of mass 9.84 kg swings down and strikes a Charpy specimen, fracturing it. Its final height is 40.2 cms less than its starting height. What is the impact energy of the specimen?

The energy lost by the pendulum to the specimen is

$$\begin{aligned} mgh &= 9.84 \times 9.81 \times 0.402 \\ &= 38.8\,\text{J} \end{aligned}$$

3.9 The surface of an aluminium alloy tensile specimen is intersected by a 1 mm crack. At what stress would the crack cause fracture? (The fracture toughness K_{Ic} of this particular aluminium alloy was 30 MPa $m^{0.5}$.)

From eq (3.3) or (3.4)

$$\sigma_f = \sqrt{\frac{2}{\pi a}} K_c$$

where a is the length of the crack

Then

$$\sigma_f = \sqrt{\frac{2}{\pi 10^{-3}}}\,30\,\text{MPa}$$

$$= 757\ \text{MPa}$$

The specimen will probably have failed for other reasons before this
(UTS for this alloy ~500–600 MPa.)

3.10 Steady state creep rates obey the following relation:

$$\dot{\varepsilon} = A\,\sigma^n e^{-}\frac{Q}{RT}$$

where $\dot{\varepsilon}$ is the strain rate, A and n are constants, Q is an 'activation' energy, R is the Gas constant and T the temperature in K.

Use the following data, derived from creep tests on solder specimens, to determine A, n and Q and thus predict the steady state creep rate at a stress of 6 MPa and temperature of 25 °C.

T (°C)	σ(MPa)	$\dot{\varepsilon}$ (s^{-1})
20	5	3.54×10^{-6}
20	10	6.11×10^{-5}
40	5	2.97×10^{-5}

(The Gas constant R = 8.31 J mol^{-1}.)

$$\dot{\varepsilon} = A\sigma^n e^{-}\frac{Q}{RT}$$

$$3.54 \times 10^{-6} = A\,5^n e^{-}\frac{Q}{R293} \quad (1)$$

$$6.11 \times 10^{-5} = A\,10^n e^{-}\frac{Q}{R293} \quad (2)$$

$$2.97 \times 10^{-5} = A\,5^n e^{-}\frac{Q}{R313} \quad (3)$$

(1) and (2)

$$2^n = 17.26$$

and

$$n = \frac{\ln(17.26)}{\ln(2)} = 4.1$$

(1) and (3)

$$8.39 = e\frac{Q}{R}\left(\frac{1}{293} - \frac{1}{313}\right)$$

and

$$Q = 81.1\ \text{kJ mol}^{-1}$$

(1) :

$$A = 1.41 \times 10^6$$

Using n, A and Q thus defined, $\dot{\varepsilon}$ at 25 °C and 6 MPa

$$= 1.41 \times 10^6 \times 6^{4.1} \times e^{-\frac{81000}{831 \times 298}}$$

$$= 1.3 \times 10^{-5}\ s^{-1}$$

Note that the units of A don't bear thinking about: the important thing is to stick with one set of units for $\dot{\varepsilon}$, σ and Q.

Chapter 4

4.1 What would be the cost of the material in a kilometre of copper wire of diameter 0.2 mm? (The density of copper is 8.9 Mg m^{-3}. Take the cost of copper from Table 2.5.)

The volume of a kilometre of wire of diameter 0.2 mm (assumed cylindrical)

$$= \pi \times (10^{-4})^2 \times 10^3$$

$$= 3.14 \times 10^{-5}\ m^3$$

The weight $= 8.9 \times 10^3 \times 3.14 \times 10^{-5}$

$= 0.28$ kg

The cost $= 0.28 \times 2700 \times 10^{-3}$

$= 76$ ¢ (US)

4.2 Common electrical solder consists of 62 wt% tin and 38 wt% lead. What is the composition in at%? (The Relative Atomic Masses of tin and lead are 119 and 207 respectively.)

$$\text{The atomic \% of tin} = \frac{\frac{62}{119}}{\frac{62}{119} + \frac{38}{207}} = 74\% \text{ with } \therefore 26 \text{ at \% lead}$$

4.3 How many phases co-exist in Sn–50 wt% Pb at 250 °C, 200 °C, 150 °C?
in Pb–10 wt% Sn at 250 °C, 200 °C, 150 °C?

From Fig. 4.10, in Sn–50 wt% Pb
at 250 °C there is one phase (liquid)
at 200 °C there are two phases (liquid and β)
at 150 °C there are two (solid) phases (α and β)

in Pb–10 wt% Sn
at 250 °C there is one phase (β)
at 200 °C there is one phase (β)
at 150 °C there are two phases (α and β)

4.4 As liquid Sn–50 wt% Pb is cooled, what is the composition of the first solid to form?

Reading from Fig. 4.10:

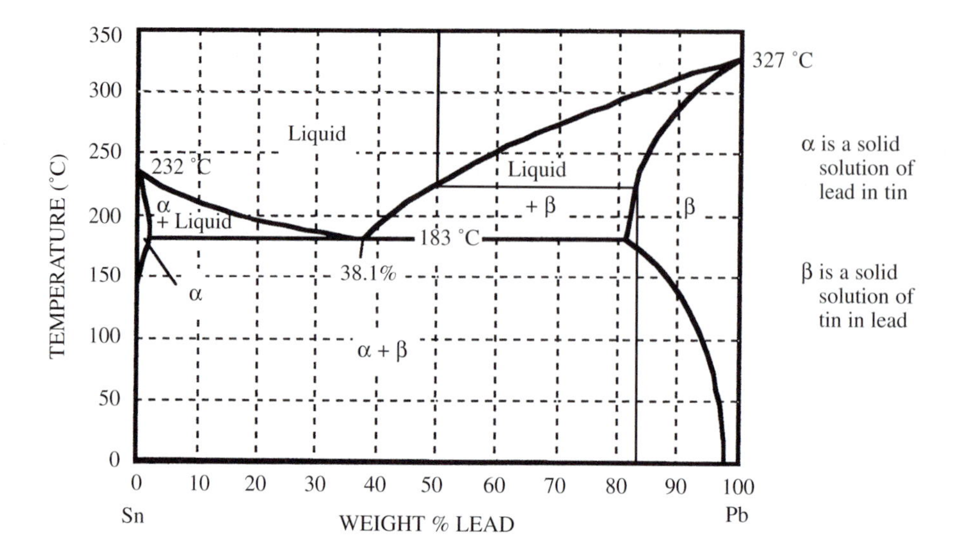

Sn–50 wt% Pb starts to solidify at 225 °C, producing solid of composition Sn–83 wt% Pb (i.e. a lot more lead-rich than the liquid alloy).

As the temperature of the alloy falls from 225 °C to the eutectic temperature, 183 °C, the composition of the liquid moves along the left hand line (called the *liquidus*) and that of the solid along the right hand line (the *solidus*). The composition of the solid (for example), at any particular temperature, is the composition of the *whole* solid, including the first solid to form, and so the composition of the previously formed solid must change during cooling. This requires solid state diffusion, which is a slow process. Unless the rate of cooling is *very* slow, there is not enough time for solid state diffusion to complete its job, and the composition of the first-formed solid remains high in lead. This effect shows up in castings, and is called *coring*.

The horizontal line connecting the liquidus to the solidus at any given temperature is called a *tie-line*. The position of the overall alloy composition along this tie-line enables us to calculate the *amounts* (as opposed to chemical compositions) of the two phases, liquid and solid (see Questions 4.7 and 4.8 below). At the eutectic temperature, 183 °C, the remaining liquid undergoes a eutectic reaction and decomposes into a fine mixture of α and β (see Fig. 4.11).

4.5 What are the solubilities of tin in lead and lead in tin at 100 °C?

Here is the phase diagram again with a bold line drawn across at 100 °C and the two solubilities arrowed:

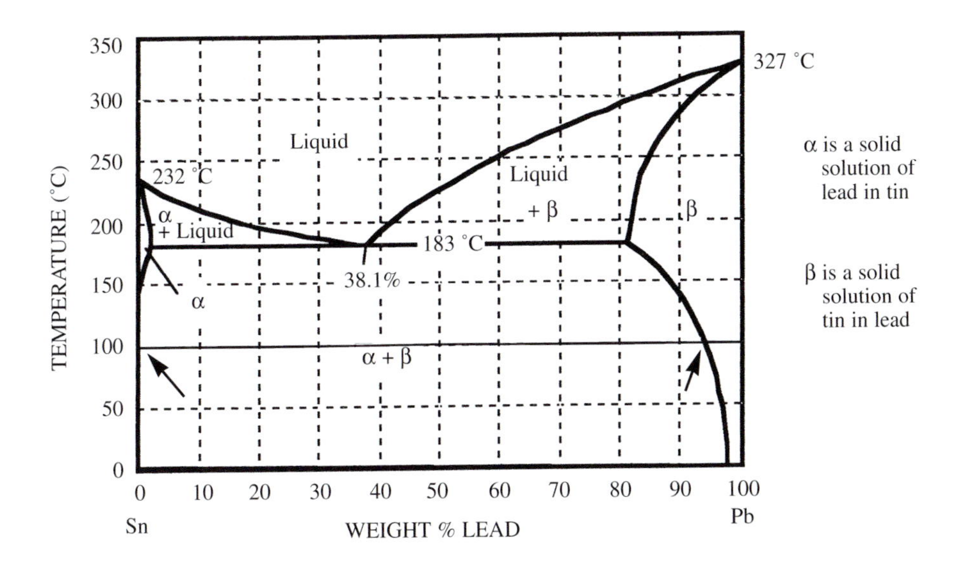

Reading off the compositions on the x-axis, the solubility of Sn in Pb is 6 wt% and that of Pb in Sn is zero, to the resolution with which I have drawn the phase diagram (solubilities are always finite, actually).

4.6 In Sn–50 wt% Pb at 100 °C, what are the proportions of the two phases?

Refer again to the figure above, which shows the lead–tin phase diagram with a line drawn across it at 100 °C. Imagine moving along this line, thus changing the overall composition of the alloy. We know that the chemical compositions of the two phases (the two solid solutions) do not change. Thus, their relative amounts *must* change.

Let us label the overall composition of the alloy x (i.e. Sn–x wt%Pb) and the compositions of the two end-points x_1 and x_2. Let us further call the fraction of the microstructure which consists of the Sn solid solution f_1 and that which consists of the Pb solid solution f_2. By noting that the total amount of Pb in the alloy is fixed, then

$$x = f_1 x_1 + f_2 x_2$$

Remembering that $$f_1 + f_2 = 1$$

then $$f_1 = \frac{x_2 - x}{x_2 - x_1}$$

and $$f_2 = \frac{x - x_1}{x_2 - x_1}$$

This is called the **lever rule**

x

x_1 x_2

for obvious reasons and is quite generally true for any two phase region in a binary (two element) phase diagram. We will use it quite a lot in Chapter 5.

Coming back to the case in point, $x_1 = 0$, $x_2 = 0.94$ and $x = 0.5$ and so the fraction of the microstructure corresponding to the tin-rich solid solution (α)

$$f_1 = \frac{0.94 - 0.5}{0.94 - 0.0} = 0.47$$

and the fraction corresponding to the lead-rich solid solution β

$$f_2 = \frac{0.5 - 0.0}{0.94 - 0.0} = 0.53$$

Thus 47% of the structure (by weight) consists of the tin-rich α phase and 53% of the lead-rich β phase.

Now we extend this approach to include the product of the eutectic reaction:

4.7 In Sn–50 wt% Pb just below the eutectic temperature, what proportion of the structure is occupied by the eutectic mixture? Of the β-phase, how much is in the eutectic mixture and how much as standalone β-phase?

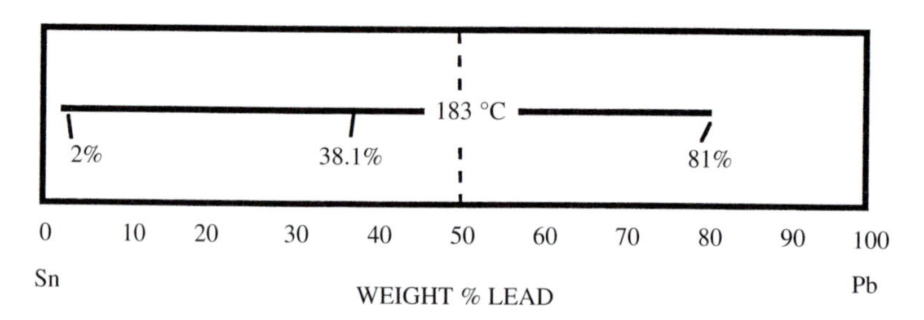

Regarding, for the moment, the eutectic mixture as a phase (which it isn't, really, of course) the 50% alloy consists of $\dfrac{0.81 - 0.5}{0.81 - 0.381} = 0.72 = 72\%$ of the eutectic mixture and 28% of the standalone β phase. The eutectic mixture consists of $\dfrac{0.81 - 0.381}{0.81 - 0.02} = 0.54 = 54\%$ Sn-rich α phase and 46% Pb-rich β phase. Multiplying by 72%, we have:

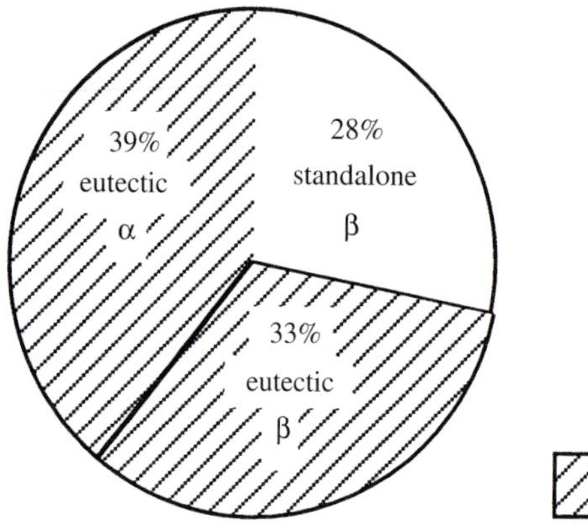

The up-market term for 'standalone' is pro-eutectic, because the phase appears 'before the eutectic reaction'.

I hope this isn't all so complicated that it has put you off! There will be another opportunity to practise with phase diagrams in Chapter 5.

4.8 Using Fig. 4.19, up to what temperature would it be safe to trust the strength of a work hardened copper wire? Does the load on the wire matter?

The level of cold work in a drawn wire is generally extremely high. From Fig. 4.19(a) 100 °C seems an absolute maximum. The stress will matter only to the extent of how close it is to the wire's strength. In fact, the wire will *creep* (see Section 3.9) before it fails.

4.9 Steel specimens with average grain sizes of 50 μm and 100 μm have yield strengths of 250 and 200 MPa respectively. To what would the grain size have to be changed to give a yield strength of 300 MPa?

(The yield strength σ_y of a metal depends on the grain diameter d via the *Hall-Petch* relationship:

$$\sigma_y = \sigma_0 + k_y d^{-0.5}$$

where σ_0 and k_y are constants.)

$$\begin{aligned} \sigma_y &= \sigma_0 + k_y d^{-0.5} \\ 250 &= \sigma_0 + k_y 50^{-0.5} \\ 200 &= \sigma_0 + k_y 100^{-0.5} \\ \therefore k_y &= 1207 \text{ MPa } \mu\text{m}^{0.5}; \quad \sigma_0 = 79 \text{ MPa} \\ 300 &= 79 + 1207\, d^{-0.5} \\ d &= 30\ \mu\text{m} \end{aligned}$$

4.10 If 1 at % gold raises the yield strength of copper by 16 MPa, what effect would you expect 2% of tin to have? (The atomic diameters of copper, gold and tin atoms in a copper environment are 0.256, 0.291 and 0.313 nm, respectively.)

Solid solution strengthening depends on the difference in sizes of the solvent (matrix) atoms and the solute atoms:

$$\frac{d\sigma_y}{dc} \;\alpha\; \left(\frac{1}{a}\frac{da}{dc}\right)^2 \tag{4.1}$$

or, to put it another (but entirely equivalent) way

$$\frac{d\sigma_y}{dc} \;\alpha\; |\Delta d|^2$$

where Δd is the difference in atomic diameters between solvent and solute atoms.
The strengthening is also proportional to the concentration of the solute.
Thus, if 1 at% gold provides copper with 16 MPa strengthening, 2 at% tin would be expected to provide $16 \times 2 \times \dfrac{(0.313 - 0.256)^2}{(0.291 - 0.256)^2} = 85$ MPa. In fact the real answer is 84 MPa.
(Note: the size of, for example, gold or tin atoms in a copper matrix is quite different from that of gold or tin atoms in elemental gold or tin.)

4.11 The strengthening phase in Cu-Cr is chromium itself (with a little copper dissolved in it). If the densities of copper and chromium are 8.9 and 7.1 Mg m^{-3} respectively, what vol % would the precipitates occupy in Cu–1 wt% Cr? If the precipitates are cubes with edge length 2 nm, what is the average separation between them?

The increase in yield strength due to the bowing mechanism (see § 4.6, section 4) is $\Delta\sigma_y = \frac{\mu b}{2\pi L}$ where μ is the shear modulus (4 × 10^{10} Pa for copper), b the magnitude of the Burgers vector (for copper 0.25 nm) and L the separation between the precipitates. Estimate the strength of Cu–1 wt% Cr assuming all the precipitates are 2 nm cubes and that they are not cut by the dislocations, and compare with the values in the table in Q3.5 (see p. 99).

$$\text{Vol\% Cr} = \frac{\frac{1}{7.1}}{\frac{1}{7.1} + \frac{99}{8.9}} = 1.25\%$$

If the average separation of the cubes is L, then $\frac{2^3}{L^3} = 0.0125$ and L = 81.6 nm.

$$\Delta\sigma_y = \frac{4 \times 10^{10} \times 0.25 \times 10^{-9}}{2\pi \times 8.6 \times 10^{-9}} = 185 \text{ MPa}$$

From the table in Question 3.5, the rise in σ_y due to the precipitates in CuCr1 is from 45 to 265 MPa i.e. $\Delta\sigma$ = 220 MPa. Thus on this (over)simple model the precipitates would need to be slightly smaller than the 2 nm quoted.

Chapter 5

(Q's 5.1–5.3: The relative atomic masses of carbon, iron and manganese are 12.01, 55.85 and 54.93 respectively.)

5.1 A steel has a carbon concentration of 0.2 wt%. (The 'wt' is often missed out in steel literature.) What is the at % of carbon?

$$\frac{\frac{0.2}{12.01}}{\frac{0.2}{12.01} + \frac{99.8}{55.85}} = 0.92 \text{ at\%}$$

5.2 The strengthening phase in plain carbon steels is Fe_3C. What is the wt% of C in Fe_3C?

$$\frac{25 \times 12.01}{75 \times 55.85 + 25 \times 12.01} = 6.7 \text{ wt\%}$$

5.3 A steel contains 0.1 wt% C and 1 wt% Mn. What are the at% of the two elements?

$$0.1 \text{ at\% C} = \frac{\frac{0.1}{12.01}}{\frac{0.1}{12.01} + \frac{1}{54.93} + \frac{98.9}{55.85}} \text{wt\%} = 0.46 \text{ wt\%}$$

$$1 \text{ at\% Mn} = \frac{\frac{1}{54.93}}{\frac{0.1}{12.01} + \frac{1}{54.93} + \frac{98.9}{55.85}} \text{wt\%} = 1.01 \text{ wt\%}$$

leaving 98.53 wt% Fe.

(Note that Mn has almost the same RAM as Fe.)

5.4 What are the names for Fe_3C, the solid solution of carbon in b.c.c. iron and the solid solution of carbon in f.c.c. iron?

Fe_3C is **cementite**.
The solid solution of carbon in b.c.c. iron is **ferrite**.
The solid solution of carbon in f.c.c. iron is **austenite**.
('Cementite' and 'ferrite' were names given by the American metallurgist Dr H.M. Howe. Cementite was first observed in cemented or 'blister' steel. Ferrite comes from Latin 'ferrum'. Professor. Roberts-Austen, a British metallurgist, was remembered via 'austenite' following a suggestion by F. Osmond, another metallurgist.)

You will find an understanding of Q's 4.2–4.7 (Chapter 4) very helpful for Q's 5.5 and 5.7–5.11. Use Figs 5.1 and 5.2.

5.5 (a) What is the solubility limit of carbon in b.c.c. iron at 727 °C?
(b) What is the solubility limit of carbon in f.c.c. iron at 1148 °C?

From Fig. 5.1 the solubility limit of carbon in b.c.c. iron at 727 °C is 0.022 wt%.
From Fig. 5.1 the solubility limit of carbon in f.c.c. iron at 1148 °C is 2.11 wt%.
These numbers have been read off the diagram. Actually, if you use a ruler, you will find the drawing does not agree very exactly. What is important is that the first figure, the solubility in b.c.c. iron—ferrite—is tiny.

5.6 (a) The lattice parameter of b.c.c. iron is 0.286 nm. Treating the atoms as touching spheres, where are the largest interstices (holes) in the structure? What is their size?

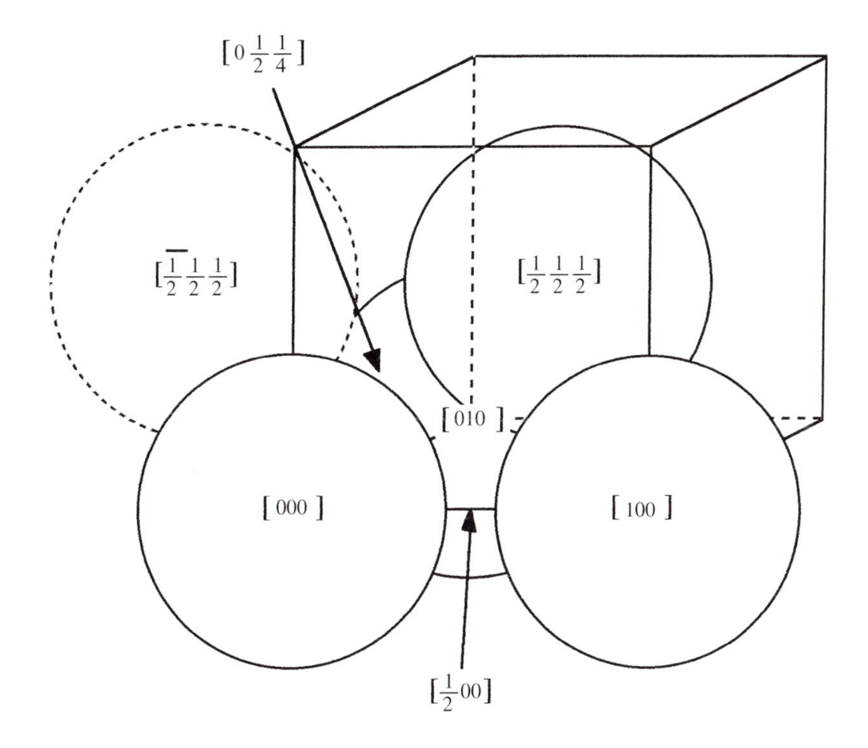

For the purposes of clarity, I have shown some only of the atoms in the b.c.c. unit cell. By inspection and a certain amount of trial and error the largest interstices are at $\left[\frac{1}{4}0\frac{1}{2}\right]$. These interstices are usually referred to as 'tetrahedral' because of the tetrahedral arrangement of atoms surrounding them. The radius of the interstice is $\frac{\sqrt{5}-\sqrt{3}}{4}a = 36$ pm.

(NB In steel the carbon atom does *not* sit here. It sits at $\left[\frac{1}{2}0\,0\right]$ (see figure above) because the neighbouring iron atoms can be pushed aside more easily.)

(b) The lattice parameter of f.c.c. iron (extrapolated to room temperature) is 0.365 nm. Where are the largest interstices and what is their size?

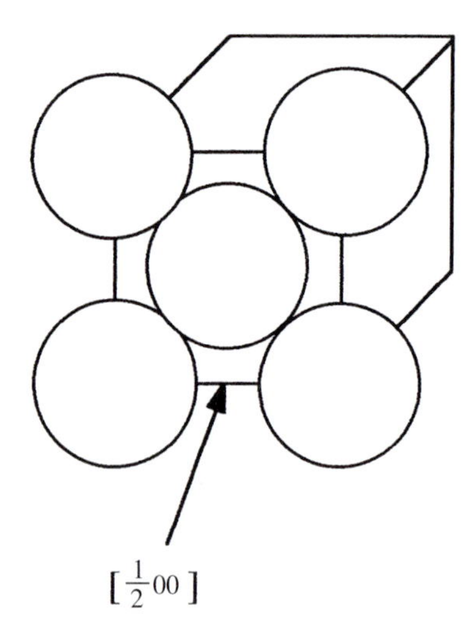

For f.c.c. the largest interstice is at $\left[\frac{1}{2}0\,0\right]\left(\equiv\left[\frac{1}{2}\frac{1}{2}\frac{1}{2}\right]\right)$. Its radius is $\frac{1}{2}\left(1-\frac{1}{\sqrt{2}}\right)a = 0.126\,a = 46$ pm.

(c) A carbon atom in iron has diameter 77 pm. What can you infer about the solubilities of carbon in b.c.c. and f.c.c. iron?

The radius of the carbon atom ($77 \div 2 = 38.5$ pm) lies between the size of the b.c.c. interstice (smaller, remember, than the 36 pm radius calculated above) and that of the f.c.c. interstice. This gives rise to a difference in solubility of ~100 times.

Q's 5.7–5.11 relate to a 0.2 wt% C mild steel.

5.7 Above what temperature must the steel be raised in order to austenitise it?

From Fig. 5.2 the boundary between the $\alpha+\gamma$ and γ regions at 0.2 at% C is about 840 °C. This is the temperature, therefore, above which the steel needs to be raised in order to convert it to austenite.

5.8 On cooling the steel (slowly) at what temperature does the 'pro-eutectoid' ferrite start to form? At what temperature does the pearlite form?

The pro-eutectoid ferrite is that which appears before the eutectoid reaction (i.e. above the eutectoid temperature during cooling). The answer to the first part is the same as that for Question 5.7: 840 °C.

The pearlite forms at 727 °C (Fig. 5.1 or 5.2).

5.9 What is the composition of the first ferrite to form? What is the composition of the ferrite when the pearlite forms? What is the composition of the ferrite at room temperature?

From Fig. 5.1 or 5.2, the boundary between the α and $\alpha + \gamma$ phase regions at 840 °C is $0.022 \times \frac{(912 - 840)}{(912 - 727)} = 0.0086$ wt%, where I have used the numbers in Fig. 5.1 rather than the drawing (see answer to Question 5.5). This is the composition of the first ferrite to form i.e. it is almost pure iron. As the temperature of the steel falls toward 727 °C the solubility of the ferrite for carbon increases to 0.022 wt%C. That ferrite which has already formed adjusts its composition via solid state diffusion. As the temperature falls towards room temperature from the eutectoid temperature (727 °C) the solubility of carbon in the b.c.c. ferrite falls again, effectively to zero. The speed of diffusion of the carbon falls also and the room temperature value requires some time before it is achieved. The same is true, superplus, of substitutional diffusion (i.e. that of all the other solutes besides carbon) and room temperature equilibrium is never truly attained.

5.10 What would you see in an etched specimen in an optical microscope? What proportion of the microstructure would be ferrite and what proportion pearlite?

The microstructure would be similar to that shown in Fig. 5.4, but the wt% carbon here is 0.2% rather than 0.4% in Fig. 5.4, i.e. half as much, and so the proportion of pearlite is half, also. Thus 3/4 of the microstructure would be ferrite and $\frac{1}{4}$ pearlite.

5.11 At 727 °C, what proportion of the pearlite consists of ferrite? What proportion of the ferrite constitutes the pro-eutectoid ferrite and what proportion is in the pearlite? What proportion of the carbon is in the ferrite and what proportion in the pearlite?

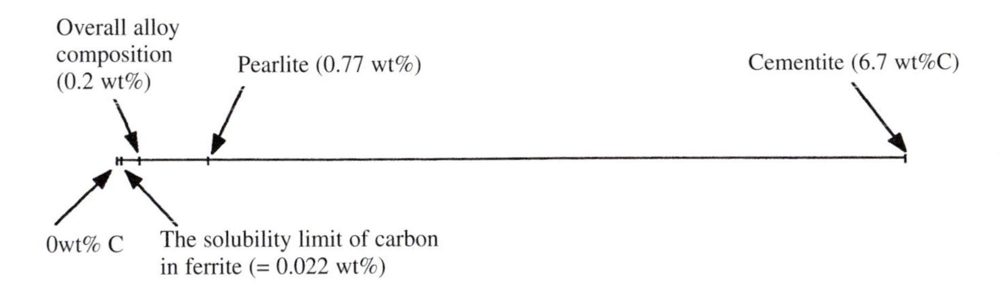

The proportion of pearlite which consists of ferrite $= \frac{6.7 - 0.77}{6.7 - 0.022} = 0.89$

The proportion of the microstructure which is pearlite is $\frac{0.2 - 0.022}{0.77 - 0.022} = 0.24$

Thus we have

ferrite	pearlite	
	ferrite	cementite
76%	21%	3%

and $\frac{76}{97} = 78\%$ of the ferrite is pro-eutectoid, leaving 22% in the pearlite. Notice how little cementite there is, considering this is where a large proportion of the strength comes from. The cementite is effectively non-deformable.

The 97% of the alloy which is ferrite contains $97 \times \frac{0.022}{0.2} = 10.7$ % of the carbon. The other 89.3% of the carbon is in the cementite, reflecting the low solubility of carbon in b.c.c. iron.

5.12 A comparison of the two parts of Fig. 5.5 suggests that the Mn content of the steel on the left is 1 wt% and that the carbon content of the steel on the right is 0.13 wt%. For a steel with both 0.13 wt% C and 1 wt% Mn, enumerate the four main contributions to the tensile strength.

The four contributions are roughly:

Ferrite	300 MPa
Solid solution hardening of ferrite by manganese	40 MPa
Grain boundaries in the ferrite	60 MPa
Pearlite	80 MPa

5.13 What types of steels would the following be made from? In each case give also the main method of fabrication.

(a) Electricity transmission tower (what you may know as a 'pylon').

(b) A disc in the low temperature part of a power generation turbine. (The disc is the central part to which the blades are attached.)

(c) A disc in the high temperature part of a turbine.

(d) The core of a transformer.

(e) The pressure vessel of a nuclear reactor.

(a) Electricity transmission tower: typically a 0.15 wt% C, 1.5 wt% Mn plain carbon steel. This steel has a 'carbon equivalent' of ~0.4 i.e. if Mn were converted to C, the steel would behave like one containing ~0.4 wt% C in terms of its propensity for martensite formation. Anything over 0.4 wt% C cannot easily be welded, because martensite forms in the weld.
Fabrication: rolled bar.

(b) Low temperature turbine disc: 'low temperature' means down to ~80 °C. This steel is picked for *toughness* (see Section 3.6). Typically $3\frac{1}{2}$Ni 1Cr Mo $\frac{1}{2}$Mo $\frac{1}{4}$V (wt%) forged to shape, quenched and tempered. (The blades attached to it would probably be 12 wt% Cr stainless steel.)

(c) High temperature disc: this would typically operate at ~565 °C. It is picked for *creep resistance* (see Section 3.9). Typically 1Cr $\frac{1}{2}$ Mo $\frac{1}{3}$ V (wt%) forged to shape, quenched to *lower bainite* (a halfway house between martensite and ferrite+carbides) and tempered to *upper bainite* (closer to ferrite+carbides).
(The blades would probably be made similarly, and of the same steel.)

(d) A transformer core is made from Fe—3-4 wt% Si rolled sheet (see Section 10.5).

(e) The pressure vessel of a nuclear reactor is designed for toughness. (The possibility of a crack forming is totally unacceptable.) It is made from a boiler steel such as A533B, which has a composition Fe–0.2 wt%C–1.4 wt%Mn–0.5 wt%Ni–0.5 wt%Mo–0.2 wt%Si–0.1 wt%Cr and a fracture toughness K_{Ic} of 210 MPa $m^{0.5}$. The steel is cast as an ingot and *ring rolled* to circular plate sections, which are then welded together on site. The plate is 215 mm thick and most of the fabrication consists of the welding.

5.14 What welding method would you use for the following jobs?:

N.B. There is not usually any single uniquely correct answer. There are generally equally valid alternative welding technique choices.

(a) The steel frame for a building

Manual metal arc or metal inert gas (MIG).

(b) Repairing some farm equipment

Transportability and simplicity are all: oxy-acetylene.

(c) Joining stainless steel tube in a chemical plant

A high quality result with tungsten inert gas (TIG).

(d) Fixing two car body panels together

Classic example of spot welding (resistance welding).

(e) Joining two thick plate sections in a pressure vessel

Submerged arc, or, provided the geometry is very simple, electroslag.

Chapter 6

6.1 The equation $\Delta G = nFE$ relates the Free Energy change ΔG associated with a reaction to the electrical potential E of a cell based on the reaction. n is the number of electrons transferred and F is called Faraday's constant. If ΔG is expressed in kJ $mole^{-1}$ and E is in Volts, work out F. For the reaction (see equation (6.2))

$$Fe + 2H^+ \rightarrow Fe^{++} + H_2 \uparrow$$

E = 0.44 V (see Table 6.1). What is ΔG?

(Avogadro's Number $N = 6.023 \times 10^{23}$ and e, the charge on the electron, $= 1.602 \times 10^{-19}$C.)

$F = N \times e$ J $mole^{-1}$ V^{-1} = 96485 = 96.485 kJ mol^{-1} V^{-1}
If $E = 0.44$ V, $\Delta G = 42.5$ kJ mol^{-1}

6.2 In a copper electropurification cell, 1 Coulomb of electricity is passed. If the copper cathode has an area of 10 mm^2, what thickness of copper is deposited? [The electronic charge $e = 1.602 \times 10^{-19}$; the lattice parameter of copper = 0.3616 nm]

1 C corresponds to 6.242×10^{18} electrons, which will reduce 3.121×10^{18} doubly charged copper ions. If the lattice parameter is 0.3615 nm, then with 4 atoms per unit cell (see Chapter 2), 1 C will produce $\frac{3.121 \times 10^{18}(0.3615 \times 10^{-9})^3}{4} = 3.686 \times 10^{-11} m^3$. The copper plate will therefore be $3.686 \times 10^{-11}/10^{-5} = 3.686\ \mu m$ thick.

6.3 Out of the following pairs of metals, which one would you expect to corrode the other?
(a) Copper and iron (b) Iron and zinc (c) Silver and nickel.
(Use the electrochemical series (Fig. 6.13).)

Always the more noble metal causes the less noble, or baser metal to corrode.
Thus (a) copper will cause iron to corrode
(b) iron will cause zinc to corrode
and (c) silver will cause nickel to corrode

According to the electrochemical series, iron should corrode aluminium, rather than the other way round, as actually happens. Why does the iron corrode? What solutions to the problem can you think of?

Aluminium is covered by a thin, impervious and perfect layer of alumina which confers on it an artificial nobility (unless the oxide film dissolves in the electrolyte). The problem may be solved by, for example, insulating electrically the iron from the aluminium. NEVER use steel bolts on aluminium!
(The *galvanic series* takes this sort of effect into account and is more directly useful for predicting corrosion than the electrochemical series.)

6.4 What type of battery (non-rechargeable) would you expect to be used in (a) a calculator (b) a camera (c) a cardiac pacemaker?

From Table 6.2, (a) dry Leclanché or alkaline, (b) mercury or silver (mercury is being phased out on environmental grounds) and (c) lithium.

6.5 I recently bought 4 AA/LR6/MN1500 batteries for the equivalent of \$5. In Birmingham I pay about 10¢ per kWh for mains electricity. Using Table 6.2 as a guide, roughly how much more expensive was my battery power than the equivalent suppied through the electricity mains? (Take the capacity of an AA alkaline battery as 2.5 Ah. (This varies considerably, depending on the mode of use.))

From Table 6.2, the power capacity of one AA alkaline battery is 3.75 Wh. Times 4 batteries = 15 Wh. In terms of mains electricity, this would cost 0.15¢. The ratio of costs is $\frac{500}{0.15} \approx 3000$.
Don't take the details of this calculation too seriously: the important thing is that battery power is a LOT more expensive than mains power. We only use batteries for convenience, or when it is impracticable to use the mains.

Chapter 7

7.1 (I am writing this sitting in my kitchen.)

Which of the following are made from ceramic?:

(a) A wall tile.

(b) The refrigerator.

(c) The electrical plug connecting the fridge to the mains.

(d) The window glass.

(e) The window frame.

(f) The brick walls.

(g) The fluorescent light tube in the ceiling.

(a) The wall tiles will be ceramic (a mixture of ceramic phases, in fact) glazed with another ceramic.

(b) The fridge will be a mixture of plastic and metal—not much ceramic here.

(c) Window glass—'glass' of any sort, actually—is ceramic.

(d) The window frame will be u-PVC (plastic), metal or wood (a natural polymer), but definitely not ceramic.

(e) Brick is a multi-phase ceramic.

(f) The fluorescent tube will be ceramic (mainly—the glass bit anyway).

My experience is that when you sit in a room and classify the surrounding materials, most of them are polymer/plastic (see Chapter 8) or ceramic.

7.2 Assuming that the ions are touching spheres, what is the percentage of free space in $BaTiO_3$? (The ionic radii of Ba^{2+}, Ti^{4+} and O^{2-} are 0.135, 0.068 and 0.140 nm respectively.)

See Fig. 7.3. The large barium and oxygen ions will touch across the face diagonals: thus the lattice parameter, a, is defined by

$$\sqrt{2}\,a = 2(r_{Ba^{2+}} + r_{O^{2-}})$$
$$a = 0.389 \text{ nm}$$

If you count up the atoms in Fig. 7.3, you will confirm for yourself that the cubic unit cell shown contains one formula unit $BaTiO_3$. The combined volumes of the atoms divided by the volume of the unit cell

$$= \frac{\frac{4}{3}\pi(0.135^3 + 0.068^3 + 3 \times 0.140^3)}{0.389^3}$$
$$= 78.3\%$$

78.3% of space is occupied. Note that this is very efficient packing and more than is possible for an element, where all the atoms are the same size (74% maximum—see Question 2.1).

7.3 What are the electrical resistivities of copper, steel, soda glass and alumina? How would they vary if the temperature rose from room temperature to 300 °C?

(From Table 2.3) ρ for copper $= 1.70 \times 10^{-8}$ Ωm, for steel (Table 5.5) 1.58×10^{-7} Ωm, for soda glass (Fig. 7.8 and Table 7.2) 10^8 Ωm and for alumina (Fig. 7.8) 10^{16} Ωm.

If the temperature were to rise from room temperature to 300 °C, the metal resistivities would rise slightly, while the ceramic resistivities would fall by a proportionately greater amount.

7.4 What potential would you have to apply across 1 mm of porcelain before it explodes (i.e. breaks down and fragments)?

From Table 7.2 the breakdown strength of porcelain is $\sim 3 \times 10^7$ V m^{-1}. Thus for 1 mm of porcelain the breakdown potential would be 30 000 V.

7.5 What is the capacitance of a 1 cm × 1 cm × 0.1 mm piece of steatite whose faces are completely coated with thin metal electrodes?
(The permittivity ε_0 of free space is 8.854×10^{-12} Fm^{-1}.)

The capacitance C of a sheet of material is $\frac{\varepsilon_R \varepsilon_0 A}{d}$. From Table 7.3 ε_R for steatite is 6.1. The capacitance of the 1 cm × 1 cm × 0.1 mm piece of steatite is therefore $\frac{6.1 \times 8.854 \times 10^{-12} \times 10^{-4}}{10^{-4}}$ = 54 pF.

7.6 The electrical conductivity of Mn_3O_4 is measured at various temperatures as follows:

T(K)	$(\Omega m)^{-1}$
200	3.63×10^{-11}
300	3.00×10^{-7}
400	1.66×10^{-5}
500	1.57×10^{-4}
600	1.37×10^{-3}
700	5.31×10^{-3}
800	1.00×10^{-2}
900	2.56×10^{-2}
1000	5.05×10^{-2}

By plotting the logarithm of the conductivity vs 1/T, determine the activation energy of the process.

T(K)	$1/T$	σ $((\Omega m)^{-1})$	ln σ
200	5.00×10^{-3}	3.63×10^{-11}	-24.04
300	3.33×10^{-3}	3.00×10^{-7}	-15.02
400	2.50×10^{-3}	1.66×10^{-5}	-11.01
500	2.00×10^{-3}	1.57×10^{-4}	-8.76
600	1.67×10^{-3}	1.37×10^{-3}	-6.59
700	1.43×10^{-3}	5.31×10^{-3}	-5.24
800	1.25×10^{-3}	1.00×10^{-2}	-4.61
900	1.11×10^{-3}	2.56×10^{-2}	-3.67
1000	1.00×10^{-3}	5.05×10^{-2}	-2.99

$E/k = 5211$ $\quad E = 7.19 \times 10^{-20}$ J K^{-1}
= 0.45 eV K^{-1}

Chapter 8

8.1 (Another question written in my kitchen!—see Q 7.1)
Which of the following are plastics?

(a) Floor tiles
(b) Washing machine
(c) The plug for the washing machine
(d) The wall socket into which the plug fits
(e) The sink
(f) The table

(a) Either plastic or ceramic. Mine are plastic.
(b) A lot of washing machine parts nowadays are plastic, including the main tub. This is part of the reason washing machines are so much quieter.
(c) Plastic.
(d) Plastic.
(e) Either metal or plastic. Not many kitchen sinks nowadays are ceramic (at least in Birmingham, England).
(f) The main frame will be metal or plastic (including wood). There may be ceramic tiles let into the top (there are in the one I am typing at).

8.2 What is the difference between a plastic and a polymer?

A polymer is the pure chemical; a plastic is what you buy. A plastic equals a polymer plus various additives—for example glass fibre to stiffen it.

8.3 What is a typical number of mers (monomer units) in a PVC molecule? (You will need to look in other books for the answer to this.)

A few hundred upwards (fairly variable).

8.4 Are the following thermoplastics, thermosets or rubbers?

(a) polythene
(b) PVC
(c) wood
(d) araldite
(e) perspex
(f) paper
(g) neoprene

Are they crystalline, amorphous or a mixture?

(a) polythene	thermoplastic	amorphous or glass/crystal mixture
(b) PVC	thermoplastic	amorphous
(c) wood	thermoplastic	amorphous
(d) araldite	thermoset	amorphous

(e) perspex	thermoplastic	amorphous
(f) paper	thermoplastic	amorphous
(g) neoprene	rubber	amorphous

8.5 If a 1 kg weight were hung from a 1 m long 1mm^2 cross-section PVC wire, by how much would the wire instantaneously extend? (You may find Table 8.4 useful; g, the acceleration due to gravity, is 9.81 ms^{-2}.)

The stress on the wire $= \frac{9.81}{10^{-6}} = 9.81$ MPa. From Table 8.4, Young's modulus E for PVC $= 3000$ MPa. The strain ε is therefore $\frac{9.81}{3000} = 0.33\%$ and the extension is 3.3 mm.

8.6 The DC resistivity of nylon is measured over a range of temperatures with the following results:

T (°C)	ρ (Ωm)
40	3.16×10^{13}
50	4.37×10^{12}
60	6.31×10^{11}
70	8.71×10^{10}
80	1.51×10^{10}
90	3.09×10^{9}
100	6.49×10^{8}
110	1.86×10^{8}

What is the activation energy of the conduction process?

T (°C)	ρ(Ωm)	T^{-1}(K^{-1})	ln ρ
40	3.16×10^{13}	3.195×10^{-3}	31.1
50	4.37×10^{12}	3.096×10^{-3}	29.1
60	6.31×10^{11}	3.003×10^{-3}	27.2
70	8.71×10^{10}	2.915×10^{-3}	25.2
80	1.51×10^{10}	2.833×10^{-3}	23.4
90	3.09×10^{9}	2.755×10^{-3}	21.9
100	6.49×10^{8}	2.681×10^{-3}	20.3
110	1.86×10^{8}	2.611×10^{-3}	19.0

$E/k = 20957$ $\quad E = 2.89 \times 10^{-19}$ J K^{-1}
$= 1.8$ eV K^{-1}

8.7 A 1 cm × 1 cm × 0.1 mm sheet of PS has metal electrodes evaporated onto its square faces. What would the capacitance of the PS sheet be? A 1000V potential difference is applied between the electrodes. What current would you expect to flow? If the voltage were increased, at what point would breakdown occur? (The permittivity of free space, ε_0, $= 8.854 \times 10^{-12}$ Fm^{-1}.)

The capacitance C of a parallel plate capacitor is $\frac{\varepsilon_R \varepsilon_0 A}{d}$.

From Table 8.3, the dielectric constant ϵ_R of PS at 1 MHz is 2.5.

$$\therefore \mathrm{C} = \frac{2.5 \times 8.854 \times 10^{-12} \times 10^{-4}}{10^{-4}} = 22 \text{ pF}$$

From Table 8.3 the volume resistivity of polystyrene (PS) is $>10^{14}$ Ωm. From eqn (2.3), the resistance of the PS between the two electrodes is $R = \frac{\rho L}{A}$.

$\therefore R > \frac{10^{14} 10^{-4}}{10^{-4}} = 10^{14}$ Ω. and $I < 10^{-11}$ A. At this sort of current, surface leakage and conduction through the surrounding air are likely to be equally important.

From Table 8.3, the breakdown voltage, or dielectric strength, is 1.35×10^8 Vm^{-1}. The voltage necessary to cause breakdown of this plate of PS is therefore $1.35 \times 10^8 \times 10^{-4} = 13\frac{1}{2}$ kV.

8.8 How and from what would you make

(a) a three pin plug

(b) a domestic electric socket

(c) the casing for

- a TV set
- an electric iron
- an electric drill
- a refrigerator
- a washing machine?

All of these can be made by injection moulding. The materials might well be:

(a) Nylon

(b) Nylon

(c) ABS, PP, PA, ABS, ABS (respectively)

Chapter 9

9.1 Name four semiconductors.

Silicon and germanium are elemental semiconductors. CVD 'diamond' (band-gap 2.8 eV) sometimes creeps in. Common compound semiconductors are GaAs, InSb and CdTe.

9.2 Which dopants are used for silicon?

p-type: boron (Periodic Table Group IIIA: see Fig. 1.6)
n-type: phosphorus, arsenic (Group VA)

9.3 If electronic grade silicon contains 1 ppm (part per million) Fe, how much would remain in the middle of a bar after one zone-refining pass?

From section 9.3.1 and Fig. 9.5, $c_s = k\, c_o\, (1-x)^{k-1}$ and $k_{Fe} = 10^{-5}$. For a position halfway along the bar, $x = 0.5$ and $c_s = 10^{-5} \times 10^{-6} \times 2$ or 2×10^{-11}.

9.4 What accelerating voltage is necessary to implant boron into silicon to a depth of 1 μm?

From Fig. 9.9, ~400 kV.

9.5 Why, until recently, was copper not used as an interconnect?

Because it diffused into the silicon, spoiling its electronic properties. It has been found that a thin layer of Ta or TaN prevents this.

9.6 What are typical impurity and doping levels for silicon?

Ppb and ppm. (Parts per billion, million.)

9.7 What is a negative photoresist?

One where exposure to UV renders the polymer photoresist insoluble (the normal type).

9.8 Name a positive photoresist, a negative photoresist and a potting compound.

Positive photoresist:	novolac + diazonaphthoquinone
Negative photoresist:	epoxy resin
Potting compound:	epoxy resin

Chapter 10

10.1 A long solenoid is wound with 10 turns/cm and a current of 10 A is passed through it. What is the field at its centre, away from the ends? What is the induction? What force would be exerted on another wire, perpendicular to the axis of the solenoid and also carrying 10 A? (The permeability of free space $\mu_0 = 4\pi \times 10^{-7} = 1.257 \times 10^{-6}$ H m^{-1}.)

A solenoid with 10 turns/cm would have an internal field of 10 A m^{-1}. The induction $B = \mu_0 H = 1.26 \times 10^{-5}$ T. The force per unit length on the second wire is iB N $= 10 \times 1.26 \times 10^{-5} = 1.26 \times 10^{-4}$ N m^{-1}. It is perpendicular to both the wire and the axis of the solenoid.

10.2 A bar of material with relative permeability $\mu_r = 1000$ is placed within the coil of Q10.1. What is the induction within the material? What is the susceptibility of the material?

From eqn(10.6) $B = \mu H = \mu_r \mu_0 H = \mu_r$ times what B would be *without* the material present. From the answer to Question 10.1, this is 1.26×10^{-5} T and so with the material present, the induction becomes $1000 \times 1.26 \times 10^{-5} = 0.0126$ T.

From eqn (10.7), the susceptibility $\chi = \mu_r - 1 = 999$.

10.3 The material referred to in Q's 10.1 and 2 has $H_c = 15$ A m^{-1} and saturation induction $B_s = 2.15$ T. Would you describe it as magnetically soft or hard? The current in the coil is increased until the magnet is saturated, reversed to the same (negative) value and then returned to zero. Roughly how much energy is absorbed (order of magnitude estimate)?

An H_c of 15 A m^{-1} corresponds to a soft magnetic material. Equally well, you could say that a relative permeability of 1000 corresponds to a soft magnetic material. H_c and μ_r are linked, as an experimental fact. High H_c implies low μ_r and vice versa. VERY roughly, $\log_{10}\mu_r = 5 - \log_{10}H_c$.

The material is being taken around its hysteresis curve (see Fig. 10.4). Very roughly, the area of the curve, which is the energy absorbed per cycle per unit volume of material, is $(2 \times H_c) \times (2 \times B_s)$

$= 130\ \mathrm{J\ m^{-3}}$. In fact, this is not a bad estimate. Engineers tend to give hysteresis losses in $\mathrm{W\ kg^{-1}}$, which means you would need to know the density of the magnetic material, and the characteristics of the electrical supply (frequency etc.) to convert.

10.4 Classify the following materials as diamagnets, paramagnets and ferro/ferrimagnets.

- copper
- polythene
- silicon
- iron
- aluminium
- magnetite (Fe_3O_4)

copper	diamagnet
polythene	diamagnet
silicon	diamagnet
iron	ferromagnet (!)
aluminium	paramagnet
magnetite	ferrimagnet

10.5 A piece of soft iron is dropped on the floor. Its coercivity doubles. Why is this?

Dropping something soft on the floor causes it to deform plastically, introducing dislocations. These pin magnetic domain movement and increase the coercivity. This can be quite a problem when making soft magnets. Machining operations, for example, must be followed by an annealing treatment to remove the dislocations and restore the desired magnetic properties.

10.6 Why does dissolving silicon into iron affect its electrical resistivity more than dissolving cobalt into it?

The change in resistivity comes mainly from the difference in valency between solvent (iron) and solute (silicon or cobalt). Cobalt is very similar electronically to iron, whereas silicon, in the middle of the periodic table and with potentially four valence electrons per atom, is very different. Thus silicon has a much greater effect on the resistivity of iron than does cobalt and is one of the reasons iron-silicon is used for transformers. (See Fig. 10.8 and Chapter 4 (in particular eqn (4.2)).)

10.7 Classify the following as soft and hard magnetic materials:

- iron
- steel
- iron-silicon
- iron neodymium boron
- iron silicon boron glass
- alnico
- barium ferrite
- samarium-cobalt
- magnetite (Fe_3O_4)

Mainly from Tables 10.3–5,

iron	soft
steel	hard(ish)
iron–silicon	soft
iron neodymium boron	hard
iron silicon boron glass	soft
alnico	hard
barium ferrite	hard
samarium-cobalt	hard
magnetite	soft

Chapter 11

(Superconductivity)

11.1 If a 1A current flows around a 1 m diameter circle of superconducting wire with 1 mm^2 cross-section for one year without diminishing detectably (e.g. loss < 1%), what is the maximum resistivity of the wire?

(The current i(t) flowing around a loop of wire decays according to

$$i(T) = i(0)\, e^{\frac{R}{L}t}$$

where t = time, R = electrical resistance and L = self-inductance.)

If the resistivity of the wire is ρ Ωm, the resistance of the loop is $\frac{\rho \times \pi \times 1}{10^{-6}}$ Ω.

$$0.99 = 1.00\, e^{-\frac{R}{4\pi 10^{-7}} 365 \times 24 \times 60 \times 60}$$

$$R < 4 \times 10^{-16}\ \Omega$$

and

$$\rho < 1.3 \times 10^{-22}\ \Omega\text{m}$$

11.2 If a metal needs to be cooled to 2K to superconduct, estimate the size of the superconducting band-gap.

Boltzmann's constant, k, is 8.62×10^{-5} eV K^{-1})

You might expect the band-gap to be less than ~kT. At 2 K $kT = 1.7 \times 10^{-4}$ eV and the band-gap should be of this order.

11.3 (See Fig. 11.1) What are the ceramic superconductors Y-123 and Bi-2223?

Y–123 is $YBa_2Cu_3O_{7-\delta}$ — The naming convention is fairly obvious.

Bi-2223 is $Bi_2Sr_2Ca_2Cu_3O_{10-\delta}$ — δ is a fraction.

(Optical fibres)

11.4 (See Fig. 11.3) In, for example, a monomode fibre, what are n_1 and n_2? What is the angle for total internal reflection? What % of e.g. Ge would be required to produce this change?

In a monomode fibre n_1 and n_2 typically differ by 0.3% only—e.g. $n_1 = 1.463$ and $n_2 = 1.460$. Then $\sin\theta = \frac{n_2}{n_1}$ and 90–$\theta = 6.5\,^\circ$. This would require 2–3% Ge (as Ge_2O_3).

Index

Bold indicates principal reference